Heat and Mass Transfer

Series Editors: D. Mewes and F. Mayinger

Norbert Kockmann

Transport Phenomena in Micro Process Engineering

With 214 Figures and 17 Tables

 Springer

Dr.-Ing. Norbert Kockmann
Albert-Ludwigs-Universität Freiburg
Institut für Mikrosystemtechnik IMTEK
Lehrstuhl Konstruktion von Mikrosystemen
Georges-Köhler-Allee 102
79110 Freiburg
Germany
e-mail: kockmann@imtek.de

Series Editors

Prof. Dr.-Ing. Dieter Mewes
Universität Hannover
Institut für Verfahrenstechnik
Callinstr. 36
30167 Hannover, Germany

Prof. em. Dr.-Ing. E.h. Franz Mayinger
Technische Universität München
Lehrstuhl für Thermodynamik
Boltzmannstr. 15
85748 Garching, Germany

Library of Congress Control Number: 2007935989

ISSN 1860-4846
ISBN 978-3-540-74616-4 Springer Berlin Heidelberg New York

Springer is a part of Springer Science+Business Media
springer.com

© Springer-Verlag Berlin Heidelberg 2008

Typesetting: Data supplied by the author
Production: LE-TeX Jelonek, Schmidt & Vöckler GbR, Leipzig, Germany
Cover: deblik Berlin

SPIN 12043293   7/3180/YL - 5 4 3 2 1 0   Printed on acid-free paper for Eva Christina, Paul Joey, and Lilly Sophie ...

Preface

Transport phenomena and kinetic processes are found in almost every technical application where mass or energy are transformed into other shapes and forms. Characteristic time and length scales determine process characteristics and their interaction. The unified notion of transport phenomena was introduced by Bird, Stuart, and Lightfoot in their well-known textbook [1, 2] and has a broad influence in many engineering areas. With miniaturization of process equipment, the characteristic length and time scales shift to different regimes, where they are beneficial for enhanced applications, see [3]. Conversely, they may prevent conventional process conditions, which must be reassessed. The transport phenomena in microstructures treated here concern momentum, species, mass, and energy transfer combined with entropy generation, reaction kinetics, and coupled processes. Almost all transport processes dealt with are in the continuum range and can be described by linear correlations, however, the limitations and exceptions are given. The length scale of transport processes reaches from molecular scale of gas scattering to complex turbulent flow in the millimeter range.

The present work is outlined according to the nature of the transport phenomena and kinetic processes. The first two chapters introduce the scope of micro process engineering and general process engineering to the unfamiliar reader. The second chapter also presents the fundamentals of transport processes from single molecular encounters to transport in multiphase flow. Engineering tools, such as dimensional analysis or order-of-magnitude estimations are explained, which allow treatment of complex nonlinear systems.

The third and fourth chapters cover momentum and heat transfer in microchannels with emphasis on laminar convection and the design of appropriate channel geometries. Convective transport enhancement is employed for mixing enhancement in microchannels, described in Chapter 5. Aside from the appropriate characterization of the mixing process and device, typical properties of convective laminar micromixers demonstrate their excellent performance with rapid, mixing-sensitive chemical reactions for increased selectivity or fast particle precipitation, described in Chapter 6.

Chapter 7 is unique in the field of micro process engineering topics and deals with coupled transport processes. Micro fabrication technology enables tailor-made geometries with appropriate materials to augment coupled transport phenomena. This is used for efficient generation of electrical energy by thermoelectric devices or single-phase separation by thermodiffusion. In spite of the advances presented, there is still a lot of effort required to improve devices and integrate them into suitable applications. Some suggestions for this ongoing process are referenced and summarized at the end of each chapter, as well as in Chapter 8.

This work may also be a contribution to the tenth anniversary of the IMRET conference [4] on micro reaction technology, first held in 1997 in Frankfurt, Germany. It might also be regarded as the twentieth anniversary of the first patent on microstructured devices for process engineering in the former Eastern Germany [5]. Finally, the first textbook on microchemical technology is more than one hundred years old [6, 7], and the first literature on microscope usage in chemical analysis appeared approx. 130 years ago [8]. My own first contact with micro process engineering, observing technical processes under the microscope, was a diffusion cell in crystallization, in which a doctorate colleague observed diffusion processes in crystal layers and inclusions under a temperature gradient. This concept of an observation cell was very effective and is similar to many of the concepts presented here.

Whilst writing this work, the idea of the "rhizome" introduced by Deleuze [9] often came to mind. Originally indicating the fine network of plant roots, Deleuze and Guattari [10] presented their philosophical concept to indicate a living, interwoven network, cross-linking many disciplines and fields, and mirroring the modern, complex world. The topic of this thesis is just a small area in science and engineering, but embraces and influences so many disciplines and fields that this idea may also apply here. The combination of different disciplines presents new opportunities for many fields in microfluidics and process intensification. I hope the reader is not affronted by the complexity of micro process engineering, but will take this work as a guideline for tackling future challenges. Assisting this process, Kornwachs stated [11] that engineers, while building machines, shape the technical landscape for actual necessities. Hence, engineers are becoming system designers and creators of culture, opening a new role, which they must learn to use.

This book is the outcome of many diploma works, a dissertation, many group discussions, and the collected effort of many contributors from the University of Freiburg, Department of Microsystems Engineering – IMTEK, other universities, research labs, and from the "real world". Thanks for the good atmosphere, for your encouraging support and many helpful discussions. The text was carefully proof-read by Paul Thomas who is not responsible for any remaining faults.

Finally, very special thanks go to my family, for their enduring care, support, and love throughout the years.

Freiburg   *Norbert Kockmann*
August 2007

Contents

Nomenclature

Latin Letters

Variable	Denotation	Unit
A	(surface) area or cross section	m^2
a	activity	–
a	temperature conductivity $a = \frac{\lambda}{\rho\, c_p}$	m^2/s
a_V	specific surface $= A/V$	m^2/m^3
B_{hom}	rate of homogeneous nucleation	$1/m^3\, s^1$
b	geometrical factor, channel width	m
C	constant	–
C_f	friction factor coefficient for laminar duct flow	–
C_i	ratio of heat capacity fluxes	–
c	specific velocity	m/s
c	molecular velocity	m/s
c	speed of sound	m/s
c_i	concentration of component i	–
c_p	isobaric specific heat capacity	J/kg K
c_v	isochoric specific heat capacity	J/kg K
D	diffusivity, diffusion coefficient	m^2/s
D	diameter	m
d_P	particle diameter	m
d_h	hydraulic diameter $= \frac{4 \cdot \text{cross section area}}{\text{wetted perimeter}}$	m
E	energy	J
E_{kin}	kinetic energy	J
E_{pot}	potential energy	J
e	specific energy	J/kg
F	force	N
F	cumulative number distribution (PSD)	–

Variable	Denotation	Unit
f	probability distribution function	–
f	number fraction frequency (PSD)	–
G	mass velocity $= \dot{m}/A$ (US literature)	kg/m^2 s
G	linear growth rate (PBE)	m/s
g	gravity constant	m/s^2
g	temperature jump coefficient	–
H	enthalpy	J
ΔH_R	reaction enthalpy	J/mol^3
Δh_R	specific reaction enthalpy	J/m^3
H_i	Henry coefficient	Pa
h	specific enthalpy	J/kg
Δh_V	latent heat of vaporization	J/kg
h	heat transfer coefficient (US literature)	W/m^2 K
h	geometrical factor, height	m
I	electrical current	A
J	general energy current	J/s
K	general coefficient	–
k	overall heat transfer coefficient	W/m^2 K
k	reaction rate constant for reaction of order m	1/s
k	Boltzmann constant ($1.380662 \cdot 10^{-23}$)	J/K
L	length	m
L_{ij}	general transport coefficient	–
L_p	rate of production	mol/s
l	length, characteristic	m
l_C	channel length	m
l_m	characteristic mixing length	m
l_P	length of the wetted perimeter	m
M	molar mass	kg/mol
M_m	mass of a single atom or molecule	kg
m	mass	kg
$\dot{m}$	mass flow rate	kg/s
m	molality	mol/kg
m	reaction order	–
N	number	–
N_i	ratio of transferred heat to heat capacity, number of transfer units NTU	–
n	amount of substance	mol
$\dot{n}$	molar flow rate	mol/s
n	number density concentration	1/m^4
n	rotation speed	1/s
P	power	W
p	pressure	bar, Pa $=$ N/m^2

Variable	Denotation	Unit
p_i	partial pressure	bar, $Pa = N/m^2$
Q	heat	$J = Nm$ $= kg/m^2 s^2$
$\dot{Q}$	heat flux	$W = J/s$
q	specific heat	J/kg
$\dot{q}$	specific heat flux	W/m^2
R	radius	m
R	radius of spherical particles	m
R	individual gas constant, $= R_m/M$	J/kgK
R	electrical resistance	Ω
R_i	transformation rate	$mol/m^3\ s$
R_m	universal gas constant (8.314 J/mol K)	$J/mol\ K$
r	spherical coordinate (radius)	m
r	reaction rate	$1/s$
S	entropy	J/K
S	selectivity (competitive reactions)	–
S_a	activity-based supersaturation	–
S_c	concentration-based supersaturation	–
S'	heat production potential	–
s	specific entropy	$J/kg\ K$
T	temperature	K, °C
ΔT_L	spatial temperature gradient	K/m
ΔT_t	temporal temperature gradient	K/s
t	time	s
t_m	characteristic mixing time	s
t_{m90}	characteristic mixing time with 90 % mixing quality	s
t_R	characteristic reaction time	s
t_P	mean residence time	s
U	inner energy	J
U	electric voltage	V
u	specific inner energy	J/kg
u	velocity in x-direction	m/s
V	volume	m^3
$\dot{V}$	volume flow rate	m^3/s
v	velocity in y-direction	m/s
v	specific volume	m^3/kg
W	work	J
W_t	technical work	J
W_{diss}	dissipated work or energy	J
w	width of phase lamellae	m
w	velocity in z-direction	m/s
$\vec{w}$	vector of velocity (u, v, w)	m/s

XVI   Contents

Variable	Denotation	Unit
w	specific work	J/kg
X	general transport variable	–
X	conversion	–
$\vec{x}$	position vector (x,y,z)	m
x	cartesian coordinate	m
x	vapor quality or void fraction	–
x_i	liquid concentration of component i	–
Y	vapor concentration	kg/kg
Y	reaction yield	mol/m^3
y	cartesian coordinate	m
y_i	vapor concentration of component i	–
z	cartesian coordinate, main flow direction in a channel	m
z	separation factor	–

Greek Letters

Variable	Denotation	Unit
α	heat transfer coefficient	W/m^{2}K
α_A	channel or structure aspect ratio $= h/b$	–
α_m	mixing quality	–
$\alpha_{\dot{V}}$	mixing quality based on volume flow	–
β	molar mass transfer coefficient	m/s
γ	momentum transfer coefficient	kg/m^{2}s
γ	interfacial tension	N/m
γ_i	relative amount of substance, mole fraction	–
γ	molar activity coefficient	–
δ	boundary layer thickness	m
δ	molecular spacing	m
ε	specific energy dissipation	m^2/s^3
ε	porosity	–
ζ	pressure loss coefficient of channel fitting or internals	–
ζ	slip length of rarefied gas flow	m
η	dynamic viscosity	kg/m s = N s/m^2
θ	dimensionless temperature	–
θ	contact angle	°
θ	heat exchanger efficiency	–

Variable	Denotation	Unit
κ	isentropic exponent $= c_p/c_v$	–
Λ	mean free path length	m
λ	heat conductivity	W/m K
λ	channel friction factor	–
μ	dynamic viscosity (US literature)	Pa s
μ	chemical potential	J/mol
ν	kinematic viscosity $= \eta/\rho$	m^2/s
ν_i	stoichiometric coefficient	–
ξ	ratio of bulk concentrations in binary solutions	–
π_i	mole fraction acc. to Raoult's law	Pa
ρ	density $= m/V$	kg/m^3
ρ_m	molar density $= n/V$	mol/m^3
ρ_P	particle density	kg/m^3
σ	standard deviation	depends
σ	collision cross section	m^2
σ	surface tension	N/m
$\dot\sigma$	dissipation function or local entropy production	J/kg s
τ	time	s
τ	shear stress	N/m^2
Φ	potential for diffusive mixing	1/m
Φ	dissipation function	J
ϕ	reaction affinity	J/mol
ϕ_c	gradient-based mixing scale	1/m
φ	probability	–
φ	relative moisture	–
χ	dispersion	–
ψ	growth potential	mol/l
ω	angular velocity	1/s

Dimensionless Numbers

Name	Description
$Bi = \frac{\alpha l}{\lambda}$	Biot number ($\lambda = \lambda_{solid}$)
$Bo = \frac{w l}{D_{ax}}$	Bodenstein number
$Bo = \frac{q''}{G\, h_{fg}} = \frac{\dot Q}{\dot m \cdot \Delta h_V}$	boiling number (US literature)
$Bo = \frac{d^2 g \rho}{\sigma}$	Bond number (US literature)

Name	Description
$Ca = \frac{\eta\, w}{\sigma}$	Capillary number
$DaI = \frac{t_r}{t_P}$	1st Damköhler number
$DaII = \frac{k_S}{k_D} = \frac{t_r}{t_D}$	2nd Damköhler number
$DaIII = \frac{\Delta h_R \cdot r \cdot l_C}{\rho \cdot c_p \cdot \Delta T_{fluid} \cdot \bar{w}}$	3rd Damköhler number
$DaIV = \frac{\Delta h_R \cdot r \cdot d_h^2}{\lambda \cdot \Delta T_{log}}$	4th Damköhler number
$Dn = Re\left(\frac{D}{R_c}\right)^{1/2}$	Dean number
$Ec = \frac{w^2}{c_p\, \Delta T}$	Eckert number
$Eu = \frac{\Delta p}{\rho\, w^2}$	Euler number
$Fo = \frac{a\, t}{l^2}$	Fourier number
$Fr = \frac{w^2}{d\, g}$	Froude number
$Gr = \frac{g\, \beta_p\, s^3\, \Delta T}{\nu^2}$	Grashoff number
$Gz = \frac{\bar{w} d_h^2}{D l_C} = Pe \cdot \frac{d_h}{l_C}$	Graetz number
$Kn = \frac{\Lambda}{l}$	Knudsen number
$Le = Sc/Pr = \frac{a}{D}$	Lewis number
$Ma = \frac{w}{c}$	Mach number
$ME_I = \frac{1}{Eu} \cdot \frac{d_h}{l_m}$	mixing effectiveness
$ME_{II} = \frac{Re}{Eu} \cdot \frac{d_h}{l_m}$	mixer effectiveness
$MP = \frac{d_h}{l_m}$	mixing performance
$Ne = \frac{P}{\rho\, n^3\, d^5}$	Newton number for stirred vessels
$Nu = \frac{\alpha\, l}{\lambda}$	Nußelt number ($\lambda = \lambda_{Fluid}$)
$Pe = Re\, Pr = \frac{w\, l}{a}$	Péclet number of heat transfer
$Pe' = Re\, Sc = \frac{w\, l}{D_m}$	Péclet number of diffusion
$Pr = \frac{\nu}{a}$	Prandtl number
$Ra = Gr\, Pr$	Rayleigh number

Name	Description
$\mathrm{Re} = \frac{w\,l}{v}$	Reynolds number
$\mathrm{Sc} = \mathrm{Le}\,\mathrm{Pr} = \frac{v}{D}$	Schmidt number
$\mathrm{Sh} = \frac{\beta\,l}{D}$	Sherwood number
$\mathrm{Sr} = \frac{f\,l}{w}$	Strouhal number
$\mathrm{St} = \frac{d_P^2 \rho_P C \bar{u}}{18 \eta l}$	Stokes number
$\mathrm{We} = \frac{w^2\,d\,\rho}{\sigma} = \frac{n^2\,d^3\,\rho}{\sigma}$	Weber number

Mathematical Operators

Notation	Description
D	substantial differential
d	general differential
∂	partial differential
Δ	delta, difference
$\nabla = i\frac{\partial}{\partial x} + j\frac{\partial}{\partial y} + k\frac{\partial}{\partial z}$	Nabla operator, in cartesian coordinates
$\mathrm{grad}\,\varphi = \nabla\varphi$	gradient of a scalar φ, gives a vector
$\mathrm{grad}\,\vec{w} = \nabla\vec{w}$	gradient of a vector φ, gives a tensor of 2nd order
$\mathrm{div}\,\vec{w} = \nabla\vec{w}$	divergence of $\vec{w}$, gives a scalar
$\mathrm{rot}\,\vec{w} = \nabla \times \vec{w}$	rotation of $\vec{w}$
$\begin{aligned}\Delta\varphi &= \nabla^2\varphi \\ &= \mathrm{div}\cdot(\mathrm{grad}\varphi) \\ &= \nabla\cdot(\nabla\varphi)\end{aligned}$	Laplace operator

Main Conferences and Their Abbreviation

IMRET 1:   Ehrfeld W (Ed.), *Microreaction Technology, Proc. of the 1ˢᵗ Int. Conf. on Microreaction Technology 1997*, Springer, Berlin, **1998**.

IMRET 2:   *Proc. of the 2ⁿᵈ Int. Conf. on Microreaction Technology*, AIChE National Spring Meeting, New Orleans, **1998**.

IMRET 3:   Ehrfeld W (Ed.), *Microreaction Technology: Industrial Prospects, Proc. of the 3ʳᵈ Int. Conf. on Microreaction Technology 1999, Frankfurt*, Springer, Berlin, **2000**.

IMRET 4:   *Microreaction Technology: Proc. of the 4ᵗʰ Int. Conf. on Microreaction Technology*, AIChE National Spring Meeting, Atlanta, **2000**.

IMRET 5:   Matlosz M, Ehrfeld W, Baselt P (Eds.), *Microreaction Technology, Proc. of the 5ᵗʰ Int. Conf. on Microreaction Technology 2001, Straßbourg*, Springer, Berlin, **2002**.

IMRET 6:   Baselt P, Eul U, Wegeng RS, Rinard I, Horch B (Eds.), *Proc. of the 6ᵗʰ Int. Conf. on Microreaction Technology*, AIChE National Spring Meeting, New Orleans, **2002**.

IMRET 7:   Renken A, Matlosz M (Eds.), *Proc. of the 7ᵗʰ Int. Conf. on Microreaction Technology 2003*, Lausanne, **2003**.

IMRET 8:   *Proc. of the 8ᵗʰ Int. Conf. on Microreaction Technology*, AIChE National Spring Meeting, Atlanta, **2005**.

IMRET 9:   *Proc. of the 9ᵗʰ Int. Conf. on Microreaction Technology*, Dechema, Potsdam, Germany, **2006**.

ICMM2003:   Kandlikar SG, *ASME 1ˢᵗ Int. Conf. on Microchannels and Minichannels*, Rochester NY, **2003**.

ICMM2004:   Kandlikar SG, *ASME 2ⁿᵈ Int. Conf. on Microchannels and Minichannels*, Rochester NY, **2004**.

ICNMM2005: Kandlikar SG, Kawaji M, *ASME 3ʳᵈ Int. Conf. on Nano, Micro and Minichannels*, Toronto, Canada, **2005**.

ICNMM2006: Kandlikar SG, Davis M, *ASME 4ᵗʰ Int. Conf. on Nano, Micro and Minichannels*, Limerick, Ireland, **2006**.

ICNMM2007: Kandlikar SG, Reyes Mazzoco R, *ASME 5ᵗʰ Int. Conf. on Nano, Micro and Minichannels*, Puebla, Mexico, **2007**.

1

Micro Process Engineering – An Interdisciplinary Approach

The aim of this chapter is to give an general overview of the new area of Micro Process Engineering. The important role of transport phenomena is addressed as well as some requirements for the successful application of microstructured equipment. Scaling down to micrometer dimensions changes not only the length or volume of a device, but also influences the performance of unit operations. This chapter encompasses of the entire field of *Micro Process Engineering*, the science of enhanced transport process in microstructured devices. To illustrate recent advances in this area, some key applications and research programs are presented from around the world.

This introductory chapter gives an overview of miniaturization effects and beneficial phenomena in microchannels with characteristic dimensions from 10 to 1 000 μm. Actual research projects in Germany in the Microsystems Technology program (2004-2009) are described as well as the actual EU projects. Further emphasis is placed on industrial projects in Europe, in the United States, and in Far East for laboratory or production purposes. Characteristic examples illustrate the first successful applications, the crucial drawbacks, but also the need and demand for future research and development. The integration of microstructured devices together with sensing and actuating elements, as well as within the "macro world", will be one of the key issues for future development.

1.1 Introduction and Motivation

The enhanced performance of key processes is crucial for the economical development of process industries. One major area in chemical engineering is process intensification, which benefits from the miniaturization of channels and ducts within devices, where the characteristic lengths reach into the scale of boundary layers, see [12]. The higher transport rates can be used for many different purposes such as rapid, mixing-, or temperature-sensitive reactions, temperature homogenization, or nanoparticle precipitation.

Process intensification by miniaturized equipment touches many disciplines and crosses mutual borders. Each discipline, from mechanical or chemical engineering, chemistry to microsystems engineering, has its own rich, sometimes specific wealth of knowledge, methods, and technology. Due to each fields focus on their own specific problems, the concept of "how everything works and fits together" and how it is entangled with economy, politics, and society becomes somewhat lost. Communication between the disciplines is very important and is emphasized in parts of this book. Also short-cut calculations, rules of thumb, and an integrated engineering approach with systems integration and optimization are included to generate interest for complex systems and highlight new, combining, and combined elements. Motivation competence enables to seek and find bridging and common elements between the different disciplines. The general aim is to teach and learn methods and skills, not only knowledge, which rapidly becomes outdated.

The wider impact of microstructured equipment and plants on social and cultural areas as well as people's behavior is described by Hessel et al. [13] Chapter 1. One key issue, which touches the future of economy and society, is the appropriate energy supply of single components, of social units, and of the entire economy. In their role as technical designers, engineers are at the same time social designers and shapers. From this starting point, Arno Rolf [14] concludes in his short essay on key qualifications of students, that convincing models are still missing, which respect the methods of each discipline and, at the same time, sharpen the orientation competence of the students within their discipline. The statement can also be expanded to include engineers, technicians, and researchers from many fields.

1.2 Orientation of Micro Process Engineering

Process technology and microsystems technology are both interdisciplinary engineering and science branches connecting physics, chemistry, biology, engineering arts, and management techniques to an enabling toolbox for various applications. Process engineering embraces process simulation, equipment design, and system orientated, cross-linked thinking for the multidisciplinary field [15]. Chemical process engineering covers not only the design and implementation of process routes into chemical production plants, but also equipment design with appropriate materials, their properties, and operation control layout. The aims of process engineering are the economical and safe production of the desired product(s). On the way to economical and sustainable production of chemicals, the methods and tools of process intensification play a major role and demand new pathways for simulation in the treatment of unit operations and integrated processes, see Figure 1.1.

Microsystems technology, coming from information technology and microelectronics, has now entered many fields of daily life. Silicon chips and sensors can be found in cars, washing machines, or smart cards, with various functions. Micromechanical systems started to enter technical systems in the 80s and 90s [16], enabling fluidic systems to be developed. Starting from data processing, microsystems have now integrated mechanical, optical, fluid mechanical, and chemical func-

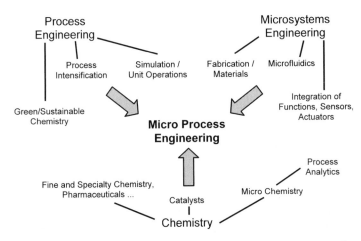

Fig. 1.1. Micro Process Engineering as enabling and focusing technology as well as interdisciplinary area for many engineering areas.

tions for tasks like sensing and analyzing, controlling larger systems, or producing suitable goods. Besides engineering skills like design, simulation, fabrication, or material knowledge, microsystems engineering also comprises a thorough physical and chemical knowledge for functional design issues and device fabrication. Also, medical and biological skills are useful for the growing application fields for therapeutics and diagnostics. A good overview of the state-of-the-art in microsystems technology is given by Gad el Hak in [17]. A recent review of fabrication technologies and process for MEMS is given by Quinn et al. [18] presenting achievable geometrical shapes, surface characteristics, and required process parameter such as solvents, temperature, and pressure. On the way to still smaller systems, microsystems technology is a major link to nanotechnology [19, 20, 21].

Chemistry gives the background orientation and a multitude of applications for both disciplines, besides fuel processing or automotive applications, see [13, 22]. Figure 1.1 gives an impression of the wide scope and complexity of all disciplines, but also illustrates the multiple interfaces, links, and common fields. The various ideas from all sides may inspire further development in the disciplines and result in increased possibilities and applications for innovations across the borders. A good overview is given by various authors in [3] with a clear and concise presentation of the fundamentals, design, fabrication and integration of micro process equipment, as well as prominent applications from around the world.

1.3 The Role of Transport Processes

Due to the reduced length scale of microstructured process equipment, the transfer lengths are short and precisely defined, and the areas are small, but high surface-to-volume ratios and tiny volumes dominate everything. This presents many opportunities for chemical and process engineering to be handled in microstructures. On an ACS symposium in 2003 [23], a recent overview was given on North American and European activities of chemical reaction engineering with microstructures.

Small channels allow short transport lengths for heat and mass transfer. This results in high transfer rates, as described for diffusive mass transfer with the mean transport length from the Einstein-Smoluchovski equation [24] Chapter 4 and 5.

$$x^2 = 2 D t \tag{1.1}$$

The transport length by diffusive mixing in gases ($D = 10^{-5}$ - 10^{-6} m²/s) and in liquids with low viscosity ($D = 10^{-9}$ - 10^{-10} m²/s) is displayed over the corresponding time in Figure 1.2. Conventional equipment has typical geometries in the range of centimeters and produces fluid structures in the range from 100 μm to 1 mm. The corresponding diffusion time is in gases slower than approx. 1 ms and in liquids in the range of 1 s. Microstructured devices with typical length scales from 100 μm to 1 mm provide fluid structures with length scales of approx. 1 μm. These small fluid structures lead to mixing times shorter than 100 μs in gases and approx. 1 ms in liquids. This is the main reason for the enhanced selectivity and high yield of chemical reactions in microreactors.

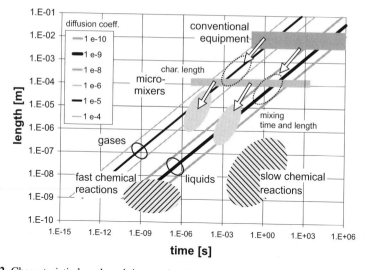

Fig. 1.2. Characteristic length and time scales for mixing in microstructured devices together with chemical reactions.

The length scaling often coincides with a time scaling of the relevant processes. In general, the shorter the length, the shorter the characteristic time for transport processes will be, and the higher the transformation frequencies. The diffusion of a species in a surrounding fluid displays this process and is described in Eq. 1.1. The typical diffusion length within one second is approx. 7 mm in gases (air) and approx. 70 μm in liquids such as water. Similarly, the conduction length can be derived from the basic balance equation for the momentum ($x_p = \sqrt{2 \, v \, t}$) and the heat transfer ($x_q = \sqrt{2 \, at}$). The characteristic time is proportional to the square of the length variation and to the transport coefficient. More details are given in Section 2.3.4.

Additionally, the typical length and time scales for fast chemical reactions like neutralizations or slow chemical reactions such as complex formation or polymerization are given as well as typical scales for micromixers to compare the processes. The mass transfer in micromixers acts on a length scale of a few microns within milliseconds or less. Different time scales are typical for partial reactions in complex chemical reaction systems. With properly designed micromixers and an adjustment of the component concentration, the selectivity of a complex reaction can be increased dramatically. The mixing time is only one criterion for an optimal chemical reaction, another is the scale of fluids residence time within the device. Within small devices, the fluids rest only briefly (seconds or less), which can be detrimental to slow reactions. A good overview of residence time, backmixing, and heat transfer in microchannels is given by Knösche in [25] with a typical example. The backmixing can be described by the Taylor-Aris model, where the molecular diffusion coefficient is replaced by an effective axial diffusion coefficient D_{ax}

$$D_{ax} = D + \frac{\bar{w}^2 d_h^2}{192 \, D} \tag{1.2}$$

and expressed by the dimensionless Bodenstein number Bo

$$Bo = \frac{\bar{w} \, L}{D_{ax}}. \tag{1.3}$$

The Bo number is similar to the dimensionless Péclet number Pe, which is defined for heat transfer problems $Pe = Re \, Pr = \bar{w} d_h / a$ and for mass transfer problems $Pe = Re \, Sc = \bar{w} d_h / D$. The Pe number plays a major role in heat and mass transfer problems, as discussed in Section 4.1.3 for heat transfer and in Section 5.3.2 for mass transfer. The backmixing described by the Bo number is not just influenced by the channel diameter and flow velocity, but also by the channel length. Fast reactions require short and small channels and a sufficiently high number of channels. More details of characteristic times and time scales are given in Section 6.1.2.

In process engineering, other characteristic times determine the process conditions. The channel length and the mean flow velocity determine the mean residence time $t_P = L_C / \bar{w}$ of a fluid element in the channel. With shorter lengths, the residence time decreases. The residence time must be designed appropriately for the actual process. A very important characteristic time in process engineering belongs to the chemical reaction: Besides the concentration c and the reaction order m, the reaction time scale depends mainly on the reaction rate constant k ($t_R \propto 1/k$). The reaction rate constant is directly influenced by the temperature. Due to the reduced mixing or residence time inside short channel equipment, a slow reaction may be incomplete at the channel outlet. Especially fast reactions, on the other side, take profit from the rapid mixing and heat exchange. Combined reactions with slower side-reactions or unstable intermediates will show a higher selectivity and higher yield in microstructured devices. Also reactions with high energy demand or release are suitable for micro devices. The concentrations can be increased to intensify the reaction, or new reaction paths can be addressed where rapid mixing plays an important role. In summary, the three characteristic mixing times t_m, of the residence in the device t_P, and of the chemical reaction t_R must be adjusted for the entire process within a chemical reactor to yield an optimal result.

With unsteady heat transfer, the characteristic time t for heating or cooling of a body is proportional to the temperature difference and the ratio of the heat capacity $m \cdot c_p$ to the heat transfer $\alpha \cdot A$ within the environment.

$$t = \frac{m\, c_p}{\alpha\, A} = \frac{\rho\, c_p\, V}{\alpha\, A} = \frac{\rho\, c_p}{\alpha\, a_V} \propto \frac{\rho\, c_p}{\alpha} \cdot d_h \tag{1.4}$$

With a smaller length scale d_h, the surface-to-volume ratio a_V increases and this characteristic relaxation time t becomes shorter. Similarly, relaxation times of other unsteady processes can be estimated.

The heat transfer in a straight channel with laminar flow is described by a constant Nußelt number Nu, see Section 4.1.3:

$$\mathrm{Nu} = \frac{\alpha\, d_h}{\lambda} = 3.65 \tag{1.5}$$

for constant wall temperature. With smaller channel diameter d_h, the heat transfer coefficient α increases. Additionally, convective effects in bent channels can increase the Nußelt number for better performance [26], but also increase the pressure loss and may be critical for particle processing due to deposited particles in the bends [27]. The fluid temperature T in the channel quickly approximates the wall temperature according to the following equation

$$T(x) \propto e^{-x/l_h} \tag{1.6}$$

with the characteristic length of

$$l_h = \frac{\dot{m}\, c_p}{3.65\, \pi\, \lambda}. \tag{1.7}$$

Combining the channel distance and the mean residence time with the mean velocity $x = \bar{w}\, t$, and solving Eq. 1.6 for the time-dependent temperature change give for the fluid temperature

$$T(t) \propto e^{-t/t_h} \tag{1.8}$$

with the characteristic time of

$$t_h = \frac{\rho c_p d_h^2}{3.65\, \pi\, \lambda}. \tag{1.9}$$

With decreasing channel diameter, the fluid temperature exponentially approaches the wall temperature. Efficient heat transfer is also important for high exothermic chemical reactions to transport the heat away from the reaction zone and to avoid hot spots. The high surface-to-volume ratio is also responsible for the fast heat transfer in microchannels. Additionally, a high surface-to-volume ratio is beneficial for surface reactions such as heterogeneous catalysis, emulsification [28], or transport-limited processes, see Groß [29] for a comprehensive description of surface characteristics.

In the case of chemical reactions in microchannels, the reactor performance is a function of the input and operating variables, the typical rates of the chemical reaction and the mixing patterns. Here, a short summary is given, more details are given in Chapter 6. In microstructured devices, the mixing patterns can be controlled very effectively. Renken and Kiwi-Minsker [30] have shown for continuously operating microstructured reactors, that heat transfer and temperature control for isothermal operation, as well as mixing and residence time characteristics, are optimal for a channel diameter of approx. 200 µm. In comparison to fixed bed reactors, the mass transfer coefficient is higher in microchannel flow, hence mass transfer limited reactions are well-suited for microchannels. They explain the design of microstructured devices utilizing characteristic parameters and dimensionless numbers. Microreactors can improve the performance of rapid and highly exo- or endothermic reactions and ensure process safety. The characteristic dimensions of microreactors are in the range from 50 to 500 µm, which produces a high specific surface and allows an effective heat and mass transfer.

The chemical reaction engineering methodology treats processes on molecular scale ($< 10^{-9}$ m) with strictly empirical, kinetic based models over eddy or particle scale transport processes to the process scale in the range of meters with state balances and dynamic models for control and optimization [31]. The appropriate numerical models range from molecular dynamics MD and statistical models like Direct Simulation Monte Carlo method (DSMC) over Direct Numerical Simulation (DNS) of the governing continuum mechanical equations to Computational Fluid Dynamics (CFD) assisted by micromixing models or empirical models for reactor devices like continuously-stirred tank reactor (CSTR), plug flow reactor (PFR), or phenomenological axial dispersion models. The scale-on interaction of transport phenomena and kinetic phenomena (chemical and biological processes) is crucial for the characteristics and performance of the devices.

In conventional process engineering, the scale-up process is quite elaborate, see Figure 1.3. Starting with laboratory experiments, the gained data are transferred to mini plants with almost the same length scale, but a more accurate representation of continuous flow processing. The pilot plant provides feedback on a semi-industrial scale for the correct design of the final production plant. Wherever possible the process conditions of temperature, pressure and concentrations are kept constant. The geometry is represented along with suitable process parameters in dimensionless numbers like the Reynolds number Re or Sherwood number Sh, which are kept constant or at least partially constant for the process transfer. A good description of this procedure is given by Zlokarnik [32].

The numbering-up procedure increases in the simplest approach just the number of channels, platelets with microstructures or devices to enlarge the capacity, see Figure 1.3. The flow distribution and correct integration must be considered and are the major critical points for a successful implementation. With more channels in a microstructured device, well designed manifolds are extremely important for performance. For example, if the flow distribution in heat exchangers varies by approx. 5 % across the channels, the performance will also be 5 % less than the design value, in most cases the optimum value, see [12]. Therefore, a combined approach was proposed by Klemm [33] and applied during the DEMiS™ project of Uhde GmbH and Degussa GmbH [34], see also Section 6.3.5.

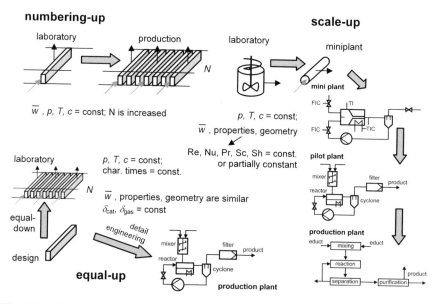

Fig. 1.3. Scaling procedures for chemical engineering purposes with microstructured devices, see also Figures 6.10 and A.4.

Beginning the process or product to be realized, the main effects and parameters are identified for a beneficial miniaturization. These key parameters, like enhanced transfer properties or integration possibilities, are transferred to the laboratory device in an equal-down step. The intended design of an industrial reactor also influences the design of the laboratory reactor. Process and reaction engineering within the constraints of economical design and fabrication guide the reactor development and dimensioning concerning the fluid dynamics, the heat and mass transfer as well as mixing and reaction kinetics. The laboratory design is equal in relation to the industrial design with respect to key geometries, fluid dynamics, mixing, reaction kinetics, and heat management. The experimental results from the laboratory reactor will clearly indicate, which dimensions cause the micro effects (e.g. the gas flow boundary layer δ_{gas}), which effects have to be transferred, and what is the shape and structure of the active surface (e.g. the catalyst layer thickness δ_{cat}), see [12]. From the laboratory results, the detailed design of the pilot reactor on industrial scale is done in the next ″equal-up″ step, see Figure 1.3. The key parameters of the large device, like channel height or wall thickness, are equal to the laboratory equipment, however the geometry of the active surfaces or the fabrication techniques may differ.

For example, the microchannels may have the same cross section or only the same height, but differ in length.

With this procedure comparable process conditions can be realized in the laboratory device to optimize many parameters and conditions quickly, as well as in the production device with high flow rates and controllable safety requirements. First experimental results from the industrial plant will give a reasonable reproducibility in comparison to the laboratory plant [34]. The equal-up design strategy and the consequent implementation of microstructures easily allows the enhancement of the throughput of devices from laboratory to production scale without the risks and costs of the conventional scale-up procedure.

In the near future microstructured plants will help to yield laboratory data and information on the process and the plant [35], which can be directly transferred to the production plant. The established scale-up procedure with advanced simulation tools combines the miniplant with the pilot plant and abridges one or two development phases. With the equal-up strategy, the sophisticated use of micro effects, even the mini plant investigations can be abridged. The simulation of process parameters and equipment geometry is similar for laboratory micro plants with microstructured equipment and production facilities. This leads to an easy scale-up procedure, a more cost efficient process development, and a shorter time-to-market period, however the goal should be to use as many microstructures as necessary, whilst being beneficial, and to build them to a scale that is useful and manageable.

1.4 Main Issues of Successful Microstructures

The benefits of microstructured devices, such as high transfer rates and small volumes, require justification for certain process conditions and substance classes in order to negate the apparent drawbacks, like small flow rates per channel or high fabrication cost. In chemical engineering, fast reactions with high heat loads, as well as mixing sensitive, complex reactions with low selectivity are suitable for application in microreactors. Roberge et al. [36] analyzed the reaction portfolio of Lonza AG, Switzerland, for the suitability of implementation in microreactors and concluded that approx. 50 % of 84 reactions would benefit from continuous processing. However, less than 20 % of all analyzed reactions were suitable for microreactors due to the presence of solids. The criteria for suitable reactions have been the reaction speed, mass transport limitations (highly selective reactions), safety issues, multiphase flow, and scale-up sensitivity [37]. Additionally, chemical reactions demanding external energy are potentially suitable for microchannel processing, like photochemical reactions (μPR project, see [38] (www. microchemtec.de), ultrasonic or microwave reactions, or electro-synthesis. In many applications, heat is transferred within the device to perform the desired process, hence, the heat transfer characteristics must be known and controlled in a suitable manner.

1.4.1 Wall heat conductivity

In microstructured heat exchangers, the heat conduction through the walls parallel to the flow direction is important for the performance and efficiency of the device. The ratio of wall thickness to channel diameter is relatively high in microchannel devices, hence, a considerable amount of heat is transferred through the wall parallel to the flow direction, which lowers the driving temperature difference and decreases the transferred heat amount, see Figure 1.4. The amount of parasitic heat flux has to be considered for highly conductive wall materials like copper, alumina or silicon, and for a low heat capacity flow of the transfer media, such as gases or low flow velocities. More detailed information and calculations can be found in [12] and in [39], where microstructured heat exchangers with a wall heat conductivity of approx. 2.5 W/m^{2}K exhibit the best performance.

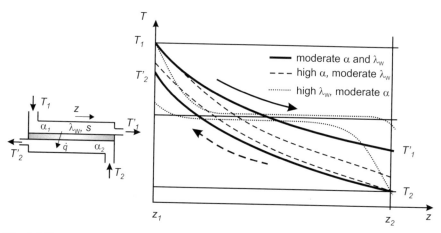

Fig. 1.4. Temperature profile along a counter-flow heat exchanger with different heat transfer coefficients and wall heat conductivity, see also Section 4.3.2.

1.4.2 Pressure loss

For high flow rates, small channels induce a high pressure loss due to the high surface-to-volume ratio. Hence, low viscosity fluids are preferred for application in microchannels due to a tolerable pressure loss. The pressure loss in a channel or device with inlet 1 and outlet 2 is determined from the energy balance, also known as Bernoulli equation, see Section 3.1.3.

$$\Delta p_{12} = p_1 - p_2 + \frac{\rho}{2}\left(w_1^2 - w_2^2\right) + g\left(y_1 - y_2\right) \tag{1.10}$$

In microchannels, the gravitation force often is negligible compared to friction forces.

$$\Delta p_{12} = p_{tot,1} - p_{tot,2} \text{ with } p_{tot} = p + \frac{\rho}{2} w^2 \qquad (1.11)$$

The total pressure p_{tot} is often used in fluid dynamics, where high velocity gradients occur. The pressure loss is calculated for a complex channel arrangement according to

$$\Delta p = \left(\lambda \frac{l}{d_h} + \zeta \right) \frac{\rho}{2} \bar{w}_{ref}^2 \qquad (1.12)$$

with constant reference velocity $\bar{w}_{ref}$. The pressure loss consists of portions from straight channels described by λ (l/d_h) and portions from bends, curves, connections, and other internals described by ζ. The pressure loss coefficient ζ is primarily defined for turbulent flow in devices and can be found in textbooks [40] or handbooks [41], Chapter 6. In laminar flow, the pressure loss in channel elements is still under investigation, see Section 3.2.4 or [42]. In general, the flow below Re = 10 can be regarded as straight laminar flow where no vortices appear and the pressure loss coefficient ζ can be neglected. For high Re numbers, especially for Re $\leq$ Re$_{crit}$, the laminar contribution can be neglected. In the transitional regimes, $10 \leq \text{Re} \leq \text{Re}_{crit}$, a square fit of laminar and turbulent values can serve as a first estimation for the pressure loss.

The hydraulic diameter $d_h = 4A/l_P$ is defined as the ratio of 4-times the channel cross section A and the wetted perimeter l_P. For laminar flow in long straight channels the channel friction factor λ is inversely proportional to the Re number in the channel.

$$\lambda = \frac{C_f}{\text{Re}} = \frac{C_f \, \eta}{\rho \, d_h \, \bar{w}} \qquad (1.13)$$

The channel friction factor λ is proportional to the dynamic viscosity η and inversely proportional to the channel diameter d_h; a smaller channel increases the friction factor. The pressure loss in a channel can be calculated from Eqs. 1.12 and 1.13:

$$\Delta p = \left(C_f \, v \, \frac{l}{d_h} + \zeta \, d_h \, \bar{w} \right) \frac{\rho \, \bar{w}}{2 \, d_h} = C_f \frac{\eta l}{2 d_h} \bar{w} + \zeta \frac{\rho}{2} \bar{w}^2. \qquad (1.14)$$

In long straight channels, the pressure loss Δp depends mainly on the first term in Eq. 1.14, hence, the pressure loss follows almost linearly the velocity, channel length, and viscosity, but is inversely proportional to the hydraulic diameter d_h. In curved channels with internals, the hydraulic diameter influences the pressure loss only marginally, but convective effects occur and determine the pressure loss coefficient ζ. Considering the Re number in a channel for the hydraulic behavior, the pressure loss can be described by

$$\Delta p = \left(C_f \frac{l}{d_h} + \zeta \, \text{Re} \right) \frac{\rho \, v^2}{2} \frac{\text{Re}}{d_h^2}. \qquad (1.15)$$

The pressure loss depends on the Re number with a linear and a quadratic part. With decreasing device length dimensions, the pressure loss is nearly proportional to the square of the Re number. Introducing the volume flow rate $\dot{V}$ through N nearly rectangular, parallel channels

$$\dot{V} \propto N \left(d_h^2 \, \bar{w} \right) \tag{1.16}$$

results in the following equation for the pressure loss

$$\Delta p = \left(C_f \, \eta \, l + \zeta \frac{\dot{V}}{N} \right) \frac{\rho}{2} \frac{\dot{V}}{N \, d_h^4}. \tag{1.17}$$

The pressure loss consists of two parts, where the laminar part (C_f) depends linearly on the volume flow rate. The convective part of internals (ζ) depends on the square of the volume flow rate. The hydraulic diameter plays a crucial role with the inverse dependency of the fourth power. Decreasing the channel diameter and the resulting higher pressure loss can be balanced by either increasing the channel number or decreasing the channel length or volume flow rate.

Eq. 1.17 only accounts for parallel channels with a single flow manifold. The determination of pressure loss and the design of more complex manifolds are described in Section 5.7 and in [43]. The viscosity influences linearly the pressure loss and is less important in short, curved channels with many internals. Within short microchannels, not only is the pressure loss low, but also precipitation or polymerization of various chemicals are possible. The latter is already introduced in industrial production [44] at Idemitsu Kosan C. Ltd., Chiba, Japan, with a capacity of 10 tons PMMA per year. Here, the emulsification and proper mixing of the immiscible reactants are the main issue in designing and operating the microstructured equipment.

For gas-flow microreactors with long, straight channels, the pressure loss is approx. 2.5 times lower than in fixed bed reactors for the same mass transfer conditions [30]. In this calculation, the surface roughness is not encountered, but may play a dominant role in microchannels, see Herwig [45]. If the surface elements reach far into the channel, the hydraulic diameter is constricted, which increases the pressure loss and leads to an earlier transition to turbulent flow, see Kandlikar et al. [46]. To include the surface roughness ε, a constricted hydraulic diameter d_{cf} is defined with the narrowest gap width instead of the mean inner distance d_t of the channel walls

$$d_{cf} = d_t - 2\,\varepsilon. \tag{1.18}$$

The corrected hydraulic diameter explains quite well the observed phenomena in rough microstructured channels [46].

1.4.3 Corrosion, fouling, and catalyst deactivation

In many applications, particle generation and processing are major steps and often regarded with caution [37] in relation to microchannels. Particles may attach to the wall, decrease the cross section, and influence the pressure loss and flow velocity. The particles sticking to the wall may attract more particles and lead to fouling and blocking of the channels. Fouling in microchannels is often besides or is neglected, but now receives more attraction, see [27] and [47]. Aside from proper channel design, an appropriate flow velocity, wall material, and suitable integration of microstructured devices into macrosized peripheral equipment is important for particulate processing. Section 6.5.6 describes the experimental investigation of re-active precipitation of an azo-pigment within a silicon micromixer integrated in a fluidic mount. The mixing channel is 1 mm long on the mixing chip; the product mixture leaves the device through the fluidic mount and a short flexible tube into a beaker. The entire residence time of the product was less than 1 sec and the mixer operated without blockage [48, 49].

Heterogeneous catalysis with high exothermic or endothermic character are suit-able, but also demand effective catalysts due to the high device performance. Catalyst deactivation or poisoning can be critical for long-term operation. Catalytically active wall material may cause chemical reactions or adsorption effects, which are negli-gible in conventional equipment such as stirred vessels. Corrosion becomes relevant in microchannels where the surface roughness influences the flow behavior and the transport characteristics. A corrosion layer of approx. 100 μm is tolerable in conven-tional systems, but fatal for microchannels at this size. The small volumes are bene-ficial for valuable goods or hazardous components to minimize the risks and costs of a plant, which also impacts on the authorization and approval process of production plants and products. A more detailed treatment can be found in Section 4.3.3.

In conclusion, with correct design for the applied process or reaction, microchan-nel devices present many opportunities for application in chemistry, power systems, or consumer products. However, their disadvantages and special requirements must be taken into consideration for successful application.

1.5 Scaling dimensions and issues

In the following, the consequences of size reduction are described for different pro-cesses arranged according to the relevance for material properties and behavior, pro-cesses, and miniaturized devices. This listing sheds light on the actual state of the art, and might be a starting point for gaining knowledge and technical skills in an emerging field of new technology and application. This picture is certainly not com-plete, but shows the complexity of the interwoven fields and gives many hints for miniaturization aspects.

1.5.1 Material properties

Typical length scales in properties and materials behavior in process technology and other areas are shown in Figure 2.1. The behavior of rarefied gases in pipes, channels, and ducts is treated with more detail in Sections 2.3.1 and 4.1.4. Multiphase flow with additional interfacial forces can be found in Sections 3.3 and 3.3. Surface effects are treated in many circumstances within this book and play a major role in transport and systems behavior.

With decreasing length, not only the material properties change, but also the operation and efficiency of unit operations is affected in various ways. The miniatu-rization of chemical equipment mostly emphasizes the length reduction of the main

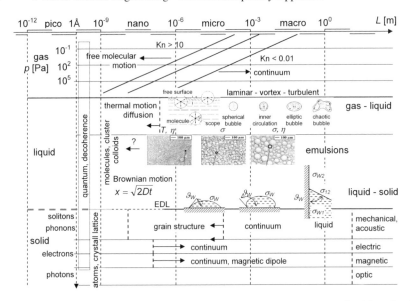

Fig. 1.5. Typical length scales of material properties and behavior in gas, liquid, and solid phase as well as interfacial phenomena. For comparison, the electron sheath of atoms has a diameter of approx. 10^{-10} m, the nucleus of an atom is approx. 10^{-15} m, electrons and quarks are approx. 10^{-18} m in diameter.

dimensions, but constant process conditions like pressure, temperature, or concentration must be obtained. In thermodynamics, these parameters are called intensive state variables, which remain unchanged when reducing the size of the system.

Continuum hypothesis

Decreasing the volume also influences the properties of fluids and the measurement of process parameters like density, velocity, or temperature. Experience treats physical quantities and fluid properties as being spread uniformly over a control volume. Very large control volumes, as in atmospheric systems, show local non-uniformities like depression systems in weather formation, see Figure 1.6 right side. Variations and fluctuations also occur in small systems, where the behavior of single molecules become important.

Measuring fluid properties with an instrument inserted in the fluid affects the volume in the direct vicinity. Normally, the particle structure of the fluid does not influence its behavior. However, with decreasing volume, molecular fluctuations disturb the smooth measurement. If only a few molecules are measured, the single molecule behavior becomes important and varies in an irregular way. Figure 1.6 illustrates the way in which a measurement of density of the fluid would depend on the sensitive volume of the instrument. The fluid can be regarded as a continuum, where the measured fluid property is constant for sensitive volumes, which are small on a macroscopic scale but large on a microscopic scale. A cube with a corner length

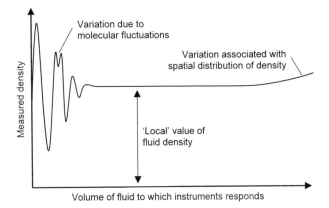

Fig. 1.6. Effect of size of sensitive volume on the density measured by an instrument, acc. to Batchelor [50]. Other macroscopic variables like velocity or temperature show a similar reaction with the observed behavior.

of 10 μm contains approx. $3 \cdot 10^{10}$ molecules of air at normal temperature and pressure. Liquids such as water contain even more molecules, which is large enough to take an average over the molecules. Hence, an instrument with a sensitive volume of 10^3 μm^3 would give a measurement of a local property. A microstructured device with dimensions and process conditions in that range can be treated with continuum methods, which belong to the toolbox of process engineers, see Section 2.4. The limits of the continuum hypothesis are shown in Section 2.3.1; a description of rarefied gases beyond the continuum hypothesis is given in Section 4.1.4.

Flame distance and explosion limits

The typical dimension of microstructured equipment is just below the extinction length or "quench distance" of many fast reactions and oxidations, which is approx. 1 mm. The measurement of the extinction length or distance is standardized in IEC 79-1A and IEC60079-1-1. It is combined with the maximum experimental safety gap MESG, where a flame is stopped to proceed further on. In Table 1.1 typical values of the extinction distance and safety gap width are summarized for oxidation of common fuel materials.

Veser [52] reports on a calm hydrogen oxidation with flame suppression in channels smaller than 500 μm. The author determines three different explosion limits for a stoichiometric H_2/O_2 mixture depending on pressure, temperature, and channel diameter. The main mechanisms are thermal quenching by wall heat conduction and "radical" quenching by kinetic effects where the mean free path of the molecules is in the range of the channel dimensions. Miesse et al. [53] observed small flame cells in flat microchannels. They regard the coupled heat and mass transfer together with the reaction kinetics as a self-organized system, which exhibits characteristic features like flame length or periodicity. For cold walls, the thermal quenching of

Table 1.1. Typical values of extinction length and safety gap width MESG in millimeters for the oxidation of fuels, according to Arpentier et al. [51]

component	extinction distance	gap width
hydrogen H_2	0.585	0.28
methane	2.03	1.14
ethane	1.50	0.91
acetylene	0.64	0.37
ethylene	1.22	0.65
butadiene	1.22	0.84
benzene	1.90	0.99
methanol	1.50	0.92
ethylene oxide	1.17	0.66
ethyl ether	1.83	0.84

the reaction is the dominant mechanism to limit combustion. They give three design rules for successful micro burner layout and complete combustion:

- the wall material should not quench the radicals to keep the gas-phase reaction undisturbed,
- the thermal insulation must be sufficient to maintain a high temperature for combustion,
- the flow and temperature management must ensure a sufficiently high bulk temperature, while preventing a melting of the wall material.

For fuel cell applications, Hessel et al. [22] Chapter 2, show in a review that reactions with strong exo-, or endothermic behavior and high-energy transfer, like explosive reactions, benefit from the high transfer rates in microchannels with precise temperature control.

1.5.2 Processes

Gradients dX/dx

For constant differences of process parameters like temperature, pressure, or concentration, a reduction of the transfer distance increases the gradient, the driving force for transport processes. Therefore, the corresponding transfer flux is increased and the equilibrium state is achieved more rapidly. Surface effects have to be considered and can be dominant opposed to volume effects such as gravity or momentum. Not only are the main transport processes (mass, momentum, and energy) increased by higher gradients, but also less predominant processes become interesting, and coupled processes gain importance [54].

The thermoelectric effect is exploited for sensing (thermocouple, NiCr-Ni, 0.04 mV/K), for cooling (Peltier element) or for the generation of electrical energy (Seebeck element, approx. 500 µV/K), see Section 7.2. The generation of a species current under a temperature gradient is called thermodiffusion, the Soret effect, see

Section 7.4. The Soret coefficient D' has an order of 10^{-12} to 10^{-14} m^2/sK for liquids and of 10^{-8} to 10^{-10} m^2/sK in gases [55] Chapter 13. The species concentration gradient due to pressure difference is called pressure diffusion, which was originally investigated for isotope separation [56]. The fabrication of the separation nozzles for pressure diffusion combined with centrifugal forces is one of the birth places of microfluidic technology and of the LIGA technology (Lithographie, Galvanik, Abformung) process [57]. Earlier work in this area can be found in [58] and [59]. The exploitation of these processes is described in more detail in Section 7.5. A summary of the historical development of microsystems fabrication can be found in [60].

Aside from coupled transport phenomena, transfer processes of species and energy benefit from shorter distances. Rapid cooling or heating can generate temporal temperature gradients of more than 10^6 K/s for sensitive products, which is described in Section 6.4. Rapid mixing leads to high concentration gradients in a short period and quick homogenization of concentration gradients, as described in Chapter 5. The selectivity of fast, parallel chemical reactions can be increased with rapid mixing, which is the major issue for microreactors aside from high heat transfer rates, see Chapter 4.

Characteristic times

Length scaling often correlates to time scaling of the relevant processes. The shorter the length, the shorter the characteristic time of transport processes will be, and the higher the frequencies of changes, pulsations or fluctuations. Characteristic times have been treated in Section 1.3, time scales are discussed in Section 6.3. Microstructured devices are well suited for rapid processes, but are also very sensitive to process changes. Within micromixers and microreactors, several characteristic times of mixing t_m, residence time in the device t_P and other process times like reaction kinetics have to be adjusted to yield an optimal result. This is one of the key issues in successful design and application of microstructured devices and plants.

Fast nucleation and controlled growth of nanoparticles

The reduced capacity and inertia within a small internal volume allows higher exchange rates for heating, cooling, and mass transfer, see also Section 1.3. Fast mixing of two nearly saturated mixtures leads often to a high supersaturation of the mixture and particle generation. The supersaturation S is measured as the ratio of partial pressure p_V to saturation pressure p_S.

In gaseous flow, the particles (liquid or solid phase) are called aerosols, like fog in humid air or spray atomization. Mixing of a warm vapor-gas mixture close to saturation with a cold gas stream leads to homogeneous condensation of small droplets. This process is also called fog generation and is a well-known phenomenon in meteorology. Fast mixing produces a homogeneous supersaturation resulting in small particles with narrow size distribution. Aerosol generation in microchannels is a promising option when small droplet generation is used as a starting step for further processes, see Section 6.4.

In liquids, the particulate flow of a suspension is often combined with crystallization. Besides exceeding solubility limits in liquids, often chemical reactions are involved in the precipitation process. Two solvable components in solution react to a third component with low solubility, which precipitates after the reaction.

The precipitation process is determined by the nucleation of the dissolved component and the growth of resultant particles. A high supersaturation generates many nuclei in the mixture, see Figure 1.6, right side. The faster the mixing process, the higher the supersaturation and the more nuclei are formed. The homogeneous nucleation of the dissolved component is followed by the growth of the particles, which is determined by the mass transfer from the bulk to the nucleus. Experimental investigations of the precipitation of barium sulfate in aqueous solution have shown that the particle size distribution is strongly dependent on the mixing intensity [61]. Industrial production of particulate products has already been introduced with pigments, see Clariant patents [62, 63]. This topic is treated in more detail in Sections 6.4 and 6.5.

In general, fast mixing in microchannels can be used for the production of nanoparticles in droplet flow, emulsions, as well as with rapid chemical reactions and succeeding precipitation in bulk flow. The main problem is the sticking and adhesion of particles on the wall, see the fouling topic in Section 4.3.3.

Distillation and rectification

Subsequent evaporation and condensation enriches the low-boiling component in the vapor phase and provides a concentration difference between gas and liquid phase. This is the fundamental mechanism of distillation and rectification, the most frequently applied separation techniques in process technology [64] Chapter 7. The coupled heat and mass transfer is performed in small laboratory equipment, such as the micro distillation apparatus as well as in huge rectification towers like 60 m the high cryogenic rectification columns of air separation units. A detailed treatment of the underlying transfer processes can be found in textbooks [65, 66] or concise handbooks [41]. Sowata and Kusakabe [67] demonstrated the vapor-liquid separation of a water-ethanol-mixture on a chip (20×40 mm^2 footprint area) with mixer-settler arrangement. Both streams get in contact in a Y-shaped mixer and the bubbly flow streams through a meandering channel. At the outlet, the two phases are separated by gravity in a separation chamber (8×8 mm^2 area). The experimental data indicates the equilibrium state for the gaseous phase and a liquid concentration of 0.65 mole fraction of methanol, which represents a stage or plate efficiency of approx. 65 % for the liquid side. Better mixing in the meandering channel may enhance the plate efficiency and lead to better separation. Hence, a proper mixing process also enhances related separation processes, due to the rapid achievement of thermodynamic equilibrium.

Fink and Hampe presented a design study of a distillation column to illustrate the setup of an entire microplant [68]. The experimental investigations show that a flow rate of approx. 20 ml/h has been realized with 2.5 theoretical separation plates. A drawback is the deteriorating influence of even small temperature changes that lead to operational disturbances and uncontrolled concentration changes [69].

In an own design study [70], a distillation chip with various geometries was fabricated and tested by evaporation and condensation of a water-ethanol mixture. The chip with a meandering separation channel is made of an anisotropic KOH-etched silicon wafer covered with two Pyrex glass lids, see Figure 1.7. From the heated side (hot water, 93 °C) the mixture partly evaporates and condenses on the opposite side of the channel. The McCabe-Thiele diagram on the right side of Figure 1.7 explains the stage-wise operation of a rectification column.

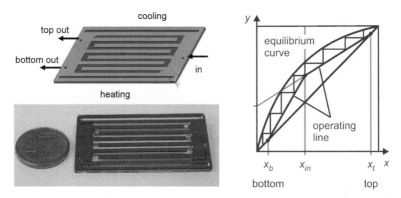

Fig. 1.7. Left top: Geometrical setup of a distillation / rectification chip (dimensions of $20 \times 40 \times 1.6$ mm^3) made of a silicon wafer with anisotropic KOH-etched channels and covered with two glass lids; bottom: image of a rectification chip with meandering separation channel; Right: Stage-wise operation of a rectification process according to the McCabe-Thiele diagram.

The operating line in the upper part of the diagram describes the enrichment section of the separation column, where the low-boiling substance enriches in the vapor phase. A theoretical distillation stage or tray is described by the horizontal line for the condensation of the rising vapor and by the vertical line for the evaporation of liquid from the tray. In the ideal case of achieving the equilibrium state, the lines touch the equilibrium curve, indicating the number of theoretical stages or trays. The graphical treatment of these operations is faster than analytical or numerical calculations and gives a better physical insight into the process. More detailed descriptions are given in [66, 71, 72].

The first experimental results indicate low mass flow rates and poor thermal insulation to the ambient. From the feasibility study, the following main issues have been identified:

- The fluid transport can not rely on gravity, other principles must be employed, i.e. the capillary transport or external force fields like centrifugal fields from equipment rotation [73].

- The contact between liquids of different concentration must be avoided, which also limits the minimum size of the gaps for evaporation, condensation, and vapor flow.
- The fabrication of suitable geometries to guide the liquids and vapor currents is a crucial issue. Besides the silicon techniques, alternative fabrication techniques like polymers or metal foil structuring should be investigated, see [69]. Additionally, the surface properties play an important role and should be used to assist the phase separation.
- An internal numbering up may help to increase the low volume flow, which is a considerable drawback of the actual devices.

These points are generally valid for multiphase flow transport problems and coupled heat and mass transfer. A distillation unit as part of a methanol production plant was presented by Tonkovich et al. [74] at the IMRET9 conference. The aim is the enhanced heat and mass transfer by contacting thin vapor and liquid films with a correspondingly reduced height of the equivalent theoretical plates (HETP). It was shown that the HETP value can be reduced from over 60 cm in a conventional distillation column to approx. 1 cm in a microchannel distillation unit. Major problems have been reported on the temperature control over short distances and appropriate vapor-liquid phase separation at the end of a stage. Defined surface characteristics, like suitable wetting angle or assisting capillary structures, can support the phase separation. The application of a hydrophilic liquid phase channel and a hydrophobic vapor phase channel are presented by Iwatsubo et al. [75] at the IMRET9 conference. The device consists of glass and PMMA substrate, where the hydrophilic nature of glass surfaces was modified by silanization to be hydrophobic. The phase separation was observed with a nitrogen-methanol/water system, which was partially evaporated. Hot and cold water streams produce the temperature gradient to achieve evaporation and condensation of the methanol/water system.

However, the potential of a high specific area provides a high transfer area and high exchange rates between the phases. A short distance between the phases and concentration differences allows a rapid attainment of the thermodynamic equilibrium and leads to smaller equipment, but the low throughput and complex fluid management must also be considered.

Absorption and desorption

Similar to the rectification and distillation process, absorption is a coupled heat and mass transfer process, where a gaseous component is dissolved in the liquid phase [71]. Desorption or "stripping" is the reverse process, where a dissolved component is released from the liquid phase into the gaseous phase. The state changes and stage-wise operation is displayed in the operation diagram with the equilibrium curve, given by Henry's law, and the operating lines for desorption and absorption.

The orientation of the operating line depends on the process conditions and is explained in [76] Chapter 1, or in [71]. In absorption processes, a component from a gaseous mixture is step-wise accumulated in a liquid and partially chemically bound in the liquid by additives. For the successful operation, a high mass transfer between the phases rapidly leads to thermodynamic equilibrium. Prominent examples are synthesis gas cleaning in the chemical industry (CO_2 removal) or CO_2 enrichment in water for the production of sparkling water.

The opposite process of absorption, the stripping of a dissolved component from a liquid (toluene in water) was investigated by Cypes and Engstrom [77]. The liquid mixture flowing in a silicon channel was treated with a superheated steam entering from a second channel through a perforated wall. After leaving the channel the bubbly flow was separated in a chamber and analyzed. The comparison between a microfabricated stripping column and a conventional packed tower indicates a transfer rate, which is nearly an order-of-magnitude higher than conventional packed towers, indicating a high specific interface. This large interface enhances the convective mass transfer process by reducing the diffusive length.

An absorption heat pump includes both absorption and desorption integrated into a process cycle, see [76] Chapter 6. A further device including both processes is the closed absorption and regeneration cycle to reuse the washing fluid. An absorption heat pump process with microstructured elements was presented by Ameel et al. [78] and Stenkamp et al. [79], but no actual applications are known to the author.

Adsorption

In adsorption, volatile molecules are arranged and attached to a solid, often a porous surface from a gaseous or liquid phase. A high surface-to-volume ratio is essential for an efficient adsorption process, however, this high ratio also can cause unwanted side effects and material interaction between the fluids and the wall. Catalytic reactions or component adsorption belong to these effects in microstructures, which are not relevant in conventional batch processing. In conventional equipment porous pellets or beads provide the porous surface for adsorption. The pellets are arranged within packed beds in larger vessels. Thus, a miniaturization of the entire plant results in a low volume, which is quickly loaded and, therefore, detrimental to the amount of adsorbed species. Smaller pellets give a higher specific area per volume, but also increase the pressure loss of the volume flow through the bed. The calculation and design procedure of the adsorption isotherm and the Knudsen regime for transport processes in pores is not treated here. Interested readers are referred to [41] Chapter 16, [65] Chapter 10, [80] Chapter 13, and [66] Chapter 12.

In small pores, the molecules interact more with the wall than with themselves. This process is described by the Knudsen diffusion. For perfect gases, the validity of this regime depends on the total pressure and the pore diameter. For analytical purposes, the faster loading results in a fast response time for chromatographic analysis. Loading, detection, and cleaning of the chromatographic separation column can be achieved more quickly than in conventional systems. This allows a fast signal response to variations of input concentration, which is very important for the proper control of modern chemical plants.

The fabrication methods and unique systems integration opportunities of microsystems technology enhance the system performance, as has been demonstrated by micro gas chromatography, see Figure 1.8. A gas chromatograph on a silicon wafer was already proposed by Winde and Heim [16] in 1988, meanwhile, miniaturized gas chromatographs are commercially available [81]. The chromatographic separation column is fabricated with a combined bulk and surface micromachining process on a silicon chip. The silicon-glass chip of the separation column has a footprint of 1 cm^2. The channel has a length of 0.86 m and a diameter of 60 μm. This column is heated to 200°Cwithin 20 seconds, using an electric heating power of 5 W. Because of the low thermal capacity, it cools rapidly back down to ambient temperature. This results in a measurement cycle shorter than 60 sec and lower carrier gas and electrical power consumption.

Fig. 1.8. Micro gas chromatograph; left: separation column on a silicon chip, covered with a glass plate; right: IC plate and analytical equipment of the micro-GC. See also Figure A.2, courtesy of SLS micro technology [81].

The development and characteristics of an ion-exchange separator and detector is described by Kang et al. [82]. The chromatographic performance exhibits a high efficiency measured by an electric conductivity detector.

Extraction

Based on the employed phases, it is distinguished between liquid-liquid and liquid-solid extraction (leaching) [41] Chapter 15, [71] Chapter 3, [72] Chapter 1. The extraction process is based on the different solubility of components A and B within the extraction fluid C. Often, the low solubility of component A in component C leads to a two-phase flow with mass transfer at the interface between the phases. The mixture $(A + B)$ is put in contact with the extraction fluid C and well mixed over a certain time to establish thermodynamic and chemical equilibrium between the phases. The residence time of the fluid mixture with the mass transfer determine the enrichment of component B in C. The succeeding phase separation leads to a purification of A and an enriched extractant C.

$$(A+B)+C \Rightarrow A+(B+C) \qquad (1.19)$$

Many applications of extraction processes can be found in the pharmaceutical industry, for various foods and detergents, vitamins, or in petrochemistry. The Nernst distribution describes the equilibrium concentration between the three phases, see Figure 1.9. Material data and properties can be found in the literature [72] Chapter 1.9 or [71].

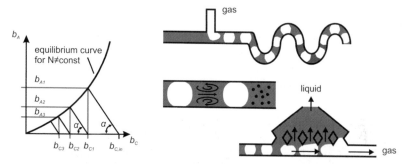

Fig. 1.9. left: Loading b_A of the component B in A over the loading b_C of component B in C; step-wise cross flow extraction with four stages, equilibrium line is given by the Nernst distribution N; right: geometrical examples of micro contacting equipment, (adopted from Tice et al. [83] and Günther et al. [84]), contacting and mixing in a serpentine channel, flow characteristics in bubbly flow, and separation by geometry and surface properties using capillary forces.

Ehrfeld et al. [85] and Klemm et al. [69] describe the extraction process in microstructured equipment. The main problem in microstructured extraction devices is the phase separation as the final step of the process. Recently, Okubo et al. [86] presented the rapid extraction process on a chip. TeGrotenhuis et al. [87] developed micro devices for transport processes in solvent extraction. Kusakabe et al. [88] investigated the flow details and transport processes within a droplet in a micro contactor for solvent extraction. Their comparison with experimental data shows a good correlation for the extraction of ethyl acetate from n-hexane to water. Specific problems to solvent extraction are the fluidic transport and phase separation after the mixing. In conventional equipment the driving forces for the fluidic transport are the density difference, gravity, external pumps as well as capillary effects. In microstructures, the pressure driven flow and capillary effects are the predominant fluid transport principles. A design study for an extraction device with microstructured elements is shown in Figure 1.10. The bottom right image shows the chip layout for contacting, mixing in a meandering channel and separation (settler) of a liquid-liquid flow. The connections to further processing are sketched, such as the pump wheel. In the following layout, the fluids are guided perpendicular to the chip plane. In the top right image, the radial arrangement of different mixer-settler chips is illustrated. On the left side, the chip blocks are displayed in a side view with intermittent pump wheels for fluid transport and a connecting pump shaft.

An example for the combination of microstructures with macro devices for fluidic transport is given by the mixer-settler chip with radial arrangement and the step-wise transport by a pump wheel as displayed in Figure 1.10. The general setup was

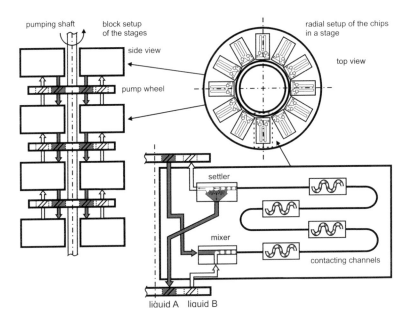

Fig. 1.10. Micro-macro integration of a mixer-settler extraction apparatus. A mixer-settler chip for two-phase flow is displayed on the right side, bottom. Twelve of these chips are radially integrated in a separation stage. Inlet and outlet connections are located near to the inner circle to be supplied by a pump wheel. This pump wheel turns between two stages and is adequately sealed to the separation stage. All pump wheels are fixed on a central shaft.

adopted from conventional extraction equipment with rotating internals for mixing and fluid transport like the Scheibel- or the Kühni-column, from Schlünder and Thuner [71] Chapter 3. For the design and fabrication, much effort is spent on packaging the chips in the radial mount and the sealing between the block and the pump wheel.

The integration of microstructures into macro equipment is necessary in order to utilize the benefits and possibilities of both areas and additional emerging effects. A suitable guideline for the design process is: Not only as small as useful for the designated application, but also as much miniaturization as necessary for the actual process.

Free convection

The heat transfer due to free or natural convection is induced by the thermal expansion and density differences of a fluid under a temperature gradient. The heat transfer between two plates with different temperatures is characterized by the Grashoff number Gr, the ratio of the volume force to the viscous force.

$$\mathrm{Gr} = \frac{g\,\beta_p\,s^3\,(T_{w1} - T_{w2})}{v^2} \tag{1.20}$$

with the thermal expansion coefficient

$$\beta_p = -\frac{1}{\rho} \left(\frac{\partial \rho}{\partial T}\right)_{p=\text{const.}} \tag{1.21}$$

Depending on the temperature difference between the walls and the distance s between the walls, the fluid will circulate in the gap. A prominent example of natural convection is the Bénard convection in a horizontal layer, which is heated from below, see [55] Chapter 14. If the layer or the gap is too thin, the viscous forces will dampen the convection and the heat is transferred solely by conduction. If the Rayleigh number Ra (= Gr Pr) is smaller than $3 \cdot 10^8$, the free convection flow is laminar and ruled by the viscous forces, see [89] Chapter 3 and [90] Chapter 5. For Ra numbers lower than 175, the heat transfer augmentation by natural convection is smaller than 1 %. In a gap filled with air or water, the influence of natural convection on the heat transfer is smaller than 1 % for a gap width smaller than 8 cm or 4 cm, respectively. Therefore, natural convection may be negligable for most circumstances in micro process technology, however, it is the major cause for heat losses of equipment to the environment. For the measurement of incline or acceleration, a sensing device with fast temporal response employs the temperature and flow field around a heated wire with free convection [91].

1.5.3 Devices

Thermal insulation length

In small devices, the thermal insulation of devices with different temperatures is very difficult. One major problem of some modules in the "backbone system" (microchemtec, Germany [38]) was the insufficient thermal insulation of the different reactors with operating temperature from below 260 K to over 470 K. Technical polymers like PEEK ($\lambda = 0.25$ W/mK) can be used as the construction material for low temperature differences up to 100 K. For higher temperature differences or very short distances, multilayer super-insulation from cryogenic technology can help to overcome the difficulties [92] Chapter 7. Although the multilayer insulations are very expensive and difficult to apply to complex shapes, they offer the best performance of all technical insulations ($\lambda \approx 10^{-8}$ W/mK). The multilayer insulation consists of evacuated alternating layers of highly reflective material (aluminum foil or copper foil) with low-conductivity spacers, like glass or polymers, see [93]. Their extremely low thermal conductivity depends on the reduction of all three modes of heat transfer: radiation, gaseous and solid conduction. Insufficient thermal insulation between small elements is the main reason, why Peterson et al. [94] give the size limits of small regenerative heat engines as a few millimeters. Small heat engines will suffer from an insufficient heat recovery in the regenerator and from difficulties to maintain the necessarily high temperature difference.

High specific area for transfer processes

When miniaturizing the length, the area is decreased proportional to the square of the length, L^2. The area plays a major role as a cross section for transport fluxes like mass, volume, or energy flow or as an active area, where transfer or transformation processes take place. For a constant flow rate like fluid velocity, species transport, or heat exchange, the total amount of the quantity is decreased proportional to L^2 with smaller length scale. For a constant driving force, the area specific flux remains constant. The pressure loss in channel flow increases with smaller cross section, see Chapter 3. The order-of-magnitude of the pressure loss increase depends on the variables, which remain constant, see Section 1.4.2. With a constant absolute flux like energy transfer (in W), the specific flux dramatically increases (in W/m^2). For example, the energy load in modern microelectronic devices increases with their performance improvement, which already leads to heat loads of above 100 W/cm^2 [95]. Finally, the small cross section of immersed bodies shows low fluidic resistance, which appears in low settlement velocities and almost Lagrangian particle behavior (massless particles).

Many transport processes take place at phase boundaries, which have the form of films, bubbles, or droplets in emulsions, dispersions, or foams. With high specific area and high gradients, the transport rates are increased and equilibrium state is reached more rapidly. Temperature and concentration differences are equalized sooner, which plays a prominent role in mixing and heat transfer. Considering simple geometric arrangements, the specific surface-to-volume ratio $a_V = A/V$ is inversely proportional to the characteristic length,

$$
\begin{aligned}
a_V = \frac{A}{V} &= \frac{4\pi r^2}{4/3\,\pi r^3} = 3/r \quad \text{for spherical bodies,} \\
&= \frac{l_P \cdot L}{A \cdot L} = 4/d_h \quad \text{for a single straight channel,} \\
&= \frac{l_P \cdot L}{A \cdot L/\varepsilon} = 4\varepsilon/d_h \quad \text{for a microreactor block with porosity } \varepsilon = A_C/A_{tot}.
\end{aligned}
\tag{1.22}
$$

The specific surface of spheres is relevant for the mass transfer around small particles or bubbles in suspensions or emulsions. Heat transfer or catalytic processes are determined by the specific surface in straight channels. The porosity ε, the ratio of the channel volume to the entire volume, also characterizes the entire device and the possible volume flow rate. The smaller the fluidic structures, the larger the specific interface and, therefore, the active area for transport processes, e.g. for heterogeneous catalytic reactions with functionalized walls. This effect will be shown in this section for major separation unit operations, together with other specific miniaturization effects.

The volume is proportional to the cubic length L^3 and is directly related with the mass via the material density. A lower mass reduces the capacity of elements in dynamic processes (see Section 1.3) and corresponding higher frequencies. Due to the small overall size, the most likely implementations of microstructured devices can be found in distributed power systems and fuel cells for automotive and mobile systems. A low volume in process equipment is useful for the handling of toxic and hazardous products, as well as for expensive, or valuable goods. With smaller plants, a distributed on-site production is possible, i.e. the production at the point of use or point of care (in medicine or diagnostics). The production-on-demand of small amounts of chemicals is much easier and allows a more flexible production, see [13] Chapter 1. At the same time, questions concerning correct and safe operation, quality control, and documentation must be answered. Due to the small internal volume, in-situ preparation of hazardous and explosive chemicals have a high potential to process in microchannels. Considerations for these potential applications are the authorization demands and adapted approval procedures, as well as the user-friendly start-up, operation, and shut-down of the devices and plants.

Filter and membrane processes

The variety of membrane processes covers a broad area ranging from food technology over potable water generation to blood and serum cleaning. Melin and Rautenbach [96] give a good introduction and overview of membrane types and specific applications. Conventional membrane processes already work with hollow fibers with a diameter of 40 to 400 µm and a high specific surface $a = A/V$ up to $20\,000$ m^2/m^3, [64]. These hollow fibers reach a specific permeate volume flow (fluid passing through the membrane) of approx. 300 m^3/h per cubic meter equipment volume. Microstructured membranes from tantalum, or palladium and their alloys are used for carbon monoxide transformation in fuel cell applications [22] Chapter 2. In microchannels, a thin membrane is used to stabilize the interface between gas and liquid phase and to control the mass transfer between the phases. A interesting field in membrane technology is the area of liquid membranes, where a liquid inside a porous structure absorbs components from one side and transfers them to the other. They exhibit good product selectivity, low pressure difference, and high transfer rates [96] Chapter 2. Besides the appropriate material properties, a crucial factor is the protection of the enclosed liquid against the pressure difference between the two membrane sides. The fabrication of the structures for liquid membranes can be assisted by micro fabrication technology.

Molecular distillation

In molecular distillation and vacuum rectification, the equipment length between the evaporating and condensing sides plays a major role, where the mean free path Λ of the molecules is larger than the distance between evaporating and condensing surface due to low operating pressure. These processes usually separate liquid mixtures, which are thermally sensitive, like vegetable oils, vitamins, or fatty acids (see [97] for the historical development and [98] Chapter 4 and [99] Chapter 10). The Knudsen regime serves for a gentle thermal separation and low product degradation. A shorter distance requires a lower process vacuum, less energy consumption, and will probably result in higher product purity. The design length is limited by the film thickness of the often high viscous liquids (approx. 1 to 5 mm). Although an application has not yet been reported, a falling film microstructured device under vacuum operation would be an interesting test object.

Cyclone

The density difference between dispersed particles and the surrounding fluid leads to separation within a force field such as gravitational or centrifugal forces. The latter are generated in a rotating system like a centrifuge or a cyclone [80]. In a mechanical centrifuge, the acceleration is proportional to the radius, hence, miniaturization will reduce the radial acceleration and, therefore, the separation efficiency. A hydro- or gas cyclone is a mechanical separation device without moving parts for dispersion of solid - gas or solid - liquid mixtures. The separation of immiscible liquids with different densities is possible, however, other separation processes like settling, filtering, or coalescing are more effective for this purpose, see [80] Chapter 6.

The vortex in a cyclone induces a high tangential velocity near the vortex center, which decreases with the radius r. The tangential velocity of a viscous vortex, like an Hamel-Oseen-vortex [100], scales with r^{-n}, where n ranges from 0.5 to 0.8, see [80] Chapter 6. The centrifugal acceleration is proportional to the square of the tangential velocity w_t divided by the radius r. The separation factor z of the cyclone is the ratio of the centrifugal force to the gravitation force and is expressed by

$$z = \frac{w_t^2}{r\,g} = \frac{\text{const.}}{r^2\,g}.$$
(1.23)

The separation factor increases in a cyclone with a smaller radius, hence, small particles are more likely to be separated by cyclones with a smaller inner diameter. In a feasibility study, Kockmann et al. [70] fabricated a micro cyclone with an inner cyclone diameter of 1 mm from PMMA and separated an aqueous SiO_2 suspension (mean diameter approx. 4.5 µm, particle density approx. 2 500 kg/m³).

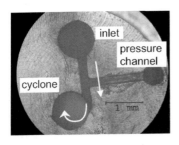 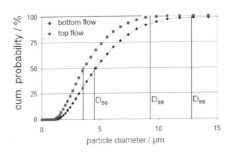

Fig. 1.11. Left: Cross section of the micro cyclone with inlet and channel for pressure measurement; Right: Cumulative probabilities of the particle diameter for the top and bottom flow of the cyclone, small particles are collected in the top flow.

The suspension tangentially enters the cyclone chamber and, driven by the centrifugal force, large particles flow to the outer radius and are collected at the bottom chamber outlet. Small particles are collected in the center and leave the cyclone through the top outlet. The separation efficiency of a cyclone can be seen in Figure 1.11 at the D_{98} value (9 µm and 13 µm for the top and bottom flow, respectively). Cascading of many micro cyclones will give a higher separation factor, but requires a higher pressure difference. The cyclone process can be extended to pressure diffusion, a coupled thermodynamic process described in Section 7.5.

In two papers, Ookawara et al. [101, 102] presented the centrifugal separation of an aqueous suspension flow in curved rectangular channels (200×150 µm^2 cross section, 30° and 180° curves, 20 mm radius). With a pressure loss of approx. 200 Pa, the typical Dean numbers Dn were between 30 and 45 with related Re numbers from 1 400 to 2 200. The experimental results and numerical evaluation by CFD simulation showed a good separation of particles larger than 15 µm from the fluid. Counteracting the centrifugal force, the Dean vortices pushed the particles inwards by friction forces and disturbed the separation flow, hence, a shorter bend may be more effective. Complementary to separation of solid particle dispersions, cyclones are employed to separate emulsions or gas-liquid dispersion, In these applications, the surface characteristics of the wall materials and the fluids have to be considered for successful design.

In the product catalogue of the German microchemtec project [38], three companies (little things factory, Ilmenau [103], Ehrfeld BTS, and mikroglas, Mainz) offer micro cyclones, which fit into their series of process equipment.

Pumping

In general, microfluidic systems typically have small flow rates, short residence time, and small hold-up in the devices. This is also valid for micropumps, which have now, after a sufficient history of development, a reasonable variety of types [104]. Like conventional pumps, the micropumps can be classified into two different classes, the displacement pumps with reciprocating or rotating elements, and the dynamic pumps using centrifugal, acoustic or electrical fields [105]. Due to the small dimensions, the flow rate is low, but electro-hydrodynamic or magneto-hydrodynamic effects can be employed.

A pump is mainly characterized by the flow rate and the induced pressure difference for micro process applications. The frequency of displacement pumps is directly coupled with the flow rate up to a certain value, where sealing losses and dynamic effects, like fluid acceleration, limit and decrease the flow rate [104]. The flow rates of a micro diaphragm pump decrease almost linearly with increasing counter pressure. Maximal flow rates of approx. 35 ml/min (0.78 kPa pressure difference) of air and 16 ml/min (4.9 kPa pressure difference) of water have been reported [105].

Electro-osmosis employs the negative charge of the wall and the positive charge of counter ions in the fluid. An electrical field generates the motion of the counter ions near the wall and, therefore, of the entire fluid in the capillary. The electro-osmotic effect increases with smaller channel diameter, which have characteristic dimension of approx. 1 µm and smaller. High pressure differences, up to 2 MPa with water, have been realized for low flow rates of 3.6 µl/min. High flow rates up to

33 ml/min have also been measured, which indicates that the main application of electro-osmotic pumps is in dosing and analytical applications [105].

The rotating gear pump of HNP Mikrosysteme GmbH [106] belongs to the displacement pumps for dosing and pumping of liquids in analytical and production applications. The pumps, with an internal volume from 3 to 48 μl, deliver a flow rate up to 17 l/h against a counter pressure of 30 to 50 bar.

Device integration

High-throughput screening is a valuable tool for catalysis research [22] Chapter 3 and chemical synthesis [107], however, the extremely low volume in microstructured equipment also affects many other processes. The research and development of new chemical products is assisted by high-throughput synthesis of classes of chemicals or reactions. Data handling and analysis is a major issue in this research. With small devices employing process conditions similar to larger equipment, pilot plants can assist process development and process route discovery with fast feedback for process parameter variations and equipment optimization.

1.6 Actual Applications and Activities

Currently, there are many applications of microstructured equipment worldwide in various technological fields. Hessel et al. give a comprehensive overview of chemical applications [13] and for power systems [22]. A good representation is found at the relevant international conferences like AIChE/DECHEMA - IMRET on microreaction technology [108, 109], ASME - ICNMM on nano, micro, and minichannels [110] or the German ACHEMA 2006 conference, which included over 50 scientific contributions on micro process engineering [111]. The ACHEMA organization team publishes trend reports including new products, technologies, and possibilities relating to the main topics of the conference under the heading of process intensification and microtechnology (www.achema.de/Trendreports.html). In the following, a short overview is given of the main research initiatives and some industrial activities in various countries and regions. The listed projects and activities are not complete and highlight just a selection.

1.6.1 European activities

In Germany, the Federal Ministry for Education and Research (www.bmbf.de) has set up the Micro Process Engineering research program for three years (2005-2007) with a total budget of approx. 16 million Euro [38]. Six projects are defined with the objective to transfer microstructured devices into industrial production. These research projects mainly target small and medium companies, which have limited access to and resources for microstructured devices, which are often barriers preventing the first step into a new technology. In 2007, the Federal Ministry for Education and

Research has launched 6 projects in regional clusters to introduce microstructured equipment in university education of chemists and chemical engineers.

Besides these research projects, several other activities in Germany target at the successful application of microstructured equipment. Specifically, modular plant setup and integration of several functions, sensors, and control equipment is the main focus of the developments. A good example is given in Figure 1.12 with the relatively new development of Syntics GmbH, Bochum, Germany (www.syntics.com).

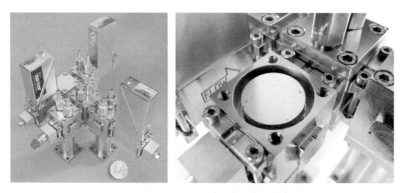

Fig. 1.12. Left: Syntics plant layout with a modular setup of flow sensors, mixers and heat exchangers. Right: Detail of a turbulence mixer with cooling chamber and housing, see also Figure A.5.

A similar modular setup on a chess-board like base plate is produced and promoted by Ehrfeld Mikrotechnik BTS (www.ehrfeld.com), see Figure 1.13. The acronym BTS stands for Bayer Technology Service, which is partner in a BMBF research project [38]. The development of the high-throughput device in Figure 1.13, right side, mirrors the trend to higher flow rates for industrial production applications.

Karlsruhe Research Center, IMVT (www.fzk.de/imvt), and IMM Mainz (www.imm-mainz.de), Germany, are long-term players in micro process engineering. Recently, both have presented successful applications of microstructured devices in industrial production. The Karlsruhe Research Center, IMVT, has developed, fabricated, and installed a combined microreactor with heat exchanger at DSM, Linz, see Figure 1.14. The microreactor has a length of 65 cm, a weight of 290 kg and allows a throughput of 1 700 kg/h of liquid chemicals. The interior consists of thousands of microchannels for mixing and heat transfer of several hundred kilowatts from the exothermic reaction. German companies fabricating and distributing microstructured equipment are named by Bayer and Kinzl in [112]; a complete list can be found under www.microchemtec.de [38].

In the EU framework program FP6, the IMPULSE project is a research initiative with 18 participants from industry, research institutes, and universities [113]. The IMPULSE project (www.impulse-project.org) is the acronym for "Integrated

Fig. 1.13. left: Modular microreaction system, sample setup for single step reaction with heterogeneous catalysis; Right: High-Flow rate Microreaction Modules: Heat exchanger and mixer for nominal throughput of 1 000 l/h (Ehrfeld Mikrotechnik BTS www.ehrfeld.com).

Fig. 1.14. Combined microreactor with heat exchanger (290 kg weight), throughput 1 700 kg/h liquid chemicals (source Karlsruhe Research Center, for DSM Linz, www.fzk.de, Press Release 13/2005).

Multiscale Process Units with Locally Structured Elements". The project aims at effective, targeted integration of innovative process equipment (microreactors, compact heat exchangers, thin-film devices and other). An analysis of their fundamental basis is leading the new, hybrid approach with multiscale design of large-scale systems with small-scale inner structuring. The project coordinator Prof. Matlosz, ENSIC, INPL-Nancy, France, gives more details on this comprehensive and ambitious project in [113].

Another research project on microreaction technology in the European framework program FP6 is NEPUMUC, an acronym for New Eco-efficient Industrial Process Using Microstructured Unit Components (www.nepumuc.info). The overall objective of the NEPUMUC project is the knowledge-based development of an intensified multi-purpose process for highly sophisticated nitration reactions using microreaction technology with focus on environmental issues and safety [114]. The consortium consists of 8 partners from France, Germany, Poland, Romania, Switzerland and The Netherlands. The novel approach for the design of the microreaction units and the entire process is based on highly innovative modeling work using different tools and strategies. New microstructured sensors and sampling devices for analysis will be developed and adapted to new types of microfluidic reactors that are designed and fabricated in glass for high-throughput nitration processes. A flexible automation system is developed to receive process data from the microfluidic sensors to monitor and control the process. The two industrial partners will perform prototype testing (parameter screenings, in-depth analysis, small-scale productions) by investigating and improving different nitration processes at different sites.

Many universities in Great Britain (London, Birmingham, Manchester, Sheffield, Glasgow among others) have several activities in the area of microsystems engineering. A milestone in the development was the Taylor report in 2002 summarizing main activities in the UK (www.mntnetwork. com). Universities in the Netherlands (Eindhoven, Twente, Delft among others) have started project activities in micro chemical engineering and fluid dynamic research. The TU Eindhoven, The Netherlands, Faculty of Chemical Engineering and Chemistry (http://www.chem.tue.nl/scr/) has strong research activities in chemical micro process engineering. The chemical industry in Switzerland has started activities in micro process engineering, like Lonza [37], Clariant [62], Roche, DSM, or Ciba. In Austria the engineering company MicroInnova, Graz (http://www.microInnova.com), develops processes and designs complete plants with microstructured devices, mainly IMM devices. Some projects are performed in cooperation with TU Graz, Prof. Marr, on particle handling and other research areas [115].

1.6.2 Activities in the US

A good and recent overview of the US activities and companies involved with fabricating and engineering of microstructured devices is given by Palo et al. [116] for process engineering. Many activities are concerned with Homeland Security and Single Person Supply, which are not treated here. An active role on energy systems and fuel processing is played by Velocys, Plain City, OH (www.velocys.com) with their own fabrication technology and many industrial projects. The company designs and fabricates chemical processing hardware for the areas of oil and gas, refining and fuels, as well as commodity and specialty chemicals. The company MesoSystems, Albuquerque, NM (www.mesosystems.com), offers a spectrum of analysis equipment for fuel processing, air purification, and mobile gas analysis. Their expertise lies in instrument automation and integration, as well as miniaturization and microfabrication among others. Exergy, LLC, Garden City, NY (www.exergyinc.com) are producing compact heat exchangers for various applications in process industry.

The Stevens Institute of Technology in Hoboken, NJ (www.stevens.edu) founded the New Jersey Center for MicroChemical Systems (NJCMCS, www. njcmcs.org) in 2002 with long-term state, federal, and institutional commitments. Research projects cover DNA chips, methanol converters and other microfluidic devices for process intensification. In a collaboration between Pacific Northwest National Laboratory PNNL, Richland, WA (www.pnl.gov) and Oregon State University (OSU), Corvallis, OR (oregonstate.edu), the Microproducts Breakthrough Institute (www.pnl.gov/microproducts) works in the areas of electronics, energy, environment, defense, health, chemicals, space, and transportation. The development of a multiple compact modular reactor system reduces the size by a factor of 10 to 100, which enables distributed processing and reduces the transportation of hazardous chemicals. Distributed heating and cooling units within a building increase the energy efficiency and decrease the operating cost of facilities.

An initiative for sampling and measuring with microstructured devices was founded in 1999 by end-users and manufacturers under the sponsorship and auspices of CPAC (Center for Process Analytical Chemistry) at the University of Washington in Seattle, the NeSSI^{TM} system for analytical purposes (www.cpac.washington.edu/ NeSSI/NeSSI.htm). Their main objectives are

- to facilitate the acceptance / implementation of modular, miniature and smart sample system technology based on ANSI/ISA SP76 standard substrate,
- to provide a technology bridge to the process for "sensor/lab-on-a-chip" microanalytical devices,
- to promote the concept of field-mounted (By-Line) smart analytical systems, and
- to lay the groundwork for an open-connectivity architecture for intrinsically safe transducer communications and industry standard communication protocols.

More information can be found on their homepage, a modular system for sensing and analyzing is shown in Figure 1.15. The company Swagelok has launched a microanalysis system, which is also shown in Figure 1.15.

Fig. 1.15. Parker Intraflow^{TM} System in the initiative of NeSSI^{TM} for analysis purposes.

1.6.3 Activities in the Far East

The Japanese initiative for Micro Chemical Process Technology (MCPT, www. mcpt.jp/ eindex.html) is probably the most well-known project in the Asian region. The project organization diagram is shown in Figure 1.16 with three projects concerning the development of microchemical plant technology, microchip technology, and systematization of the gained results. The financial budget was 947 million JPY (approx. 8.23 million Euro) for the year 2002, 1 212 million JPY (approx. 10.1 million Euro) in 2003, and 1 150 million JPY (approx. 8.65 million Euro) in 2004.

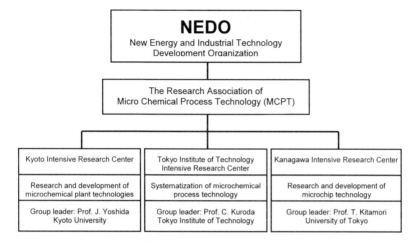

Fig. 1.16. Organization chart of the Japanese MCPT program 2002-05, see [117].

The forum to present the actual results is the Micro-chemical plant international workshop, which was held in 2006 at Kyoto University with over 180 participants from Japan, Europe, and the United States. The project program for the year 2005 included the following research and development activities by the three groups.

1. Research and development of microchemical plant technologies: Under Prof. J. Yoshida, Kyoto University, the research on micro unit operations was expanded further. Three continuously operating pilot plants using micro devices were built to confirm the suitability of micro devices for fast reactions [118]. Five more different pilot plants were planned for the year 2006. The research results were applied to production processes to give techniques for optimal unit operations design with integrated measurement and control systems.

2. Research and development of microchip technology: Under Prof. T. Kitamori, University of Tokyo, the research on microchemistry in the microscopic space of a microchip for micro analysis or production uses multi-chemical processes on a variety of analytical chips and detection methods. This included the evaluation and improvement of prototype microfluidic devices such as pumps, valves, and sensors.

3. Systematization of microchemical process technology: A research team under Prof. C. Kuroda, Tokyo Institute of Technology, in collaboration with the two other research groups named above was engaged in collecting and rearranging technological data (both existing and new data) related to common fundamental technologies of the microchemical process. It was planned to build systems to support design, operation and control of the microchemical processes to be utilized for commercialization, as well as restructuring the enhanced knowledge collaboration systems to a level where it can be utilized for simulations. The results were stored as knowledge data of intensive research laboratories.

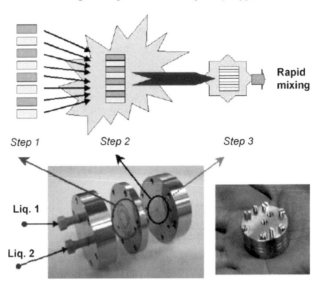

Fig. 1.17. Mixing process description by lamellae engineering and equipment design (K-type micromixer), from the Japanese MCPT program.

The success of the Japanese research project is documented in the development of microstructured devices, the implementation of many chemical reactions reaching production scale, and in numerous publications. Mae [119] give an overview of the recent activities and the philosophy behind them including illustrative examples, see Figure 1.17. The author propose a paradigm shift to advanced chemical reaction engineering, especially in areas of conventional batch production.

In the Peoples Republic of China, only a few projects are known with companies or institutes besides some research activities at universities. For example, the University of Science and Technology of China (www.hfnl.ustc.edu.cn/ustc/front_en) in Hefei harbors the National Laboratory for Physical Science at the Microscale with a Physics and Chemistry Analysis Laboratory. In cooperation with Xi'an Chemical Industrial Group, the IMM Mainz, Germany, planned and erected a production plant in China with continuous flow nitroglycerin production [120]. In autumn of 2005, the first test runs of the microstructured reactor plant were performed with a throughput of approximately 15 kg/h. The product must be produced under GMP conditions and meet high quality standards. The new plant gives higher yield, better product quality, increased safety, and reduced environmental hazards [120].

Many activities in South Korea are managed and funded by the Korean Institute of Science and Technology (KIST, www.kist.re.kr), the Korean Science and Engineering Foundation (KOSEF, www.kosef.re.kr), and the Korean Research Foundation (KRF, www.krf.or.kr). The main focus of the KIST program lies in the microfluidic analysis systems and on BioMEMS. In 2001, KOSEF and industrial partners designated a Center of Excellence in chemistry and chemical engineering, the Center for Ultra-microchemical Process Systems (CUPS, cups.kaist.ac.kr/). The nine-year funding begins with the infrastructure period for the first three years, the development period from the 4th to 6th years and will end with the take-off and independence period in the next three years.

The center embraces four groups with the topics Microfluidics (microfluidic channels, micro flow control and micromixer [121], micro heat transfer), Multichannel microreactor (fabrication, high performance materials using combinatorial catalysis and materials for micro fuel cells), Microchemical analysis and control (micro automation and optomechatronics, microsensors), and Process on chip (microsensors and self-assembly, micropatterning, and soft-lithography for protein chips). The goal of the Center for Ultra-microchemical Process Systems (CUPS) is to design and manufacture the chemical process systems with ultra-micron scale, which will potentially make low-cost mass production possible. The center applies the new technology to discover materials for customers in the life sciences, and in new chemical, biological and electronics industries, see Figure 1.18.

In the area of mechanical engineering, the research center of Micro Thermal System was established at the Seoul National University under Prof. J.S. Lee in 2001. The Micro Thermal System Research Center (µTherm, microtherm.snu.ac.kr) works on the development of micro heat exchangers, micro power devices, and the related manufacturing technology using high energy beams. The projects include the design and fabrication of micro heat exchangers and their experimental testing. One aim is the mass production of micro heat exchangers and micro power systems. Industrial partners like HANJOO Technology Co., Ltd or NEUROS Co., Ltd are sponsoring the micro thermal systems research. Another international company, Samsung Engineering, delivers micro-chemical plants (www.samsungengineering.co.kr/secl/eng/business/bus_m3.htm) for various application fields.

Moreover, other countries in the Far East are active in microsystems engineering, mainly in the MEMS area. In India, research activities on flow boiling in microchannels are found at the Indian Institute of Technology (IIT) in Roorkee, Prof. Kumar and Prof. Gupta (www.iitr.ernet.in/ departments/MI). Fluid dynamics in micro nozzles is investigated at the Indian Institute of Science in Bangalore, Prof. B.N. Raghunandan. The institute cooperates with foreign organizations like KOSEF in Korea or universities in Japan and the UK. In Australia, some activities in microsystems engineering are located at the Commonwealth Scientific and Industrial Research Organisation (SCIRO www.csiro.au) in Highett, Victoria. For example, the description and experimental results of a micro fuel cell can be found under (www.solve.csiro.au/1104/article6.htm). At the University of Melbourne (www.rmit.edu.au/SECE), a research group under Dr. Rosengarten works on microchannels and their coupling with thermoelectric structures for cooling and sensing purposes. The Australian CRC for microtechnology is a joint venture aimed at producing microfluidic devices with design, modeling, manufacturing, and mass production of microsystems to handle fluids. The devices contain active and passive microstructures to control the flow and mixing of the fluids for physical, chemical, biochemical, and microbiological reactions.

Basic technology for UCP &their application

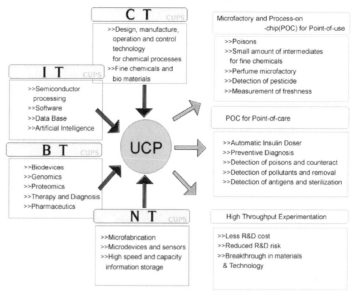

Fig. 1.18. Technologies and Applications of the Korean Center of Excellence Ultra-microchemical Processing UCP (acc. to cups.kaist.ac.kr/cups/intro_tech.htm).

Summarizing the international activities in Micro Process Engineering, there are many research and industrial activities worldwide, but in general there are still many barriers to overcome.

1.7 Barriers and Challenges

The major drawbacks of microstructured equipment are low flow rates and high pressure loss, which conflict with industrial mass production. Therefore, most of the potential applications are seen in fine and specialty chemistry or in pharmaceutical production. Unlike conventional applications, the high surface to volume ratio leads to amplified phenomena like material interaction or species adsorption with undesired consequences.

High performance equipment like microstructured devices are very sensitive to small changes of process parameters or conditions like surface structure, material composition, flow characteristics, or ambient conditions. For example, the performance of catalytic microreactors is severely deteriorated by catalyst deactivation. Corrosion can diminish the flow rates through the devices and the mechanical stability of microstructures. Fouling and clogging also decrease the flow rates, increase the pressure drop, lead to maldistribution of flow, or even block the entire device [27]. Proper cleaning and mitigation techniques must be developed for the main applications, which can be transferred to new application fields. Alternatively, low cost disposable devices can operate for a certain period and then be substituted with new devices.

The micro fabrication technology adopted from microelectronics allows, under certain conditions, a cost effective mass production of devices with a dense integration of various elements. The integration of microstructured devices into the environment also needs conventional fabrication technologies and should be considered during device design. Many micro fabrication techniques allow very complex geometries for combination, switching, and proper arrangement of the various process steps. Recently, lithographically structured metal sheets from offset printing technology were proposed for cost-efficient mass production of microstructured reactors [122].

Strict fabrication tolerances and effective sealing prevent leaks between channels, but must remain in the scope of economic fabrication. Proper design of flow manifolds avoids the maldistribution of the flow into the numerous channels. The design must also allow for the malfunction of flow distributors by fouling, material deposition, or corrosion. The coupling of microstructures to the conventional environment is also a critical point, which needs to be solved for each application and specific environment.

Microstructured devices are continuously operated and may lead to acceptance problems in areas where batch processes are dominant. Existing vessels and plants are economically depreciated and operate efficiently with well known protocols. Therefore, it is intended to introduce microstructured equipment into the education and curricula to teach students how to work with continuous flow devices and plants [123]. The starting point is the education of chemists, chemical technologists, chemical engineers, or mechanical engineers for heat and mass transfer phenomena with or without chemical reactions. In 2007, the German Federal Ministry of Education and Research (bmbf) has launched a training program of six regional clusters at universities and research institutes to spread micro process engineering tools into education.

The lack of long-term experience with microstructured devices is one of the major barriers for industrial application in the chemical industry, but is being addressed in current research projects in Germany [38], in Europe, as well as in the MCPT initiative in Japan.

In spite of the many research activities in recent years, the development of microstructured devices for various applications still presents many challenges for basic and applied research. The interplay of fluid mechanics, heat transfer, mass transfer and chemical reactions in microchannels is not understood well enough to model or simulate complex microreactors correctly. The numerical diffusion due to high Schmidt numbers $Sc = v/D$ and relatively coarse CFD grids "smear out" the related concentration fields and over-estimate the diffusion length, see also Section 3.2.3. This effect leads to high mass transfer rates and well-mixed channel flow, which does not match actual systems. To yield adequate mass transfer results compared to fluid dynamic results, the CFD element size must be smaller by the factor of $Sc^{1/2}$. In conventional technology, mixing models for turbulent mixing cover this gap, see [124]. For mixing processes in microchannels, adequate models are missing since experimental and numerical data are scarce. Mixing in laminar flow is quite complex and needs a high mathematical effort [125, 126]. A simple model for convective mixing in T-shaped micromixers and combined meandering micromixers was proposed by Engler [127, 128]. For larger geometries and composed microstructured devices the numerical simulation of transport phenomena is too elaborate and time-consuming for CFD calculations. Alternatively, analytical modeling helps to reduce the complexity and leads to lumped element modeling [129] or to bond graph methods [130].

However, not all unit operations are suitable for miniaturization. Mechanical unit operations often include mechanical gears and motions with friction and wear, which cause many problems in microsystems. Therefore, pure mechanical processes like pressing or crushing are not feasible in micro process technology. As demonstrated in Section 1.5, unit operations and potential applications must be screened for successful applications of microstructured devices, which is an enduring process and still far from being finished.

The integration of various sensors and actuators into microstructured devices increases the potential for process control. Löbbecke describes in [131] the integration of optical in-line measurement devices (Raman, UV/Vis-spectroscopy) and micro-sized sensors for temperature, pressure, or flow rate. The calorimetry of a chemical reaction is determined with Seebeck/Peltier elements integrated in a microreactor. A modular setup is helpful for the integration of the various devices and also offers an interface to the macrosized environment. The design of microstructured high-performance devices for high flow rates in production applications demands an information process chain from simulation, design, and layout through fabrication issues to application-specific requirements [132]. Finally, industrial production is different from analyzing substances, where most of the actual microchannel devices are applied. High flow rates and low pressure loss combined with reliable operation are essential to achieve high performance with low effort [133].

Concluding this section, micro process engineering at the interface of (at least) chemistry, process engineering, and microsystems engineering provides good opportunities to introduce new possibilities in many application fields. Multiple channel devices demand new design methodology to increase channel number and provide safe operation conditions. In chemistry education, continuous processing should be taught from the beginning with process and chemical engineering methods. In mechanical engineering education, microchannels present the opportunity to grasp the essentials for fluid flow, heat transfer, boiling, evaporation, and condensation within continuously operating equipment. A few examples illustrate the rapid development of the last years, but there is still a lot of research and engineering work to be done at the interface of the various disciplines to understand fundamental processes. Hopefully, this will lead to a paradigm change in many industrial applications. The following chapters illustrate some methods for successful design, fabrication and operation of microstructured devices and will hopefully provide a guideline for future research and development at the leading edge of science and engineering.

2

Fundamentals, Balances, and Transport Processes

Transfer processes and balances of the transport parameters are essential for any engineering design process as well as physical or chemical calculations. Beginning with simple atomic encounters from statistical mechanics, macroscopic balance equations and transport properties are derived for the following chapters. Furthermore, the toolbox of engineering calculation methods is presented as an orientation and introduction to limits and opportunities.

2.1 Introduction

Process technology and microsystems technology are both interdisciplinary engineering and natural science branches connecting physics, chemistry, biology, engineering arts, and management techniques to an rich and enabling toolbox for various applications. Microsystems engineering and technology, coming from information technology and miniaturization of data processing devices, has now entered many aspects of our daily life such as automotive or medical devices. Chemistry in *miniaturized* or *microstructured equipment* is an emerging discipline combining microsystems technology and chemical engineering, whilst also an old discipline of chemical analytics. More than a century ago in 1895, the first textbook on microchemical technology was published by Behrens [6, 7]. In 1911 Prof. Emich from Graz published his textbook "Lehrbuch der Mikrochemie" [134]. Prof. Pregl from the same working group in Graz was honored with the Nobel prize in 1923 for his work on microchemical analysis. Today, chemistry and chemical processing in microstructured devices has its own textbooks [3, 13] and has entered encyclopedias [69] and classical textbooks like Fitzer and Fritz "Technische Chemie, Einführung in die chemische Reaktionstechnik", updated in 2005 by Emig and Klemm [33].

Dealing with very small geometrical structures is a well-known area of study in process engineering. Adsorption technology and chemical reactions on catalytic surfaces are based on the flow and adhesion processes in nanoscale pores, [135] Chapter 4. Transformations and transfer processes on a molecular scale are called "micro processes" in contrast to a "macro process" where convection plays the dominant role. Some typical length scales for process technology, chemistry, and micro technology are given in Figure 2.1.

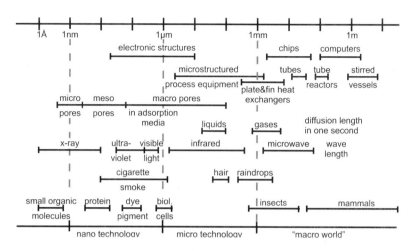

Fig. 2.1. Characteristic length of important processes and equipment in chemical engineering and microsystems technology. The top and bottom line also indicate the different terms for micro processes in both disciplines, adopted from [69].

In the process industry, there are several applications of structures with typical dimensions below 1 mm, such as compact plate & fin heat exchangers or structured packings in separation columns for enhanced heat and mass transfer. The elementary set-up of microstructured and conventional equipment is comparable and displayed in Figure 2.2.

Process plants consist of process units combining various equipment types such as heat exchangers or vessels with internal structures. The basic geometrical elements of the internal structures in conventional technology are the tube, the plate, and the film, on, in, or across which the transport processes and transformations occur. These elements form the active area for a certain process within a device. The layout of process equipment and process steps follow this scheme from small elementary active areas ("micro process") over the process space of the device ("macro process") to the balancing of the complete process.

The parallel arrangement of microstructured channels or elements is called internal numbering-up, which is the most frequently used way to increase the volume throughput of an apparatus. The parallelization of microstructured devices is called external numbering-up, applied to bypass flow distribution problems within the equipment. A relatively new concept is the equal-up concept, the parallelization of similar effects [69], see Section 6.3. The numbering-up and equal-up concepts facilitate the scale-up process from laboratory equipment to production equipment, but still face the unique problem of flow distribution in manifolds, see Section 5.7.

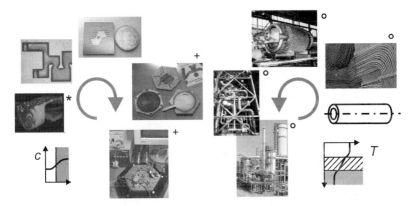

Fig. 2.2. Setup of microstructured and conventional equipment from single microstructures over combined elements to an entire plant, see also Figure A.1. The principle of the active area can be applied in both cases. (Sources: * from [136]; [+] courtesy of FhG-ICT, Pfinztal, Germany; [o] courtesy of Linde AG, Wiesbaden, Germany).

2.2 Unit Operations and Process Design

Generally, the design procedure of chemical devices and processes is different and should be distinguished. Due to the close relation between geometry, material, and inherent chemical or physical processes in microstructured equipment, however, both processes will be treated together to facilitate a better understanding. With increasing knowledge and a variety of applications gained about a process, further classification would be reasonable in the future.

2.2.1 Process simulation, scale-up, and equal-up

Since the 1960s the fundamental balance equations have been solved with the help of increasing computer performance. In the 1970s, process designers were able to quickly determine entire process chains with the introduction of unit operations, their mathematical modeling of balance equations, and, concurrently, gathering information about the essential device parameters. The aim of process simulation is to attain information for the understanding of the quantitative (and qualitative) properties of the process and systems behavior. The process engineer determines design parameters for the reliable concept development of processes for industrial plants. The employed process models, with model parameters, are an accepted and approved tool in process design, which is reinforced by experimental data from larger production facilities [132].

Besides the application of mathematical and simulation tools, the appropriate, critical judgment and interpretation of the results are important for the reasonable and responsible work of an engineer. Experimental data for the evaluation are gained from model plants in the laboratory, miniplants, pilot plants, and eventually, from industrial production plants modified for the specific purpose. The conventional process development in the chemical industry (without microstructured equipment) begins at the laboratory scale (< 1 m and < 1 kg/h), followed by miniplants (< 3 m and < 20 kg/h), laboratory pilot plants (< 10 m), and production pilot plants (< 30 m and < 2 t/h) before reaching the final production plant in large scale. Zlokarnik [32] describes this scale-up procedure in detail; the economy-of-scale is described in Section 6.3.4.

Meanwhile, with advanced scale-up knowledge, laboratory and production pilot plants are combined to save development time and investment costs. In some companies, microstructured devices help to yield laboratory data and information on the process and the plant [35], which can then be directly transferred to the production plant.

In conjunction with the integration of microstructured devices in laboratory and production plants, various implementation strategies have been developed to link the processes in single structures with the characteristics of the entire device. The numbering-up process takes the results from processes in a single device and extrapolates them to the entire production unit by multiplying it by the number of devices. Parameters like pressure, temperature, and concentration are kept constant. This simple approach has various disadvantages, such as the inhomogeneous flow distribution or parallel numbering-up of the measurement and control systems.

Alternatively, the equal-up strategy was developed to transfer a chemical process from the laboratory scale on industrial production scale, see Figure 2.3. With the equal-up method [69], the application of microstructured equipment in production plants can also simplify the scale-up process to an operational capacity using various fabrication techniques. The entire equal-up strategy is described by Becker et al. [34] in more detail and embraces the following steps.

Starting with the process or product to be realized, the main effects and parameters are identified where a miniaturization could be beneficial. These key parameters, such as enhanced transfer properties or integration possibilities, are transferred to the laboratory device in an equal-down step. The likely design of an industrial reactor also influences the design of the laboratory reactor. Process and reaction engineering guide the reactor development within the constraints of economical design and fabrication. Emig and Klemm [33] give a sophisticated description of the dimensioning of microstructured reactors concerning fluid dynamics, heat and mass transfer, as well as mixing and reaction kinetics, see also Section 6.3 for a detailed treatment of reactor design. Here, a short overview is given for introduction. The laboratory design is equal to the industrial design in relation to the governing micro effects including key geometries, fluid dynamics, mixing, reaction engineering, and heat management. It has to be clearly determined for the laboratory reactor, which dimensions cause the micro effects, which effects should be transferred, and what the shape and structure of the active area will be.

Using the laboratory results, the detail design of the pilot reactor on an industrial scale is done in the next, equal-up step. The key parameters of the large device, such as channel width, catalyst wall thickness and modified residence time, are equal to the laboratory equipment, however, the channel length and number or the fabrication

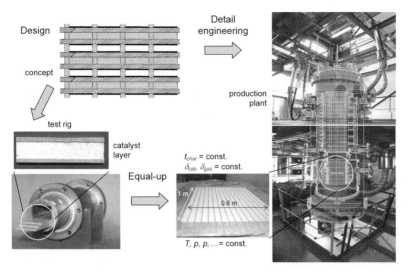

Fig. 2.3. Equipment design of a catalytic microchannel reactor with the equal-up and equal-down method according to [34, 33]. The modified residence time, catalyst thickness, channel height, and the process conditions remain constant. See also Figure A.3, courtesy of Degussa GmbH, Hanau and Uhde GmbH, Dortmund.

techniques may differ. For example, the microchannels have the same cross section or the same height, but differ in length, see Figure 2.3. Hence, the process has almost the same geometrical conditions in the the laboratory device than in the production plant, which determine the performance.

2.2.2 Method of process, equipment, and plant design

The process and plant design embraces engineering activities from the pilot study through basic and detail engineering to equipment design, fabrication, assembly, and plant start-up, see Figures 2.3 and 2.4. This process makes use of professional project management tools.

The process design begins with the pilot study, which may originate from internal ideas and initiatives, customer requests, market research, or the re-engineering of an existing plant or product. The pilot study indicates the feasibility of the project, identifies solution paths, and specifies measurable and testable variables for plant description. It embraces the analysis of available, experimental data, the determination and analysis of material properties, and the process-simulation software like AS-PEN [137, 138], MATLAB [139, 140], or POLYMATH [141, 142]. The pilot study gives a starting point for the basic engineering in the main study, which helps develop a general, functional process structure and generates the process flow sheet with main functional information, see also Figure 2.5. Commencement of the basic engineering involves the research and development group and the technical department of an engineering company.

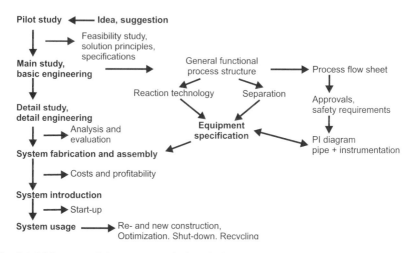

Fig. 2.4. Main steps of the process and plant design, including the basic steps of equipment design with engineering documentation (see Figure 6.7), and of the plant construction and entire plant lifetime cycle, acc. to [64]. The design of microstructured devices is described in Section 6.3.

The following documents are prepared and elaborated during basic engineering.

- Process flow diagram (PFD) with main control circuits;
- Mass and energy balances with process performance, consumption values for resources and utilities, media list with hazard category, and emissions of various kinds (waste gas or water, noise, and other);
- Process description and specification, start-up and shut-down procedure;
- Equipment list with major devices, fabrication, and materials;
- Dimensioning of major equipment, for example, with heat exchangers: area, number of passages, channel cross-section, length, distance and pitch, or in the case of separation columns, the type and geometry of inserts, special requirements for operation, maintenance and handling, see also Section 6.3;
- Measurement and control list, valves, control circuits, safety requirements;
- Auxiliary and peripheral systems, such as vessels, drains, vents, steam system, cooling water, electric supply.

Not all of these issues are important for plant design with microstructured equipment, however, the list gives a fairly complete overview for a comprehensive process design. As the main result of basic design, the process flow diagram indicates the important currents of mass, energy, and signal. It is used to specify the parameters of the process equipment and the control scheme. During the design of the process equipment and the preparation of the plant control system, the pipe and instrumentation diagram (PI diagram) is developed, which is the major information source for the process. During the detail engineering, an ongoing evaluation process brings all elements together resulting in the design documentation for the entire plant, from which the equipment is designed, fabricated, and assembled. At this stage, initial investment cost are roughly determined and the profitability of the plant can be ascertained. With the initial filling, priming and start-up of the plant, the evaluation of the specification requirements begins. The system is introduced to the operators and is used for the first production runs, however, further optimizations or adjustments may be necessary.

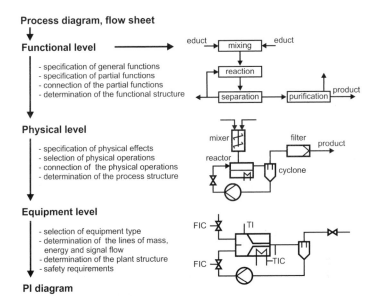

Fig. 2.5. Information and design levels in the process design procedure, adopted from Blass [64]. More information on process engineering symbols can be found in DIN ISO 10 628.

The information levels and project steps during basic and detail engineering are illustrated in Figure 2.5. On the functional level, the specification of the plant elements and the equipment functions are defined. The functional structure is displayed in the general flow sheet of the process similar to a black box diagram. The contents of the individual process steps in the black boxes are specified on the physical level, where the process structure is defined with all the integrated unit operations. The information is displayed in the process diagram, which contains the main currents of mass, energy and signals (main control circuits). On the equipment level, the process is defined in detail using a PI diagram. Here, detailed information is given for the plant setup and any further equipment specification, design, and fabrication.

Process design diagrams use standardized symbols, see Figure 2.6, in which symbols of typical micromixers have been added. The symbols represent commonly used devices in conventional, and especially, micro process engineering. They are classified in five groups, which are by no means complete.

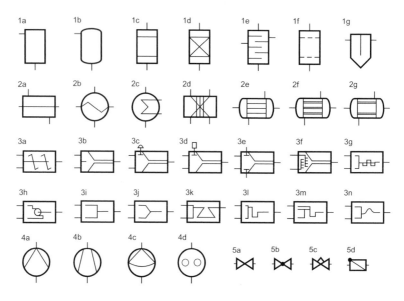

Fig. 2.6. Unit operations and proposal for micro device symbols, for more information see the text. Group 1: Vessels and separation equipment; Group 2: Heat exchangers; Group 3: Mixing devices; Group 4: Pumps; Group 5: Valves.

- Group 1 Vessels: a) general vessel; b) pressure vessel; c, d) vessel with fixed bed; e) separation column; f) filter; g) mechanical separator;
- Group 2 Heat exchangers and reactors: a) double tube heat exchanger; b, c) general heat exchangers; d) plate heat exchanger; e) shell & tube heat exchanger; f, g) reactors with internal heat exchanger
- Group 3 Mixers: a) general static mixer; b) general micro mixer; c) active micro mixer with pneumatic drive; d) active micro mixer with electrical drive; e) active micro mixer with electro-kinetic drive; f) multilamination mixer; g) split-and-recombine mixer; h) cyclone mixer; i) T-shaped mixer; j) Y-shaped mixer; k) micro contactor and arbitrarily shaped micro mixer; l) micro contactor and meandering mixing channel; m) injection mixer; n) contactor with short mixing channel; still unspecified: herringbone, superfocus, K-type, collision, chaotic advection, and other, which are named in the literature
- Group 4 Pumps: a) general centrifugal pump; b) gas compressor; c) membrane pump; d) gear pump;
- Group 5 Valves: a) general valve; b) ball valve; c) control valve; d) check valve or fluidic diode.

The symbols in Figure 2.6 indicate what is possible in microstructures: mass transfer in contacting and mixing with a short residence time and narrow distribution, high heat transfer and precise temperature control. Together with the wide palette of micromachining, surface modifications, complex geometries, different materials, and integrated sensors allow the design of highly functionable devices and integrated processes. Most applications of these devices are in the field of lab-on-a-chip and micro total analysis systems. However, there are still many unit operations, commonly used in conventional process engineering, which have not been successfully transferred into microstructured devices. Coupled transport phenomena, such as thermodiffusion or pressure diffusion are explained in Chapter 7.

The various unit operations can be categorized according to the physical forces (mechanical and electro-magnetic unit operations) or by the molecular driving forces (thermal unit operations), see Table 2.1. Mixing can be treated as a major unit operation, which is fundamental for many other processes and can adopt many forms such as homogenizing, dispersing, or suspending; mixing can occur with or without chemical reactions, or as a precursor for chemical reactions, in combustion, or polymerization. An overview of possible unit operations is given in Table 2.1.

Table 2.1. Main mixing and separation unit operations grouped according to their driving force

Unit operation	molecular / thermal	mechanical/ext. force	electro-magnetic
Mixing and aggregation, Combination, Control of segregation.	diffusion[1] dissolving[2] extracting[2c] desorption[2c]	spraying[2] aeration[2] stirring[2] active mixing[1,2] dosing[1,2]	electro-phoretic mixing[1] mixing with magnetic beads[2]
Separation	thermodiffusion[1] pressure diffusion[1] counter-current diffusion[1] condensation[2a] evaporation[2a] crystallization[2a] distillation/ rectification[2a] drying[2b] absorption[2c] adsorption[2c] extraction[2c] ion exchange[2c] membrane processes.	sedimentation[2] cycloning[2] centrifugation[2] pressure diffusion[1,2] (ultracentrifuge) filtration osmosis gas permeation classification sorting	electro deposition[2] magneto deposition[2] electro filtration electro dialysis electro osmosis electrophoresis magneto-striction

superscripts for *employed phases*:
1) single phase
2) multiphase
a) with own co-phase
b) own + additional co-phase
c) additional co-phase

More detailed operation descriptions and further reading in [41, 64, 72, 143]; Additional thermal unit operations, which are closely related to these listed:
Partial condensation[2a]; Flash evaporation[2a], vacuum evaporation[2a];
Freeze-drying[2a], radiation drying[2a], super heated steam drying[2a];
Outside/secondary steam distillation[2b]; Molecular distillation[2a];
Extractive rectification[2b], azeotropic rectification[2b];

Chromatographic adsorption;

Membrane processes: permeation, pervaporation, dialysis, osmosis and reverse osmosis, micro- and ultra-filtration;

Combination with chemical reactions in reactive distillation and reactive extraction. This list does not claim to be complete, especially the separation processes from analytics, which are only schematically shown. More information on separation and preconcentration methods in micro-analytical systems are given by Kutter and Fintschenko [144]. The development of further operations over the next few years will be enabled by enhanced fabrication and integration possibilities.

A further major process step, the transport of fluids, is not listed in the above table, but is partially included in the unit operation description. Active devices for fluid transport are pumps and compressors, which show a wide variety of possibilities depending on the fluid, its viscosity, the necessary pressure increase, and the volume flow. Pressure-driven flow is predominant in micro process engineering applications, however, other pumping mechanisms are employed in lab-on-a-chip applications, such as electro-osmotic or magneto-hydrodynamic flow. Electro-kinetic flow is also used to propel liquids on analytical chips and is described in Section 7.3.

Chemical reactions are more heterogeneous than the afore-mentioned presented unit operations and their working principles. There exists some segmentation proposals similar to the unit operations, which follow physical or physico-chemical aspects, such as heat release (exo- or endothermic), rate constants (fast, slow), type of initiation (photo, electro, acoustic, ...), or the component phases. A large group of multiple reactions are parallel reaction, which are subdivided in concurrent and consecutive, see Section 6.3.7. In complex, industrial reaction systems, the kinetics often are unknown, and the yield and selectivity are mixing-controlled, see Section 6.1.

A large field of chemical reactions deals with catalytic transformations, for example, approximately 90 to 95 % of industrial organic reactions need catalysts. Catalytic reactions are classified into homogeneous and heterogeneous catalysis. In homogeneous catalysis the catalytic active material is present in the same phase as the reactants, such as enzymatic reactions in biology, where all the components are present in the liquid phase. These reactions are determined by mass transfer between the interacting components. In heterogeneous catalysism the catalyst is often present in solid phase, bound at the wall or in porous form. Here, the mass transfer between the bulk with the mixed reactants and the supporting wall determines the transfer rates, see Section 6.2.

2.2.3 Design principles for process equipment

To assist the design procedure of chemical processes and devices, design rules and principles have been developed to incorporate the methodological knowledge and experience. According to Pahl and Beitz [145], young engineers educated in design methodology will surprisingly fast be a complete staff member without firstly acquiring detailed experience. Put simply, design should follow the principal rules that the process or device must be simple, safe, and a well-defined and clear function.

Secondly, design principles help engineers to focus on the main points and avoid typical pitfalls and traps. In the following, some important design principles for process engineering are presented in a concise form.

The *principle of an active area* indicates a platform for driving forces in molecular and thermal processes. It describes for the transfer processes with linear correlations between the flux and the driving force, also called the kinetic approach, see [146] Chapter 1. The processes take place in basic geometrical elements, such as vessel, the tube, a channel, pipe, pores, or plates, which are combined to form the process space in the chemical equipment. Within these elements the fluid itself forms geometrical shapes like beads, drops, bubbles, films, or thin layers, which determine the transfer processes, and which are confined by the geometry, see Figure 2.7.

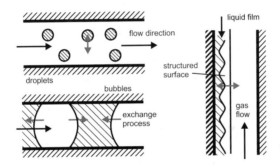

Fig. 2.7. Geometrical elements for transport processes between two phases: droplets, bubbles and a falling (rinsing) film on a structured surface.

The three phases of a pure substance allow the following combinations for phase mixtures of a carrier fluid and a dissolved phase: gas - gas, gas - liquid (droplets and aerosols), gas - solid (aerosols), liquid - gas (bubbles and foam), liquid - liquid (miscible or immiscible, emulsion), and liquid - solid (suspension). An illustration of the properties of the phases and interfaces is given in Figure 1.5 dependent on the length scale.

The *principle of technical enhancement* and *process intensification*, when compared to the natural driving forces, gives the opportunity to control the transfer rates and state conditions in a way, which is optimal for the desired results. The *principle of a selective phase* enlarges the process space by adding a new component, which enforces a new equilibrium within the process (drying, extraction, stripping etc.). The *principle of flow guidance* in the equipment and process space (co-current, counter-current, cross-current, mixed arrangement or recirculated flow), in addition to various switching possibilities (series, parallel, cascading), allows for the effective exploitation of the existing driving force. The heuristic application of these principles can benefit systems design and will encourage within the fields of process engineering and microsystems engineering research and development [64].

Control of transport processes is essential for design and proper operation of process equipment and relies on several principles. To understand and apply these principles is an essential factor for the successful design and efficient operation of devices and plants. Material properties and device geometry is determined during the basic and detail design stages depending on the fabrication technology. Fluid properties are given by the process, as do the process conditions, which are also often part of the design process and can be optimized with the application of microstructured equipment. The specific transfer area is mostly controlled by the equipment design, but can be manipulated by process conditions, such as fouling or partially flooding. The driving forces of transfer processes are gradients of process parameters, which are controlled during operation. Precise information about process parameters is essential for the effective control of a device and the process, where integrated sensors play a crucial role. Hence, there are many opportunities for integrated microstructures and sensors to optimize existing processes or to develop new pathways for chemical and process engineering.

2.3 Balances and Transport Equations

Process engineering calculations and process equipment design begin with the conservation and balance equations of mass, species, momentum, energy, as well as the definition of the entropy and its application in exergy analysis. The conservation laws of mass (continuity equations) and energy (first law of thermodynamics) are valid in the scope of chemical processes dealt within this work. They can be described in words with the following schema.

$$\begin{bmatrix} \text{System change} \\ \text{with time} \end{bmatrix} = \begin{bmatrix} \text{Incoming} \\ \text{Flow} \end{bmatrix} - \begin{bmatrix} \text{Outgoing} \\ \text{Flow} \end{bmatrix} + \begin{bmatrix} \text{Source or} \\ \text{Sink} \end{bmatrix} \qquad (2.1)$$

The source or sink of the flow parameter depends on the system and the parameter itself and is described later, together with other possible simplifications.

2.3.1 Statistical mechanics and mean free path

Before introducing the continuum and balance equations of the various parameters, a short excursion to the molecular origin of these equations begins with the derivation of the Boltzmann transport equation for thermodynamic equilibrium. In a perfect gas, the molecules are regarded as hard spheres interacting only in very short encounters with other molecules or with the boundary (wall, surface, or other limiting elements). At the molecular scale, the ratio of the mean molecular spacing δ and the mean molecular diameter σ is an important parameter. Gases with the condition

$$\frac{\delta}{\sigma} > 7 \qquad (2.2)$$

are referred to as dilute gases [147]. If the condition in Eq. 2.2 is not satisfied, the gas is regarded as being a dense gas. Air, at standard conditions, is regarded as a dilute gas, but pressurized air with 0.6 MPa is already considered as a dense gas, as explained later. An important value in this case is the number of atoms or molecules in a volume element under standard conditions, the Loschmidt number $N_L = 2.686 \cdot 10^{25}$ m_N^{-3} at 0 °C and 101.315 kPa, and the Avogadro number $N_A = 6.022 \cdot 10^{23}$ $1/mol^{-1}$, which indicates the fixed number of atoms in a mole.

It can be assumed that the probability of a molecule moving in a certain direction is equal for all three space coordinates, see also [148] p. 148. This can be expressed by the constant ratio of the derivative of the probability distribution function (PDF) $f(w)$ to the function itself and the velocity component w.

$$\frac{f'(w)}{w\,f(w)} = \frac{d\ln f(w)}{w\,dw} = -2\gamma \ \Rightarrow \ln f(w) = c_i - \gamma w^2 \tag{2.3}$$

For convenience, the integration constant is set to -2γ and determined with the kinetic energy of the molecules

$$\gamma = \frac{M_m}{2\,k\,T} = \frac{M}{2\,R_m T} \tag{2.4}$$

with the mass of a single molecule $M_m = Mk/R_m$, the Boltzmann constant $k = 1.38 \cdot 10^{-23}$ J/K, and the universal gas constant $R_m = 8.314$ J/kmol K. The integration constant c_i in Eq. 2.3 is determined by normalizing the sum of the probability to unity. The integration gives the probability distribution for one velocity component w, which stands for the other components u and v as well.

$$f(w) = \left(\frac{M}{2\pi\,R_m T}\right)^{\frac{1}{2}} \exp\left(-\frac{M}{2\,R_m T}w^2\right) \tag{2.5}$$

This is the Maxwell velocity distribution [149] of a perfect gas in thermodynamic equilibrium. The integration over a sphere in all three space coordinates gives the probability of the absolute velocity value c, independent from the direction

$$\varphi(c)\,dc = 4\pi\,c^2 F(c)\,dc$$

$$\Rightarrow \ \varphi(c) = 4\pi\,c^2 \left(\frac{1}{2\pi RT}\right)^{\frac{3}{2}} \exp\left(-\frac{c^2}{2RT}\right) \tag{2.6}$$

with the individual gas constant $R = R_m/M$. Both probability distributions are displayed in Figure 2.8 for air ($R = 287$J/kg K) with four different temperatures. For higher temperatures, the gas molecules move faster, and the distribution becomes wider.

The most probable velocity of a molecule is determined to

$$\bar{c}_{mp} = \sqrt{2\,R\,T}, \tag{2.7}$$

where the probability density distribution has its maximum. The mean velocity from the distribution is

$$\bar{c}_M = \sqrt{(8/\pi)\,R\,T}; \tag{2.8}$$

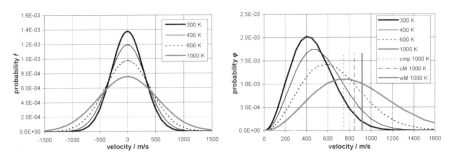

Fig. 2.8. The Maxwell velocity distribution f for air with four different temperatures (300, 400, 600, and 1 000 K) is displayed in the left diagram. The right diagram shows the absolute velocity distribution φ for air. The mean velocities are indicated for air at 1 000 K.

the mean velocity from the kinetic energy of a molecule is given by

$$\bar{w}_M = \sqrt{3\,R\,T}, \tag{2.9}$$

the mean kinetic molecular velocity.

The geometrical situation of a binary encounter is displayed in Figure 2.9. The molecules are treated as hard spheres with a diameter of d and a collision cross section $\sigma = \pi \cdot d^2$.

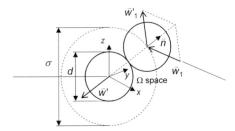

Fig. 2.9. Typical geometrical situation of the encounter of two identical molecules with a hard sphere model, see Sommerfeld [148]. A more detailed treatment of different encounters is given by Reif [150] Chapter 14.

With the number of molecules in a unit volume N_A, the number of encounters between the molecules and the mean time between these encounters can be determined. Multiplied with the mean velocity, an estimation of the mean free path of a molecule, the average length between two collisions, can be derived:

$$\Lambda = \frac{kT}{\sqrt{2}\,p\,\sigma} \tag{2.10}$$

The mean free path can also be expressed by

$$\Lambda = \frac{1}{\sqrt{2}\, n\, \sigma} \qquad (2.11)$$

with the number of particles n in a control volume, which is also given by $n = N/V = \delta^{-3}$. With the mean velocity of the particles between two collision $\bar{c}_M$, the characteristic time between two collisions is calculated by

$$t_c = \frac{\Lambda}{\bar{c}_M} = \frac{k}{4\, p\, \sigma} \sqrt{\frac{\pi\, T}{R}} \qquad (2.12)$$

The collision time is proportional to the square root of the temperature and inverse proportional to the pressure. Another characteristic length gives the limit of the continuum regime of material and fluid behavior. The continuum assumption in fluid mechanics is only valid for sufficient molecules within a given volume to achieve a stable estimate of the macroscopic flow properties, see Figure 1.6. As proposed by Bird [147], the ratio of the characteristic length scale L to the mean molecular spacing δ should satisfy

$$\frac{L}{\delta} > 100 \qquad (2.13)$$

to achieve a statistically stable estimate of the macroscopic properties. In other words, there should be at least 100 molecular spaces along each face of the sample volume, giving a total of 1 million molecules within the control volume.

The mean free path Λ divided by the characteristic length L of the system gives the dimensionless Knudsen number Kn

$$\mathrm{Kn} = \frac{\Lambda}{L}, \qquad (2.14)$$

which is used later to estimate the influence of the molecular mobility on the fluid behavior inside microstructures. The flow regimes of gases concerning rarefaction effects can be classified according to the range of the Kn number. Generally, rarefied gas behavior is classified into four different regimes, see Figure 2.10.

1. For $\mathrm{Kn} < 10^{-2}$, the continuum and thermodynamic equilibrium assumptions are appropriate and flow situations can be described by conventional no-slip boundary conditions. Some authors have suggested that the breakdown in the thermodynamic equilibrium assumption is already discernible at Knudsen numbers as low as $\mathrm{Kn} = 10^{-3}$, see for example [152].
2. In the range $10^{-2} < \mathrm{Kn} < 10^{-1}$ (commonly referred to as the slip-flow regime), the Navier-Stokes equations remain valid provided tangential slip-velocity and temperature-jump boundary conditions are implemented at the walls of the flow domain.
3. In the range $10^{-1} < \mathrm{Kn} < 10$ (transition flow regime), the continuum and thermodynamic equilibrium assumptions of the Navier-Stokes equations begin to break down and alternative analysis methods using the Burnett equations [153], particle-based DSMC (direct simulation Monte Carlo) approaches, or molecular dynamics (MD) simulations must be employed [147]. The difficulty in analyzing

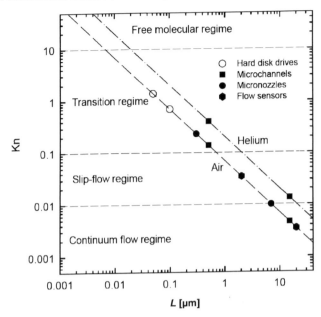

Fig. 2.10. Characteristic length scales of typical microfluidic components and the corresponding Knudsen number at standard atmospheric conditions and for two different gases, according to Barber and Emerson [151].

the transition regime from a continuum perspective arises from the fact that the stress-strain relationship for the fluid becomes non-linear within a distance of approximately one mean free path from the wall (the so-called Knudsen layer). The influence of curved surfaces or corrugated walls are still under discussion, see Sone [154] or Vasudevaiah and Balamurugang [155]. The so-called Crooke's radiometer or lightmill, a fascinating physical toy, employs rarefied gas scattering on differently textured surfaces under a temperature gradient to generate rotational motion of a wheel. In a short article series [156, 157, 158], Jones speculated about micro-spacecraft propulsion in the atmospheric mesosphere using defined gas scattering in the rarefied regime. Fundamental processes of wall and surface scattering are described by Goodman and Wachman [159] for various cases. Modern simulation methods allow the detailed treatment of gas-wall interactions. For example, Nedea et al. [160] propose a hybrid method consisting of molecular dynamics (MD) close to the wall and direct simulation Monte-Carlo (DSMC) in the bulk flow to determine the density distribution and transport properties of a gas close to a wall.

4. Finally, for Kn > 10, the situation can be described as free molecular flow. Under such conditions, the mean-free path of the molecules is far greater than the characteristic length scale and, consequently, molecules are reflected from a solid surface and travel, on average, many length scales before colliding with other molecules. The intermolecular collisions are thus negligible in comparison to

collisions between gas molecules and the walls of the flow domain. Passian et al. [161] performed exact measurements with a modified AFM cantilever and attached corrugated Si plate (500×263 μm^2 area, 2 μm thick with 2 μm square holes) for thermal transpiration in the free molecular regime of different gases. They found a 10 times higher force peak as in the continuous regime within a pressure range from 5 to 15 kPa. The highest force was measured for helium, air, nitrogen, and oxygen were significantly lower, and carbon dioxide caused the lowest force with approx. 40 % of helium.

The above limits are empirical and should be used as a guide. Additionally, the characteristic length scale of the device or flow problems are difficult to define. Hence, often the gradient of a macroscopic quantity is used for the definition, for example, of the density.

$$L = \frac{\rho}{|\nabla\rho|} \tag{2.15}$$

In Figure 2.10, the gas flow regimes in typical microfluidic devices are displayed for their characteristic length scale and the Kn number. Most devices operate in the slip-flow regime or early transition flow regime.

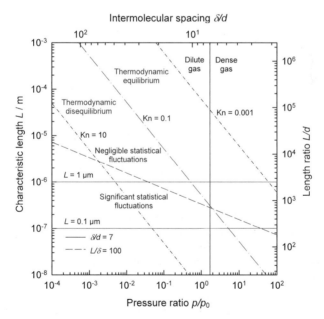

Fig. 2.11. Limiting criteria for the application of the Navier-Stokes equations to air with $\sigma = 4\cdot10^{-10}$ m, according to Barber and Emerson [151].

An alternative method to display the transition between continuum behavior and molecular behavior is shown in Figure 2.11. The conditions of a hard sphere gas are illustrated with a molecular diameter $d = 4\cdot10^{-10}$ m, which is approximately the molecular diameter of air. The left-hand ordinate represents the characteristic length scale L, while the right-hand ordinate shows the length scale normalized with the molecular diameter, L/d. The bottom abscissa represents the density normalized with a reference pressure p/p_0, which is equivalent to the normalized number density n/n_0 or density ratio ρ/ρ_0. Finally, the top axis represents the average distance between the molecules normalized with the molecular diameter δ/d.

Figure 2.11 shows the applicability limits of the Navier-Stokes equations given by $\delta/\sigma = 7$, $L/\sigma = 100$, and $Kn = \Lambda/L = 10^{-1}$. Air at standard ambient conditions has a pressure of $101\,315$ N/m² (101.315 kPa), a number density of $2.68666 \cdot 10^{25}$ m⁻³, a density ratio of unity (by definition) and $\delta/\sigma = 8.5$. With these values, air can be assumed to be a dilute gas, but is close to the upper limit of the dilute gas assumption.

The $L/\delta = 100$ line in Figure 2.11 represents the limit of molecular chaos. The continuum approximation is only valid if there are sufficient molecules within a given volume to achieve a statistically-stable estimate of the macroscopic flow properties. The remaining line with $Kn = \Lambda/L = 10^{-1}$ gives the validity limit of the thermodynamic equilibrium assumption. Since the mean-free path is inversely proportional to the number density n, the gradient of the thermodynamic equilibrium line (on a logarithmic graph) is three times steeper than the molecular chaos line.

According to Figure 2.11, the thermodynamic equilibrium assumption first fails if the flow dimensions are reduced in size, followed by a failure in the continuum assumption for a dilute gas. For a dense gas, the continuum assumption first fails, followed by a failure in the thermodynamic equilibrium assumption. Hence, special care should be taken for the limiting conditions of the Navier-Stokes equations. The usual flow classification system based solely on the magnitude of the local Knudsen number, is only one parameter to completely describe the system. More details on the treatment of gas flows in the transition regime are given in Barber and Emerson [151] with a comparison of recent simulation results. The review of current rarefied gas flow boundary conditions provides motivation for detailed experimental studies, which will give more insight and a better understanding of this problem.

2.3.2 The Boltzmann equation and balance equations

A closer look at the probability distribution f for the location $\vec{x} = f(x, y, z)$ and the velocity space of a particle $\vec{w} = f(u, v, w)$ varying with time, gives a better picture of the forces and energy distribution in a perfect gas. The integration of the probability density function (PDF) over the velocity space results in the number of particles in the control volume [162], p. 4.

$$\int f(t, \vec{x}, \vec{w})\, d\vec{w} = N(t, \vec{x}) \tag{2.16}$$

The integration of the PDF over the velocity space $\vec{w}$, divided by the entire mass m of the control volume, results in the fluid density ρ.

$$\rho(t, \vec{x}) = \frac{1}{m} \int f(t, \vec{x}, \vec{w})\, d\vec{w} \tag{2.17}$$

In his book on kinetic theory, Sone [154] discusses more properties of the PDF f. The integration of a state variable multiplied with the PDF over the velocity space gives the mean value of this variable. The total derivative of the PDF in an external field (for example, a gravitational field) is determined by the collisions of the molecules in a control volume, specifically, the current of gain and loss due to the molecule collisions.

$$\frac{\mathrm{D}f}{\mathrm{D}t} = \frac{\partial f}{\partial t} + \vec{w} \cdot \frac{\partial f}{\partial \vec{x}} + \frac{\vec{F}}{m} \frac{\partial f}{\partial \vec{w}} = J_{\text{gain}} - J_{\text{loss}} = \Delta J_{\text{coll}} \qquad (2.18)$$

This is the Maxwell-Boltzmann transport equation, an integro-differential equation. The determination of the loss and gain current, the so-called collision integral, and the construction of the PDF are the main problems in solving Eq. 2.18, see [163, 164]. The collision of two molecules with the relative velocity w_{rel} changes their probability f and f_1 before the collision to the probability f' and f_1' after the collision. Integration over a volume element and the velocity space of both molecules after the encounter determines the right side of Eq. 2.18, the collision integral [148] p. 263.

$$\Delta J_{\text{coll}} = \iint \left(f' f'_1 - f f_1 \right) \sigma \, w_{\text{rel}} \, \mathrm{d}^2 \Omega \, \mathrm{d}^3 \vec{w}_1 \qquad (2.19)$$

The difference between the PDF before and after the encounter of both particles is multiplied by the collision diameter σ and the relative velocity between both particles $\vec{w}_{rel}$. The product is integrated over the space around the reference particle $\mathrm{d}\Omega$ and the velocity space $\mathrm{d}\vec{w}_1$ of the approaching particle. Kaper and Ferziger [164] give other derivations of the collision term in Eq. 2.19.

The velocity distribution of a gas follows the Maxwell distribution, described in Eq. 2.5, if the collision integral vanishes in the equilibrium state. To describe a gas reaching the equilibrium state, Ludwig Boltzmann, [153] Chapter 4, defined a function $\mathcal{H}$ from the probability function f according to

$$\mathcal{H} = \int f \cdot \ln f \mathrm{d}w > 0, \qquad (2.20)$$

which is always positive. This is the so-called Boltzmann $\mathcal{H}$-theorem, which describes the development of irreversible processes from reversible single events like molecular encounters. The "Eta"-theorem is denoted with the capital letter $\mathcal{H}$ whose typing is nearly identical with the eighth letter of the Roman alphabet, see Brush [165]. According to Flamm in the introduction of Boltzmann's collected works [166], the symbol $\mathcal{H}$ was introduced by Burbury in 1890 instead of the symbol E used by Boltzmann himself. Burbury intended to avoid the misleading usage of the symbol E, which is also used for the energy. In 1937, Chapman assumed that Boltzmann had introduced the symbol $\mathcal{H}$, however he adopted the symbol H. To avoid these discussions, Cercignani [167] introduced the symbol $\mathcal{H}$ without naming the letter, which is also adopted here.

The $\mathcal{H}$ theorem is regarded as a paradoxon, because it leads from reversible encounters described by the probability distribution f to the irreversible change of states. The influence of the $\mathcal{H}$ theorem is still under discussion. The function $\mathcal{H}$

is also connected to the entropy S, whose temporal derivative is always negative for irreversible processes $dS/dt < 0$.

The collision term in Eq. 2.19 is also used and implemented in the numerical Lattice methods, see Krafzyk [168], and Sukop and Thorne [169]. To give a clear distinction, lattice methods are simulation methods based on cellular automata and do not solve the Boltzmann transport equation directly, see Section 2.4.5. The cellular automata contain a transport term comparable to the right side of Eq. 2.18.

The first moments of the PDF f_J, for which the collision integral will vanish in the event of local equilibrium, are also called the collision invariants and will be used to yield first solutions of Eq. 2.18. With these collision-invariant variables f_J, the Boltzmann equation can be simplified to

$$\frac{\partial}{\partial t}(\rho\, f_J) + \frac{\partial}{\partial x}(\rho\, f_J\, \vec{w}) + \frac{\rho\, \vec{F}}{m}\frac{\partial f_J}{\partial \vec{w}} = 0. \tag{2.21}$$

For $f_J = 1$ the mass conservation equation is derived from the first two terms on the left side of the above equation.

$$\frac{\partial \rho}{\partial t} + \operatorname{div}(\rho\, \vec{w}) = 0 \tag{2.22}$$

For the velocity components $f_J = u, v, w$, the momentum conservation equation for all three space coordinates is derived.

$$\rho\left(\frac{\partial \vec{w}}{\partial t} + \vec{w}\cdot\operatorname{div}\vec{w}\right) + \operatorname{grad} p + \operatorname{div}(\operatorname{grad}\vec{w}) - \frac{\rho\, \vec{F}}{m} = 0 \tag{2.23}$$

This equation is the basis for the Navier-Stokes equation of fluid dynamics, see Section 3.1 for further treatment. The collision invariant $f_J = 1/2\cdot\vec{w}^2$, which describes the continuity of the kinetic energy during the collision process, leads to the following equation.

$$\frac{\partial}{\partial t}\left(\frac{\rho}{2}\overline{\vec{w}^2}\right) + \operatorname{div}\left(\frac{\rho}{2}\overline{\vec{w}^2\cdot\vec{w}}\right) - \frac{\rho\, \vec{F}}{m}\cdot\overline{\vec{w}} = 0 \tag{2.24}$$

With the following simplifications for the velocity mean values

$$\overline{w^2} = \frac{3p}{\rho} + \overline{w}^2 \quad \text{and}$$

$$\overline{w^2\cdot\vec{w}} = \frac{2}{\rho}(-\lambda\,\operatorname{grad} T),$$

the energy equation can be written with the specific energy e of the system

$$\rho\left[\frac{\partial e}{\partial t} + \vec{w}\cdot\operatorname{grad} e\right] + \operatorname{div}(-\lambda\,\operatorname{grad} T) + p\,\operatorname{div}\vec{w} + \dot{e}_q = 0, \tag{2.25}$$

where the term $\dot{e}_q$ describes the shear and deformation tensors in the energy equation, which can be added with the viscous dissipation. A more detailed discussion on the energy equation is given by Sommerfeld [148], pp. 266. At this point, the illustration of the link between the micro state Boltzmann equation, Eq. 2.18 and the macro state balance equations, Eqs. 2.22 to 2.25 is satisfactory and gives good insight of molecular processes. The macroscopic equations are valid, if enough molecules act together to receive a smoothed signal from the collisions, see Eq. 2.13. A in-depth discussion of special micro-scale effects can be found in Sone [154]. In his book on the Boltzmann equation, Babovsky [170] illustrates the derivation of hydrodynamic equations from asymptotic solutions.

2.3.3 Macroscopic balance equations

The macroscopic calculations and balance equations for process equipment can be formulated on various levels as shown in Figure 2.12. The equipment level is the fundamental calculation level for plant design and layout. Equipment elements, such as tubes or trays, can be balanced on the level of a differential length, which refers to the "micro process" description. The third and base level is represented by differential elements (in Figure 2.12 for all three dimensions with flow in z-direction), which allows the continuum approach and an integration over the balance space.

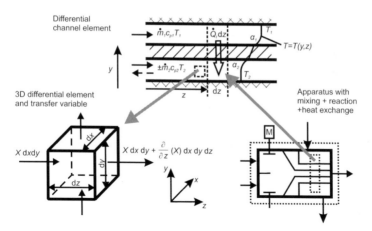

Fig. 2.12. Overview of the various balancing volumes in process engineering (from left to right): 3D differential element for general calculations, 1D differential element depending on the geometry of the active area (channel, tube, plate, film, interface, see Figure 2.7), and complete equipment for process balances, here, an active mixer according to Figure 2.6. The hierarchy of the balancing volumes is similar to the setup of devices and plants from single elements, see Figure 2.2.

In Figure 2.12, the balanced parameters in a differential element are given by X, which stands for the mass, species, momentum, or energy. The general balance equation with temporal and spatial derivatives of a system and a differential element with the volume V is written as:

$$V \cdot \frac{\partial}{\partial t} X = w \cdot \left[X \, dy \, dx - \left(X \, dy \, dx + \frac{\partial}{\partial z} X \, dz \, dy \, dx \right) \right] \tag{2.26}$$

with X as the general balanced value, see Figure 2.12.

In chemical engineering, the values of process parameters from balance equations are often displayed in tables combined with process flow diagrams. This combined display gives a comprehensive overview and is the result of basic engineering design. Blass [64] describes the design and development of chemical processes in more detail, which includes the usage of balance sheets, process flow diagrams, and pipe and instrumentation (PI) diagrams, see Figure 2.13.

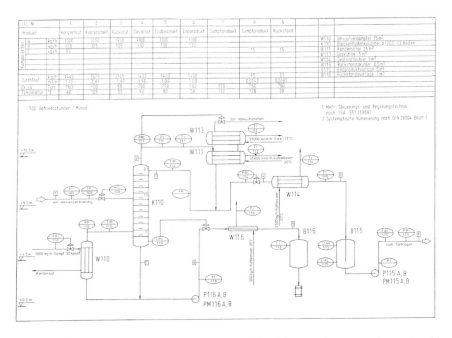

Fig. 2.13. PI diagram of a distillation column, peripheral heat exchangers and vessels with mass flow balance in tabular form, according to DIN ISO 10 628 [64].

The rather complex balance equations can be simplified when considering the actual process. In a steady process, the temporal derivative vanishes, $\partial / \partial t = 0$. In systems with high velocities, the convective transport is dominant compared to conductive and diffusive fluxes. For systems with dominant chemical reactions, only a change of substance needs to be considered. Incompressible flow is not a factor for density changes. Also, simple pipe and channel flows without chemical reactions or technical work are often found in technical systems. Other limitations on gas flow and transport processes are later in this section (2.3.1 and 2.3.6).

Within the systems treated in this chapter no mass is generated or destroyed, hence, no sink or source appears in the balance equation 2.1. The mass balance for a process element is written as

$$\sum m_{in} = \sum m_{out} - \sum m_{conversion} \tag{2.27}$$

For time dependent mass flow rates, Eq. (2.1) can be written as

$$\frac{\partial m}{\partial t} = \dot{m}_{in} - \dot{m}_{out} \tag{2.28}$$

If the fluid density is replaced by the concentration of the relevant species, the species balance of a system can be derived from the mass balance. For an arbitrary process volume, this equation can be written as

$$\sum \dot{n}_{in} = \sum \dot{n}_{out} - \sum \dot{n}_{conv} \tag{2.29}$$

without chemical reactions. Regarding the species transport, the balance value X for a differential element is $X = \rho\, c_i\, w_i$, which results in

$$\frac{\partial (\rho\, c_i\, A)}{\partial t} = -\frac{\partial (\rho\, c_i\, w\, A)}{\partial z}. \tag{2.30}$$

The transport process of species will be used in the chapters on mixing and chemical reactions.

The momentum of a moving fluid can be expressed as the product of the mass and the flow velocity. The integral of the momentum over the volume results in a net force of the fluid on the volume boundary or on the equipment. In general, the momentum balance of a device can be written as

$$\sum J_{in} = \sum J_{out} - \sum J_{loss} \tag{2.31}$$

The momentum loss can be interpreted as the viscous momentum loss, which is expressed as pressure loss along the channel or device flow. For a non-viscous fluid flow through an arbitrary channel with an external force, the momentum change can be expressed by the momentum, the pressure, and the gravity part:

$$\frac{\partial (m\, w)}{\partial t} = (\dot{m}\, w)_{in} - (\dot{m}\, w)_{out} + (p\, A_i)_{in} - (p\, A_i)_{out} + m\, g + F_z \tag{2.32}$$

The detailed derivation of the momentum equation, the Navier-Stokes equation, can be found in Section 3.1.1. Additional forces in microfluidic applications result from with surface effects in multiphase flow, such as bubbles and droplets, see also Section 3.3 on multiphase flow.

Energy can take various forms in process technology. Similar to the mass, the energy itself is conserved according to the first law of thermodynamics for open systems.

$$\sum \dot{E}_{in} = \sum \dot{E}_{out} - \sum \dot{E}_{diss} \tag{2.33}$$

The energy dissipation takes into account that energy conversion from one form into another is accompanied with natural losses. These losses are characterized by the entropy generation during a process according to the second law of thermodynamics, which is treated in Eq. 2.39.

In the case of a channel with a constant cross section, without chemical reactions and technical work consumed or produced, the entire energy of the fluid can be expressed according to the first law of thermodynamics with kinetic and potential energy.

$$m\,e = \rho\,A\left(u + \frac{1}{2}w^2 + g\,z\right) \tag{2.34}$$

For open systems, the inner energy u is replaced by the enthalpy $h = u + p/\rho$ including the displacement work. Herwig [171] discussed the combination of the inner energy with the displacement work and advised to be cautious with the physical interpretation of the enthalpy and the displacement work. In flow processes, the total enthalpy $h_t = u + p/\rho + w^2/2 - g\,z$ is used with the kinetic and potential energy part [76] p. 66. As the energy dissipation comes from the shear stress and the velocity gradient $\varepsilon = \partial\,(\tau\,w)/\partial z$ (here only the z-direction), the enthalpy form of the energy balance can be written as

$$\rho\frac{d\,h}{d\,t} = \frac{d\,p}{d\,t} + \varepsilon - \operatorname{div}\dot{q}. \tag{2.35}$$

The caloric equation of state gives the correlation between the inner energy or enthalpy and the temperature,

$$d\,u = c_v\,d\,T; \quad d\,h = c_p\,d\,T, \tag{2.36}$$

the energy equation can be rewritten as

$$\rho\,c_p\frac{d\,T}{d\,t} = \frac{d\,p}{d\,t} + \varepsilon - \operatorname{div}\left(\lambda\,\operatorname{grad} T\right). \tag{2.37}$$

The solution of this equation gives the temperature distribution for the actual process. Besides the conduction and convection, heat may also be transported by radiation, which is proportional to T^4 and the emission properties of a surface, however, radiation heat transfer is not treated here, interested readers are referred to [89] Chapter 5, and [172] Chapter 6.

A chemical reaction within the system influences not only the species equation, but the reaction enthalpy Δh_R must also be considered in the energy balance due to the apparent heat consumption or release.

$$\Delta h_R = h_p - h_r \tag{2.38}$$

This net heat balance is calculated from the enthalpy of the reactants h_r and the products h_p, see also Bejan [173] p. 371 and Section 4.1.1. In Section 6.1.1, the reaction enthalpy Δh_R is treated together with heat transfer and microreactor design.

Dividing the transferred heat by the temperature, a new state variable, the entropy s, is derived for further characterization of states and processes. With the thermodynamic correlations

$$ds = \frac{dq}{T},$$
$$T\,ds = du + p\,dv \qquad (2.39)$$
$$= dh - v\,dp,$$

the entropy correlation can be derived from Eq. 2.35

$$\frac{ds}{dt} = \frac{1}{T}\left\{\frac{\varepsilon}{\rho} - \frac{1}{\rho}\mathrm{div}\dot{\vec{q}}\right\}. \qquad (2.40)$$

The dissipation function ε is the friction loss per volume and time unit. In adiabatic systems $dq = 0$, entropy always increases $ds > 0$ by dissipation and irreversible processes, such as pressure loss or concentration homogenization by mixing, see Szargut et al. [174].

The entropy production is a major indication of the efficiency of a process. Generally, the efficiency of a process is described by the ratio of the profit or outcome to the effort required for the process. The effort for a process comes from the input, such as energy, pressure loss, or heat loss. Furthermore, the effort may be defined from the fabrication process for the device, the materials used, the effort to maintain correct operation, or reliability issues. The profit from a process depends on the task. It may be the volume flow rate or throughput, the energy transfer, the mixing intensity, or the conversion rate, selectivity or space-time yield of a chemical reactor. Due to the large variety of devices in micro process engineering, the application task is also very important. A mixing device for bio-analytical purposes possesses different specifications than a mixer for chemical reactions in industrial production. A micromixer for a fast chemical reaction may be unsuitable for a slow reaction and lead to insufficient results, therefore, the application of the device must be correctly specified, as well as the efficiency. These various applications may come from analysis systems (μTAS), from the lab-on-a-chip field, from process or material screening, from energy conversion, or from chemical synthesis and production.

For a detailed description of entropy analysis and systems optimizations, the interested reader is referred to the textbook of Bejan [175]. Basically, various control strategies are applicable for energy savings and entropy minimization [176]. For continuous flow systems, high pressure losses and unnecessary throttling should be avoided. In mixing, high solution concentrations as well as high dilutions lead to high separation effort. Heat transfer devices with high temperature gradients induce high entropy production and can be damaged in extreme cases. Linnhoff proposed an optimization method for energy integration, called pinch analysis [177], based on the temperature gradient and enthalpy content of fluid streams in a process.

2.3.4 Elementary transport processes and their description

The balance equations are derived from equilibrium states and do not reflect the transport processes between the single states itself. If the gradients are not too high (valid in this case), the transport processes can be described with linear correlations and phenomenological coefficients. With higher gradients, transport coefficients of higher order approximation must be applied, see Section 2.3.6. The entire change of a state variable is described by the transport processes of conduction in the immobile phase (solids or resting fluids), convection in the fluid phase (gases and liquids), and by the generation or depletion in the control volume.

$$
\begin{bmatrix} \text{Total flow} \\ \text{density} \end{bmatrix} = \begin{bmatrix} \text{Conduction} \\ \text{over the} \\ \text{boundary} \end{bmatrix} + \begin{bmatrix} \text{Convection} \\ \text{over the} \\ \text{boundary} \end{bmatrix} + \begin{bmatrix} \text{Source} \\ \text{or} \\ \text{Sink} \end{bmatrix} \tag{2.41}
$$

The last part of the above equation can be complemented by accumulation and generation of the quantity. The single parts can not be added directly. Depending on the geometrical setup of the particular device, single serial resistances or parallel conductivities can be added directly. The transport correlation of a single quantity can be expressed in the fundamental form

$$
J = L \cdot A \cdot X. \tag{2.42}
$$

In short: the flux (intensity) depends on the driving force (parameter gradient or difference) multiplied with the active surface (cross section normal to the flux), and the intensity (transport coefficient). The main transport phenomena in process engineering are given in Table 2.2 supplemented by the electrical transport. In 1960, Bird, Steward, and Lightfoot [1] established the logic for the unified study of momentum, heat, and mass transfer in continua. An updated version of their classical textbook was published in 2002 [2]. Brodkey and Hershey give an elementary introduction of basic concepts [178] and applications [179] of transport phenomena for undergraduate level. A modern analysis of transport phenomena is presented by Deen [180] and Kraume [181] for chemical engineering.

Conductive transport is driven by a parameter gradient; convective transport is always accompanied by a volume flow rate with a certain mean velocity. Transfer flow is introduced for complex geometrical and flow situations where a boundary layer with thickness δ determines the transport process and a transport coefficient includes all conductive elements. The geometrical situation of the flow, the temperature and concentration distribution at the channel wall is displayed in Figure 2.14, left side.

Transfer coefficients: $\beta = \frac{D}{\delta}$, [m/s], $\alpha = \frac{\lambda}{\delta}$, [W/m^{2}K], $a = \frac{\lambda}{\rho\,c_p}$, [m^2/s], $\gamma = \frac{\eta}{\delta}$, [kg/m^{2}s], with the boundary layer thickness δ

The coefficients for conductive transport of species, momentum, and heat energy, $D, \nu = \eta/\rho$, and $a = \lambda/\rho c_p$, respectively, have the same physical unit of [m^2/s] and can be combined to dimensionless numbers. The dynamic behavior of viscous fluids is governed by fluid velocity $\bar{w}$, geometry, and viscosity, displayed in the Reynolds number Re $= \bar{w} d_h/\nu$. For the derivation of dimensionless numbers, see also Section 2.4.1. Species transfer in convective flow is influenced by fluid velocity $\bar{w}$ and diffusion coefficient D, where the Re number is accompanied by the Schmidt number Sc $= \nu/D$. The Sc number indicates the ratio between momentum and species

Table 2.2. Transport mechanisms and transport properties

Flow process	mass	species	momentum	heat	electrons
Conduction	porous membrane $\dot{m} = k\,\Delta p$	$\dot{n}_i = -DA\nabla c_i$	$J = F$ $= -\eta A\nabla w$ $= -\nu\rho\nabla w$	$\dot{Q} = -\lambda A\nabla T$ $= -a\rho c_p A\nabla T$	$I = \dot{E}$ $= \frac{1}{R_{el}}\nabla U$
Convection	$\dot{m} = \rho\,Aw$	$\dot{n}_i = \rho c_i\,Aw$	$J = F$ $= \frac{\rho}{2} A\,w^2$	$\dot{Q} = \rho\,c_p\,\Delta TA\,w$	$P = U\cdot I$
Transfer flow	$\dot{m} = \rho\,A\,w$	$\dot{n}_i = \beta\,\rho\,A\,\Delta c_i$	$J = F$ $= \gamma A\,\Delta\,w$	$\dot{Q} = \alpha A\,\Delta\,T$	Flow of electrolytes [182]

transfer. For Sc $= O(1)$, as with gases, momentum transfer and fluid dynamics are in the same order-of-magnitude. In this case, concentration gradients behave similar to velocity gradients. For high Sc numbers, such as for liquids, concentration gradients continue much longer than velocity gradients, which is very important for mixing issues.

Convective heat transfer is governed by fluid velocity $\bar{w}$ and temperature conductivity or diffusivity a, which produces, combined with the viscosity, the Prandtl number Pr $= \nu/a$. Low Pr numbers indicate low thermal conductivity, as in oils or organic liquids. The Pr numbers of air or water have the order of $O(1)$, permitting a heat transfer as fast as momentum transfer. Coupled heat and mass or species transfer, such as convective condensation of aerosol droplets, are described by temperature conductivity and diffusion coefficient. Both coefficients are combined to the Lewis number Le $= a/D$, see also Section 5.3.1. A high Le number allows droplet generation from vapor cooling, while for low Le numbers, vapor will directly condense at the wall.

Both single-phase transport and transport processes at interfaces occur often in process engineering. For the fluid/fluid exchange of mass, species and energy, there are several models to describe the transport process. The *two-film model* is one of the most common models [181, 180], which gives good analytical results and shows a physically complete picture, see Figure 2.14, right side.

For a liquid A and a fluid B, which may be gaseous or liquid with a low miscibility in A, the species transport equations are written in the following form

$$J_i^A = \beta_i^A\,A\left(c_{i,\text{bulk}}^A - c_{i,\text{interf}}^A\right), \tag{2.43}$$

$$J_i^B = \beta_i^B\,A\left(c_{i,\text{interf}}^B - c_{i,\text{bulk}}^B\right), \tag{2.44}$$

with the mass transfer coefficient $\beta_i = D_{Ai}/\delta_i$. If fluid B is a perfect gas, the concentration can be replaced by the partial pressure and the perfect gas law can substitute the transfer coefficient due to the fast diffusion in gases.

For the absorption of a gas into a liquid, Henry's law describes the solubility of the gas in the liquid with the help of the Henry coefficient H_i [65] p. 535.

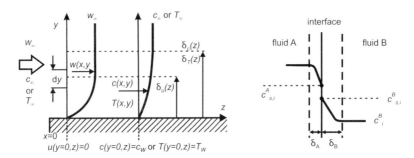

Fig. 2.14. left: Transport process in a differential element in channel flow with velocity w, temperature T, and concentration c profiles at the wall; right: Two-film model with two different interface conditions leading to different boundary layer thickness δ [135] pp. 93. Detailed measurements by Wothe [183] indicate a uniformly continuous concentration distribution over the interface of two immiscible liquids, which requires additional investigations.

$$p_{i,g} = H_i \, c_{i,i} \tag{2.45}$$

For the evaporation or condensation of a binary mixture (distillation, rectification), the relationship between the concentrations at the interface is described by Raoult's law [65] p. 278,

$$p_i = p_t \, y_i = \pi_i \, x_i \tag{2.46}$$

with the partial pressure p_i, the vapor concentration y_i, and the liquid concentration x_i. The concentration difference at the interface of emulsions can be described with the Nernst distribution for extraction applications [65], p. 634. An example of an extraction process with microstructured devices is given in Section 1.5.2 and Figure 1.9. The above equations and correlations support the primary calculations of mass transfer in multiphase flow reactors and devices.

2.3.5 Molecular velocities and macroscopic fluid properties

Aside from the balance equations, some macroscopic gas properties can be derived from the kinetic theory and molecular behavior. The derivation of these properties follows the outline of Sommerfeld [148]. Boon and Yip give in their monograph a comprehensive picture of molecular behavior and fluid dynamics [184]. The viscosity of a fluid describes the internal friction by molecular encounters and reflection under a velocity gradient. The linear velocity distribution and the geometry of a molecular encounter is shown in Figure 2.15, where a molecule hits a plane with the angle ϑ.

The fluctuations of molecules within a fluid contribute to its momentum transfer. The momentum of molecules striking a plane from the upper half sphere is expressed with the molecular mass M_m, the mean molecular velocity $\bar{c}_M$, the striking angle ϑ and the mean free path Λ by

$$\bar{\mathbf{p}}_u = M_m \bar{c}_M^+ (+\Lambda \cos \vartheta), \tag{2.47}$$

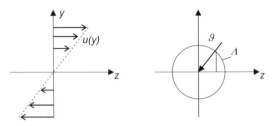

Fig. 2.15. Velocity gradient $w(y)$ along the z coordinate (left), and molecules hitting a plane under the angle of ϑ.

and similarly for the lower half sphere by

$$\bar{\mathbf{p}}_l = M_m \bar{c}_M^- (-\Lambda \cos \vartheta). \tag{2.48}$$

The Taylor expansion of the momentum difference yields an expression with the velocity gradient

$$\bar{\mathbf{p}}_l - \bar{\mathbf{p}}_u = \bar{\mathbf{p}} \approx -2 M_m \Lambda \cos \vartheta \left(\frac{\partial w}{\partial y} \right). \tag{2.49}$$

The integration over the entire surface introduces the number of particles N in the control volume to the entire momentum change at point 0

$$\bar{\mathbf{p}} = \mp \frac{M_m N \Lambda \bar{c}_M}{3} \left(\frac{\partial w}{\partial y} \right)_0. \tag{2.50}$$

A comparison of the coefficients in Stokes'/Newton's law for the linear relationship between the velocity gradient and the shear stress

$$\sigma_{yx} = -\eta \left(\frac{\partial w}{\partial y} \right) \tag{2.51}$$

with Eq. 2.50 gives the correlation for the viscosity

$$\eta = \frac{M_m N \Lambda \bar{c}_M}{3}. \tag{2.52}$$

The term $M_m N$ in Eq. 2.52 represents the density ρ of the gas

$$\eta = \frac{1}{3} \rho \Lambda \bar{c}_M. \tag{2.53}$$

The factor 1/3 is only a rough value from simplified calculations. A more exact treatment of the molecular encounters in the Boltzmann transport equation yields a better correlation for the viscosity according to Sommerfeld [148] p. 280:

$$\eta = \frac{5\pi}{32} \rho \Lambda \bar{c}_M = 0.4909 \rho \Lambda \bar{c}_M, \tag{2.54}$$

and according to Chapman and Cowling [153] Chapter 12:

$$\eta = 0.499 \rho \Lambda \bar{c}_M. \tag{2.55}$$

This result is regarded as the most precise correlation of viscosity from the kinetic theory of gases. Thermodynamic correlations for a perfect gas $p = N R_m T$, the mean velocity $\bar{c}_M$ from Eq. 2.8 and the mean free path Λ from Eq. 2.10 give the dynamic viscosity of gases

$$\eta = \frac{M_m}{\sqrt{6 \pi \sigma^2}} \sqrt{R T}. \tag{2.56}$$

This equation clearly indicates the independence of dynamic viscosity from pressure over a wide range. The viscosity increases with the square root of the temperature, converse to the typical viscosity behavior of liquids, which decreases with increasing temperature. In a similar way, the heat conductivity of perfect gases can be derived from the change of kinetic energy $E_{kin} = 3/2\, k\, T$ for a monatomic gas under a temperature gradient $\partial T / \partial y$.

$$E_l - E_u \approx -2 \frac{2}{3}\, k \cdot \Lambda \cos \vartheta \left(\frac{\partial T}{\partial y} \right)_0 \tag{2.57}$$

The integration over a volume element yields for the energy transport

$$\Delta E = -\frac{N k \Lambda \bar{c}_M}{2} \left(\frac{\partial T}{\partial y} \right)_0, \tag{2.58}$$

which can be compared to Fourier's law of heat transport $\dot{q} = -\lambda \partial T / \partial y$ to determine the heat conductivity

$$\lambda = \frac{N k}{2} \Lambda \bar{c}_M = \frac{k}{\pi^{3/2} \sigma^2} \sqrt{R T}. \tag{2.59}$$

The heat conductivity is also independent from the pressure over a wide range and also proportional to $\sqrt{T}$ like the dynamic viscosity. A comparison of both coefficients in Eqs. 2.52 and 2.59 for a monatomic gas gives the following correlation

$$\frac{\lambda}{\eta} = \sqrt{\frac{6}{\pi} \frac{k}{M_m}}. \tag{2.60}$$

Similarly, the diffusion coefficient can be derived for a species in a surrounding gas. Reif [150] and Bird et al. [2] give the following correlation for the diffusivity of a monatomic species A in a surrounding gas B with the same mass and scattering cross section:

$$D_{AB} = \frac{1}{3} \Lambda \bar{c}_M. \tag{2.61}$$

The mean velocity of the molecules is determined according to Eq. 2.8. These correlations belong to the major achievements of kinetic theory and are an early demonstration of the atomic nature of matter, here of gases. In a similar way, the heat capacity of a gas or the diffusion coefficient of gaseous molecules can be attained. Kittel derived in his textbook [185] the electrical conductivity in an electron gas from the Boltzmann transport equation. A general introduction and summary of molecular thermodynamics and transport phenomena is given by Peters [186] with emphasis on space and time scales.

2.3.6 Limits of linear transport properties

The fundamental transport equations described in Section 2.3.4 include transport properties for diffusion, heat conduction, or viscous motion of a fluid, which are regarded as a linear relationship between the force and the corresponding flux, see Table 2.3.4. In small structures, these linear coefficients may be limited due to the influence of the surface, which plays a major role for gases, see Sections 2.3.1 and 1.5.1 as well as Ferziger and Kaper [164] for additional reading. For liquids, the limits of the linear correlation are reached in far smaller scales, which are outside the scope of this book and most applications in micro process technology. Interested readers are referred to Frenkel [187].

Surface effects become more important in microstructures due to the higher surface-to-volume ratio. The viscosity of a liquid is reported to be higher in the vicinity of a surface. Visco-elastic effects due to the surface influence are described in [187], Chapter 6. The solid surface of a dielectric material induces, due to its electrical charge, an adsorbed layer of molecules from the liquid, the Kelvin-Helmholtz electro-kinetic double layer EDL. This layer extends approximately three molecular layers from the wall into the bulk fluid and induces a charge orientation in the adjacent fluid. This layer of mobile, charged molecules is called the Stern layer. In very small channels ($d_h \approx 100$ nm) and for relatively large molecules, this layer may influence the flow and the transport processes at the wall increasing the wall friction. In these cases, an order-of-magnitude estimation helps quantify the influence of this effect. The EDL also influences the local Nußelt number Nu. In micropolar flow, for example, the Nu number is approx. 7 % smaller than in laminar flow, see [188] Chapter 32.

Capillary forces may block microchannels, but can also be used for filling and directed transport of liquids in microstructures. Many aspects of interfacial transport processes are currently under investigation, a good overview is given by [189]. Importantly in process engineering, the vapor pressure depends not only on the temperature, but also on the surface shape. With curved surfaces (bubbles with a concave surface, droplets with a convex surface, capillary filling), the vapor pressure p_s depends on the surface curvature, see Figure 2.16 and Mersmann [190] Chapter 1.

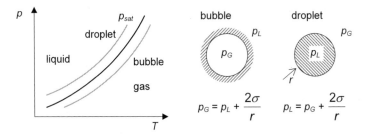

Fig. 2.16. Relationship between the vapor pressure and the surface curvature. In small bubbles, the gas dissolves sooner into the surrounding liquid. From a droplet, the liquid evaporates sooner into the surrounding vapor [190].

The vapor pressure is determined with the Gibbs-Thompson equation at strongly curved surfaces

$$\ln\frac{(p_s)_r}{p_s} = \frac{2\sigma}{R\,T\,\rho_L\,r}. \tag{2.62}$$

For concave surfaces (bubbles), the vapor pressure is low, with convex surfaces (droplets), the vapor pressure is higher than at a plane surface. Hence, during evaporation, initially small bubbles are formed, whereas small droplets occur first during condensation. For water and most organic liquids, the pressure difference is lower than 1 % for bubbles or droplets smaller than 100 nm in diameter. For example, a water bubble with a diameter of 40 nm has a 5 % lower vapor pressure than a planar surface. A detailed discussion on fluid properties and equilibrium conditions of curved interfaces is given by Mitrovic [191], who describes the energy jump across the interface as an external force.

For high force gradients, the linear correlations must be enlarged by terms of higher order, which includes the relaxation time of fast processes. This is important for high energy densities, such as laser processing or fast chemical reactions (explosions). The Fourier equation, for example, must be expanded in nonlinear terms for heat transfer processes with high energy fluxes or high temperature gradients [172]. The linear relationship is valid for temperature gradients up to a few thousand Kelvin per cm in the order of $O(100 \text{ K/mm})$, which is also 0.1 K/μm in microsystems. Here, the more accurate Maxwell-Catteneo equation is used to describe the heat flux vector [192].

$$\vec{q} = -\lambda \text{ grad} T - t_q \frac{\partial\,\vec{q}}{\partial t} \tag{2.63}$$

The measurement of the temporal propagation of heat waves gives the order-of-magnitude of the relaxation time t_q, which is $O(10^{-12} \text{ s})$ for metals, $O(10^{-11} \text{ s})$ for liquids, and approx. $O(10^{-9} \text{ s})$ for gases.

The anisotropic conduction in crystals becomes important in microstructures smaller than 1 μm. Recently, Chang and co-workers [193] observed an asymmetrical heat conductivity of carbon nanotubes with a thin metal layer on the outside. The effect is explained by the heat transfer over phase borders, which induce an irreversible effect. Besides quantum mechanical description, these processes can also be described by Rational Irreversible Thermodynamics [194] or by Extended Irreversible Thermodynamics [195], where further application limits of linear correlations can be found.

2.4 Modeling, Calculation Methods, and Simulation

Various levels of knowledge exist for technical processes. The basis is a global, but generally correct interpretation of the observed phenomena. On a second level, a detailed analysis is often obtained within the frame of measurement rules or empirical correlations, which deliver a reliable solution for certain phenomena in praxis. Thirdly, as most exact knowledge, a theory is elaborated, which shows the physical background of the investigated process [15] Chapter 1. Calculation in process engineering can be classified according to this knowledge and the need for practical application to exact mathematical calculations with complete knowledge of the correlations and solutions of the differential equations (ideal case), calculation with simplified models, where the grade of simplification determines the accuracy (numerical simulations like CFD or FEM are always simplifications and approximations), inter- and extrapolation of experimental results, or, finally, application of dimensional analysis and similarity theory for general laws and for scaling up or down.

Besides the knowledge and application of mathematical and engineering tools, accurate and critical judgment and interpretation of results is essential for the reasonable and responsible work of an engineer. The main sources of error in a calculation process are:

- incomplete or erroneous information of states and behavior of technical systems. This plays an important role in complex systems, where emergent behavior cannot be exactly predicted.
- numerical procedures, which are always general approximation methods and only approximate the exact solution, thereby possessing an inherent margin for error. These errors must be estimated by the numerical code, such as residuals or final mass or energy balance. Numerical results can be validated with benchmark problems against analytical results or experimental data.
- the internal treatment of numbers by programs and routines that can lead to errors, for example by truncations, which may influence the calculation of complex, iterative systems.

Hence, a multifaceted approach from different sides, including analytical and numerical calculations, as well as experimental validation, will assist the comprehensive treatment of microstructured devices.

2.4.1 Physical variables and dimensional analysis

Physical state variables are defined by a numerical value and a physical dimension. The dimension is the generalization of the physical parameter. For example, the mechanical theory can be explained with the help of three main dimensions: length (meter, L), mass (kilogram, M), and time (second, T). Units derived from these values are the volume (L^3), the density (ML^{-3}), the velocity (LT^{-1}), or the acceleration (LT^{-2}). Other basic units are the temperature (Kelvin, Θ), the species amount (mol, N), or the electrical current (ampere, I). The important physical parameters of a process can be combined to dimensionless numbers. They allow for a simple treatment of complex systems and to elaborate similarities from already solved problems.

There are four ways to derive the dimensionless numbers from the problem description [196]:

1. dimensionless groups from the dimensionless differential equations, for example, the Re number from the dimensionless Navier-Stokes equations,

2. the analysis of the ratio of important forces, apparent energy currents, and characteristic times. For example, the ratio of inertia forces to viscous forces leads to the Reynolds number Re,
3. the reduction of the variables by transformation to dimensionless variables, like the dimensionless time of unsteady heat conduction, which is also the Fourier number Fo,
4. and with the Π-theorem of Buckingham applied to the set of important parameters describing the physical process.

A systematic procedure to derive dimensionless numbers is given by Wetzler [197], who lists approx. 400 numbers to describe various physical processes. An important method to achieve dimensionless numbers of a physical problem is Buckingham's Π-theorem [196]. The theorem states that every equation with correct dimensions can be written as a correlation of a complete set of dimensionless parameters or numbers.

$$\underset{\text{physical variables}}{f(x_1, x_2, \dots x_n) = 0} \Rightarrow \underset{\text{dimensionless numbers}}{F(\Pi_1, \Pi_2, \dots \Pi_s)} \tag{2.64}$$

The dimensionless numbers are the potential product of the physical variables

$$\Pi = x_1^{a_1} \cdot x_2^{a_2} \dots x_n^{a_n} \tag{2.65}$$

with the exponents a_i of the physical variables x_i. The general procedure for creating dimensionless numbers is to collect a list of the important physical variables x_i, determining the physical problem, and prepare the matrix of basic dimensions for the variables x_i. The matrix is arranged to the conditional equations for the exponents a_i. One or more independent variables x_i are reasonably selected with a_i, using known dimensionless numbers for first simplifications [196]. Potential equations, such as Eq. 2.65, are set up to determine the exponents with the help of experimental data, analytical, or numerical models.

This procedure gives many possibilities for the dimensionless characterization of a physical problem. The gained dimensionless groups can be simplified, if the main variable appears only in one number or group. Using simple numbers (simplex) of similar variables, such as length or time ratios, help at the beginning to find preliminary dimensionless numbers.

2.4.2 Similarity laws and scaling laws

Based on the dimensionless numbers, the similarity of processes plays an important role in process engineering. The benefits of dimensionless numbers and groups are the possibility of scaling, the reduction of variables, the independence from the unit system (SI or British Units), and the similarity of processes with the same values for dimensionless numbers. One major drawback is their limited application range. The dimensionless numbers and their combinations are only valid for the model system, for which they are defined. For example, the Re number number for pipe flow is somewhat different from the Re of a flow over a plate, and should not be confused.

Also the correlations between the dimensionless numbers are only valid for a certain range. The friction factor for laminar channel flow is inversely proportional to the Re number, which is not valid for turbulent channel flow ($Re > Re_{crit}$). The influence parameters must be chosen carefully and adjusted to suit the actual problem. The dimensionless numbers do not offer any specific physical insight, but allow a more structured description of the process.

The principle idea of scaling is the concept that all physical rules are independent from dimension system being used. The application of the scaling laws employs the geometric similarity of the systems and the dynamic similarity of the systems. The relative values of temperature, pressure, velocity, and other parameter in a system should be the same on both scales. Additionally, boundary conditions should be the same, such as constant temperature or zero velocity at the wall.

Miniaturization is assisted greatly by using dimensional analysis and the transfer of knowledge into smaller dimensions, see also Section 3.1.4. In microstructured equipment the so-called numbering-up process partly replaces the scale-up process, which describes the data transfer from model systems to real plant dimensions.

2.4.3 Order-of-magnitude analysis

To indicate scaling correlations, the Trimmer brackets [188], Chapter 2, were introduced to give an overview of the consequences of a length variation. However, for more complex or coupled processes, this method requires more effort to display the relationship between length and other parameters. An alternative method of yielding information regarding the order-of-magnitude for the relevant effects is the scale analysis described by Bejan [90]. Scale analysis goes beyond dimensional analysis and is a relatively simple method for gaining information about system behavior.

To estimate the order-of-magnitude for the relevant quantities, the governing equations and the geometrical situation must be known. This also must be done for an "exact" analysis and is the first step of an engineering analysis. The basic equations, often in the form of partial differential equations, are transformed to equations with geometrical and state parameters. For example, the gradient is expressed as a difference ratio.

$$\frac{\partial T}{\partial t} \rightarrow \frac{\Delta T}{t_c} \quad \text{or} \quad \frac{\partial^2 T}{\partial x^2} \rightarrow \frac{\Delta T}{L_c^2} \tag{2.66}$$

Here, t_c and L_c define the characteristic time and length for the actual problem, respectively. This simplification helps to solve the governing equations for parameters, such as characteristic time or penetration depth. The relationship between two dominant parameters can be found from one equation. If there are more than two relevant parameters, the following rules can be applied to estimate the order-of-magnitude [90]:

- Summation: If $O(a) > O(b)$ and $c = a + b$, the order of c is $O(c) = O(a)$.
  If $O(a) = O(b)$ and $c = a + b$, the order of c is $O(c) \approx O(a) \approx O(b)$. The same holds for the difference and the negation.
- Product: For $c = a \times b$, the order of c is $O(c) = O(a) \times O(b)$.

- Ratio: For $c = a / b$, the order of c is $O(c) = O(a) / O(b)$.

According to Bejan [90], scale analysis is widely employed in heat transfer and one of the first steps in engineering analysis. It is a simple method for gaining a first impression and "house number" results without the effort of the "exact" analysis, which is also dependent on the model prerequisites.

2.4.4 Lumped element modeling

For complex systems and network arrangements, analytical treatment is still possible, however, this amount of elements cannot be handled in the conventional way. The numerical treatment of complex systems has advanced during the last decades. The treatment of complex and coupled systems is established, but nevertheless, the computational effort is extremely high for dealing with transport processes in fluidic networks, if possible at all. From electronic engineering, systems with millions of individual, but analytically describable, elements are known and are treated with lumped element programs like P-Spice [198, 199].

In the context of process engineering, continuum modeling and analytical modeling are preferred to bridge the gap between micro processes, equipment transfer processes and complete process design. Analytical modeling of process engineering phenomena is based on physical models with adjustable parameters. A complete calculation comprises the hierarchy of

- physical level for the process principles of unit operations,
- equipment level for the technical and mechanical processes,
- process structure for controlling and regulating automated programs, and, finally,
- the validation level for economical calculations, see also Sections 2.2.2 and 6.3.

The risk and flexibility of model assumptions and simplifications determine the system modeling. There must be a compromise between the mathematical and physical accuracy and the actual technical complexity. The mathematical models have to be appropriate for experimental test units or real plants, from which comparison data is gathered.

The analogy of transport processes and the corresponding equations, see Table 2.2, help to translate complex flow networks or heat management problems into mathematical descriptions, which can be handled by electronic simulation programs. The individual processes of mass, species, heat and energy transfer have their equivalent processes in the electrical current transport, like the resistance, the capacity (storage), and the inductivity (kinetic energy). This method is described in more detail in Schaedel's textbook [200] regarding fluidic networks and lumped element modeling. Other application in microsystems technology can be found in [198] Chapter 6, [199] Chapter 12, and [201] Chapter 4. The pressure loss in channel manifolds is simulated with the help of analytical equations [43] to achieve uniform flow distribution to each mixing element.

2.4.5 Numerical simulation and analytical modeling

Essentially, one can distinguish between continuum modeling of transport processes in microstructures [202] Chapter 4, [203] Chapter 2, statistical methods like the Monte Carlo Methods DSMC [204] Chapter 10, [205] Chapter 7 + 8, and cellular automata calculations like Lattice Boltzmann Methods LBM [168, 206]. A good introduction to cellular automata and lattice methods is given by Sukop and Thorne [169] with special emphasis on multicomponent and multiphase transport.

A relatively new approach is to treat chemical engineering problems, such as diffusion and reactions with stochastic methods. An introduction is given by Doraiswamy and Kulkarni [207]. An overview of stochastic methods and possible applications can be found in Gardiner's handbook [208], where chemical reactions and diffusive transport is described with stochastic functions. Additionally, spectral methods approximate the unknowns with the help of truncated Fourier series or Chebychev polynomials on the entire domain [209].

The problems in process engineering demand the solution of the conservation and balance equations of convection, diffusion, and/or reaction problems in various geometries under certain boundary conditions, which also requires solving ordinary (ODE) and partial differential equations (PDE) on discrete domains. The mathematical equations describing the physical problems appearing in this chapter and throughout the book are often first and second order derivatives of a state parameter according to the time t and location, represented by one space coordinate x.

The general form of a linear, homogeneous PDE of second order of a state parameter X with two independent variables z and t is given by

$$a\frac{\partial^2 X}{\partial t^2} + 2h\frac{\partial^2 X}{\partial t \partial z} + b\frac{\partial^2 X}{\partial z^2} + 2f\frac{\partial X}{\partial t} + 2g\frac{\partial X}{\partial z} + eX = 0 \qquad (2.67)$$

For example, Fick's second law for diffusion (parameter concentration c) or Fourier's law for heat conduction (parameter temperature T) is obtained with $b = 1$; a, h, g, and $e = 0$; $f = 1/(2D)$; and $ab - h^2 = 0$.

$$\frac{\partial c}{\partial t} = D\frac{\partial^2 c}{\partial z^2} \qquad (2.68)$$

This PDE is also called parabolic, because Eq. 2.67 with the condition $ab - h^2 = 0$ is similar to the equation describing a parabola for $\partial/\partial x \rightarrow x$ and $\partial/\partial y \rightarrow y$, see Stephenson [210] Chapter 2. The PDE is called elliptic for $ab - h^2 > 0$ and hyperbolic for $ab - h^2 < 0$, because Eq. 2.67 resembles a correlation for an ellipse or hyperbola, respectively. This classification is adopted for linear equations in three or more independent variables, like Laplace's equation (steady heat conduction) in three space coordinates, which is of elliptic type.

$$\mathrm{div}\,(\mathrm{grad}T) = \Delta T = \frac{\partial^2 T}{\partial x^2} + \frac{\partial^2 T}{\partial y^2} + \frac{\partial^2 T}{\partial z^2} = 0 \qquad (2.69)$$

For further treatment of ODE's and PDE's, the reader is referred to relevant mathematical textbooks, for example Braun [211]. For simple geometries and flow situations, analytical solutions exist for fluid dynamics, heat and mass transfer in laminar flow [212] and for general engineering applications [213]. A thorough treatment of heat and mass transfer problems with a good mathematical background is given by Kays et al. [214].

The majority of problems in this thesis can be described with continuum models, but must be solved within complex geometries, with complex boundary conditions, or for transient situations. Here, numerical methods assist the calculation process enormously, which are disseminated and user-friendly elaborated for various application. In the following, only a short overview of the common methods can be given, the interested reader is referred to the cited literature, and is also challenged to make their own experience and judgments.

The Finite-Element-Method (FEM) was initially derived from solving structure-mechanical problems using simple piecewise functions, but is also applied in fluid mechanics [209]. The application of FEM to fluid dynamics and heat and mass transfer problems is described by Li [215]. The commercial FEM code COMSOL, formerly FEMLab, becomes more important in numerical treatment of microfluidic problems, such as electro-kinetic flow [140] or laminar flow. The Finite-Difference-Method (FD) generates approximations for the derivatives of the unknowns at each grid point with the help of truncated Taylor series expansions [216]. An open-source code for mathematical functions and finite-difference programming is SciLab (www.scilab.org), which is used by Polifke and Kopitz [217] to solve numerically transient heat transfer problems.

Originally developed as a special FD formulation, the Finite-Volume-Method (volume-of-fluid VOF) solves the algebraic equations using an iterative method, which have been derived from the integral equations governing transport processes, see Ferziger and Peric [216]. This method directly displays the real physical processes, which are very important in thermo-fluid dynamics where fluxes over the boundaries of control volumes play an important role. Variable material properties can be realized and allow a more realistic simulation. The control volumes may have arbitrary shapes, which make it easier to complex geometries. The conservation of transport values for the fluxes is better ensured than in the finite difference methods, but may lead also to errors for coarse meshes [217].

2.5 Future Directions of Micro Process Engineering Research

Only a few physical principles and their potential for miniaturization have been discussed in this and the preceding chapter. Schubert [218] provides a comprehensive list of all known physical principles. Many of these principles are translated for design purposes in mechanical and process engineering by Koller [219] including many examples, which could be a starting point for further investigations. One motivation for process engineering can be summarized by the concept of process intensification. Additionally, processes in microstructured devices open new routes in mixing, transformation, downstream processing, separation, and purification with the help of system integration.

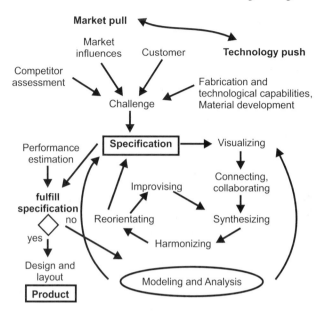

Fig. 2.17. The creativity process as one of the first steps in the design process, partially adopted from nbm [220].

An emerging market, such as micro process engineering, is governed by a technological push from new possibilities from fabrication and materials, and a market pull from the customer and application needs, see Figure 2.17. In chemical technology and production, long-term business does not allow simple restructuring of complex production plants with lifespans of decades. The role of unit operations should be extended to integrated processes in microstructures to provide a comprehensive toolbox for treating various tasks with process engineering. Chemical synthesis can break new ground with microstructured devices allowing high concentrations, new solvents, and fast reactions. Innovative reactor design adjusted to chemical reactions, optimized flow, parallel and high throughput testing, new reaction routes closer to the molecule, and their transformation (time, rate, concentration) are the new applications of microstructured equipment in the chemical industry, in energy conversion, fuel cells, and life science.

The design process with emphasis on the creativity and problem-solving process is displayed in Figure 2.17, according to [220]. The design of chemical processes is described in Section 2.2.2, while the design process of microstructured equipment is treated in Section 6.3. The initiative for a project, of developing a new product, process, or device, may come from outside, from the market or the customer (market pull). On the other hand, internal sources, such as new fabrication capabilities, new materials and combinations (technology push), or the need to enhance an existing product or process gives a strong need or challenge to initiate a project. A good starting point is to visualize the object by existing CAD drawings, blackboard outlines, or manual sketches. The Mind Map method is a very helpful tool for visualization and organization [221] of complex systems and processes. Brain storming, brain writing, or other improvised creativity techniques can yield new items or solution possibilities. A systematic problem-solving process is the key to the successful design of processes, equipment, and devices. Brainstorming in groups is regarded as an effective method to generate new ideas. A psychological study [222] showed that communication within a group, listening to and waiting for others, disturbs the creative thinking process of a single person and can hinder or even prevent the generation of ideas. Single persons, who can think undisturbed, are much more creative for a certain period. The writing down of individual ideas and subsequent communication and exchange of the ideas show much better results than just group work. Hence, a combination of individuals and group work will lead to maximum results.

Check lists [223, 224], heuristic rules [64], morphological charts [225], or the TRIZ method [226] are other methodologies for finding new ideas or to assist the design process. The TRIZ method consists of 40 general rules or laws from engineering experience, which have been set up from an extensive patent review by Altschuller. Examples of these 40 rules can be found in [227] for microsystem engineering and microelectronics and for process engineering in [228]. New and existing items are included to fulfill new functions. When collaborating, new functions can emerge that are very helpful for meeting the challenge. In synthesizing various objects, new items are generated, which must be harmonized to fit together, to produce a sound solution.

In modeling and analyzing new items and systems, their performance and abilities are checked, which may already lead to a design and layout solution. If the required functions are not obtained, a reorientation will give new viewpoints to possible solutions. Improvising may bridge unconnected tasks and solutions to give further options. Figure 2.17 makes the highly iterative process slightly clearer, but is only one description. In reality, the creative process is very complex and unique.

In the near future, research needs to address aspects, such as liquid flow and transportation or two phase flow in capillary structures with the intelligent use of minor effects, fluctuations, surface effects and coupled processes. High throughput devices like mixers, pre-mixers, reactors, or heat exchangers should be addressed by the application of cost effective fabrication technologies and integrated processes like reactors with mixers and heat exchangers. The interplay of function and structure, of materials, related fabrication technologies, and properties must be introduced into the design process for microstructured equipment and devices. A first step toward an integrated product development process is given by Marz [229] in his PhD thesis. The integration of microstructures into macro equipment must be carried out to take advantage of the benefits and opportunities of both fields and additional emerging effects.

3

Momentum Transfer

The goal of this chapter is to illustrate the fundamentals of momentum transfer to determine the flow in microchannels and microstructured devices for chemical engineering applications. Different flow conditions are treated for microchannel flow, such as rarefied gases, compressible flow, long channels, or developing flow. Emphasis is given to curved flow with enhanced transfer characteristics. The consequences for enhanced mixing and heat transfer are displayed. Multiphase flow phenomena are discussed briefly, however, their investigation should be enlarged in the future.

3.1 Momentum Transport of Single-Phase Flow

In single-phase flow, the fluid motion in microchannels is determined by wall friction, inner viscous forces and inertial forces. The governing equations, which describe the fluid motions are given in the following.

3.1.1 The momentum equation and force balance

The macroscopic balance equations for process equipment can be formulated on various levels, as shown in Figure 2.12. The macroscopic differential equations for mass, species, momentum, and energy are given here without derivation. For specific details, please refer to advanced textbooks in fluid mechanics [50, 230]. The following equations are based on the conservation of mass, species, momentum, and energy in infinitesimally small control volume elements. These balance equations only form a constitutive set of equations (number of unknowns = number of equations) after the mechanical stress/strain correlations and thermodynamic equations of state have been introduced. The relations between stress and strain $\tau = f(\eta, \mathrm{div}\vec{w})$ or between heat flux and temperature gradient $\dot{\vec{q}} = f(\lambda, \mathrm{grad}T)$ are chosen as linear isotropic relations with constant coefficients η (viscosity) and λ (thermal conductivity). This denotes Newtonian fluid behavior and thermodynamic processes close to thermodynamic equilibrium, where the state variables are properly defined.

The balanced parameter is the fluid mass in a constant cross section or a cubic differential element, given as the density ρ. The mass balance can be written as

$$\frac{D\rho}{Dt} = \frac{\partial \rho}{\partial t} + \text{div}\,(\rho\,\vec{w}) = 0. \tag{3.1}$$

This equation is also known as continuity equation [231] p. 2, and is one of the fundamental equations in fluid dynamics. For an incompressible fluid, the continuity equation can be simplified for a one-dimensional case to

$$\frac{\partial\,(\rho\,A)}{\partial\,t} = 0 = -\frac{\partial\,(\rho\,w\,A)}{\partial\,z}. \tag{3.2}$$

This continuity equation will help to simplify other balance equations.

The momentum of a moving fluid can be expressed as the product of the mass and the flow velocity. The integral of the momentum over a volume element results in a net fluid force on the volume boundary or on the equipment wall. In general, the momentum balance of a device or element can be written as

$$\sum J_{\text{in}} = \sum J_{\text{out}} - \sum J_{\text{loss}}. \tag{3.3}$$

The momentum loss can be interpreted as the viscous momentum loss, which is expressed as pressure loss along the channel or over the device. For viscous fluid flow through an arbitrary channel with an external force, the momentum change or force balance can be expressed by the momentum, pressure, and gravity part:

$$\frac{\partial\,(m\,w)}{\partial t} = (\dot{m}\,w)_{\text{in}} - (\dot{m}\,w)_{\text{out}} + (p\,A_i)_{\text{in}} - (p\,A_i)_{\text{out}} + m\,g + F_z \tag{3.4}$$

For one-dimensional differential channel elements with viscous flow, the momentum balance can be written as

$$\frac{\partial\,(\rho\,A\,w)}{\partial t} = -\frac{\partial\,(\rho\,w\,A\,w)}{\partial z} - \frac{\partial\,(p\,A)}{\partial z} + p\frac{\partial A}{\partial z} - \tau\,L_c - \rho\,A\,g, \tag{3.5}$$

where L_c means the perimeter of the differential channel element. With the continuity equation, Eq. 3.2, the above equation can be simplified to

$$\rho\,A\frac{\partial\,w}{\partial t} = -\rho\,w\,A\frac{\partial\,w}{\partial z} - A\frac{\partial\,p}{\partial z} - \tau\,L_c - \rho\,A\,g. \tag{3.6}$$

This equation can also be derived for all space coordinates independently from the coordinate system in a general form, see Oertel [232] or Schlichting [233].

$$\rho\,\frac{D\vec{w}}{Dt} = \rho\left(\frac{\partial}{\partial t} + \vec{w}\cdot\text{div}\right)\vec{w} = -\text{grad}\,p + \eta\cdot\Delta\vec{w} + \vec{k}. \tag{3.7}$$

with $\vec{k}$ as the sum term for external forces. The term $\rho\,(\vec{w}\cdot\text{div})\,\vec{w}$ is also called the inertial term and describes the convection of the velocity gradient. It is the only nonlinear term in the Navier-Stokes equations, see Foias et al. [234]. The Navier-Stokes equations are among the few equations in mathematical physics, for which the nonlinearity arises not from the physical attributes of the system, but rather from the mathematical (kinematic) aspects of the problem. For turbulent flow, Hoyer [235] stated that the Navier-Stokes equations incompletely describe the motion of viscous fluids because the Boltzmann transport equation is only valid in steady states, see also Section 2.3.2. As explained in the work of Geier [236], additional terms must be introduced to take the convective transport of shear stress into account for three-dimensional turbulent flow. This issue is still under discussion and not within the scope of this work.

Using the correlations between pressure p and shear stress τ as well as between tension σ and strain, the momentum equation or Newton's law is written as

$$\rho \frac{D\vec{w}}{Dt} = -\text{grad } p + \text{Div}\left[\eta\left(2\,\text{grad }\vec{\mathbf{w}} - \frac{2}{3}\vec{\delta}\,\text{div }\vec{w}\right)\right] + \rho\vec{g} \qquad (3.8)$$

with the velocity gradient tensor

$$\text{grad }\vec{\mathbf{w}} = \frac{1}{2}\left[\text{grad }\vec{w} + (\text{grad }\vec{w})^{\text{T}}\right] \qquad (3.9)$$

and the Kronecker unit tensor $\vec{\delta}$. For a detailed description of the derivation see Batchelor [50] Chapter 3 or Ferziger and Perić [216].

The mixed terms in Navier-Stokes equations lead to complex correlations compared to heat conduction or diffusion. For simple flow conditions or geometries, some terms can be neglected, and analytical solutions can be given for Navier-Stokes equations. Nevertheless, an exact mathematical proof is still missing for the existence of solutions describing three-dimensional turbulence. It is supposed in some works that Navier-Stokes equations are only valid for two-dimensional turbulence, see for example [234]. Lattice Boltzmann methods, which include higher moments of the Boltzmann equation, may represent better solutions for turbulent flow regimes, see [236]. An complete treatment of kinetic processes, from reversible micro-events, such as molecular hits and encounters, over the first emergence of irreversible processes, Boltzmann transport equation, chemical kinetics, and the derivation of the hydrodynamic equations is given by Gorban and Karlin in [237], where model reductions and regime transitions are extensively treated and some solution methods are presented.

3.1.2 The energy equation for fluid dynamics

In order to derive the Bernoulli equation, the energy equation must be addressed at first. A more detailed treatment with heat energy is given in Chapter 4. The energy equation or first law of thermodynamics is given by

$$\rho \frac{D\,h^+}{Dt} = -\text{div }\dot{\vec{q}} + \rho\vec{w}\cdot\vec{g} + \widehat{D} + \frac{\partial p}{\partial t} \qquad (3.10)$$

with the total enthalpy $h^+ = h + \rho \vec{w}^2/2$ and a group of diffusion terms $\widehat{D}$, see Herwig [230]. The conservative equations hold from an Eulerian viewpoint in a fixed coordinate system (reference frame), hence, the substantial derivatives include local and convective derivatives $(\mathrm{D}/\mathrm{D}t = \partial/\partial t + \vec{w} \cdot \mathrm{div})$. Alternatively, the Langrangian viewpoint follows a massless particle or body within the flow and denotes the constitutive equations in this framework. Due to channel and duct flow problems, this viewpoint is not treated here. More specific details are given by Gersten and Herwig [238] and cited literature therein.

The total energy conservation is expressed in Eq. 3.10 as the sum of mechanical and thermal parts. A dissipation term Φ occurs explicitly, if both parts are balanced individually, which describes the redistribution of energy between mechanical and thermal parts. This can be written for the mechanical energy from Eq. 3.8

$$\frac{\rho}{2} \frac{\mathrm{D}\vec{w}}{\mathrm{D}t} = \rho \vec{w} \cdot \vec{\mathrm{g}} + \widehat{D} + \frac{\partial p}{\partial t} - \frac{\mathrm{D}p}{\mathrm{D}t} - \Phi \tag{3.11}$$

and for the thermal (inner) energy from Eq. 3.10

$$\rho \frac{\mathrm{D}h}{\mathrm{D}t} = -\mathrm{div}\,\vec{q} + \frac{\mathrm{D}p}{\mathrm{D}t} + \Phi. \tag{3.12}$$

The dissipation Φ of mechanical energy is accompanied by entropy production and can be expressed by

$$\Phi = \tau : \mathrm{grad}\,\vec{\mathbf{w}} + \quad \text{with} \quad \tau = \eta \left(2\,\mathrm{grad}\,\vec{\mathbf{w}} - \frac{2}{3}\vec{\delta}\,\mathrm{div}\,\vec{w} \right), \tag{3.13}$$

see Eq. 3.8. Hence, dissipation is the inner product of the viscous stress tensor τ and velocity gradient tensor $\mathrm{grad}\,\vec{\mathbf{w}}$. The boundary conditions of the above differential equations are

- no wall slip, i.e. zero normal and parallel velocity at the wall (if no suction or blowing occurs, and for gases with Kn < 0.01),
- no temperature jump at the wall, i.e. the temperature and/or temperature gradients can be described at the wall.

For constant density ρ, non-viscous flow, and a cubic differential element for all three directions, the momentum balance can be written as

$$\frac{\partial \vec{w}}{\partial t} = -(\vec{w}\,\mathrm{div})\,\vec{w} - \frac{\mathrm{grad}\,p}{\rho} + \vec{\mathrm{g}} \tag{3.14}$$

This equation is also called the Euler equation for non-viscous flow and is one of the basic equations of hydrodynamics [231] p. 4. In the following chapter, the Bernoulli equation will be derived for long and narrow channels. Additional forces in microchannel flow may occur with surface effects in multiphase flow, such as bubbles and droplets, see also Section 3.3 and 3.3.4.

3.1.3 Basic equations for long, small channels

In microstructured devices, the channels are often long, and their stream-wise dimensions are much larger than their cross-sectional dimensions. In conventional devices, these channels are often accompanied with high Re numbers. The geometrical setup of a micro-scale slender channel is displayed in Figure 3.1 with rectangular cross section, which is often the case for microstructures due to the available fabrication technology.

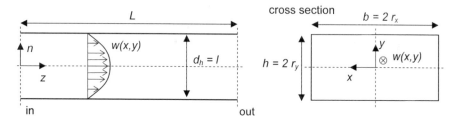

Fig. 3.1. Typical setup for a slender microchannel with rectangular cross section, here, for the y-component only.

To derive the non-dimensional correlations, the following dimensionless variables are introduced with reference values for slender channels with rectangular cross section (width b, depth h, length L, hydraulic diameter $d_h = l$)

$$z^* = \frac{z}{l\mathrm{Re}}, \qquad y^* = \frac{y}{l}, \qquad w^* = \frac{w}{w_r}, \qquad v^* = \frac{v\mathrm{Re}}{w_r}, \qquad p^* = \frac{p - p_r}{\rho w_r^2},$$

$$T^* = \frac{T - T_r}{\Delta T}, \qquad \eta^* = \frac{\eta}{\eta_r}, \qquad \lambda^* = \frac{\lambda}{\lambda_r}, \qquad \Phi^* = \frac{\Phi l^2 \mathrm{Re}^2}{w_r^2}. \tag{3.15}$$

The equations describing a steady and incompressible flow in long channels with variable viscosity and thermal conductivity are

- the continuity equation, from Eq. 3.1

$$\frac{\partial v^*}{\partial y^*} + \frac{\partial w^*}{\partial z^*} = 0, \tag{3.16}$$

- the z-momentum equation, from Eq. 3.5

$$v^* \frac{\partial w^*}{\partial y^*} + w^* \frac{\partial w^*}{\partial z^*} = -\frac{\partial p}{\partial z^*} + \frac{\partial}{\partial y^*} \left[\eta^* \frac{\partial w^*}{\partial y^*} \right] + \frac{1}{\mathrm{Re}^2} \frac{\partial}{\partial z^*} \left[\eta^* \frac{\partial w^*}{\partial z^*} \right], \tag{3.17}$$

- the y-momentum equation, from Eq. 3.5

$$\frac{1}{\mathrm{Re}^2} \left[v^* \frac{\partial v^*}{\partial y^*} + w^* \frac{\partial v^*}{\partial z^*} \right] = -\frac{\partial p}{\partial y^*} +$$

$$+ \frac{1}{\mathrm{Re}^2} \left[\frac{\partial}{\partial y^*} \left(\eta^* \frac{\partial v^*}{\partial y^*} \right) + \frac{\partial}{\partial z^*} \eta^* \frac{\partial v^*}{\partial z^*} \right] \tag{3.18}$$

- and the thermal energy from Eq. 3.10

$$v^* \frac{\partial T^*}{\partial y^*} + w^* \frac{\partial T^*}{\partial z^*} = \frac{1}{Pr} \frac{\partial}{\partial y^*} \left[\lambda^* \frac{\partial T^*}{\partial y^*} \right] +$$

$$+ \frac{1}{PrRe^2} \frac{\partial}{\partial z^*} \left[\lambda^* \frac{\partial T^*}{\partial z^*} \right] + \frac{Ec}{Re^2} \Phi^*, \qquad (3.19)$$

with the dimensionless groups of the Reynolds number $Re = \rho w_r l / \eta_r$, the Prandtl number $Pr = \eta_r / \lambda_r c_p = \nu / a$, and the Eckert number $Ec = w_r^2 / c_p \Delta T$.

In flow regimes with high Re numbers, the terms with $1/Re^2$ are very small and can be ignored. In microflows where the Re number is in the order of $O(1)$, the full Navier-Stokes equations must be considered and the influence of u and v velocities is not negligible. Only if the Re number is well above unity, can the slender channel equations (Eqs. 3.16 to 3.19) be taken into account. When considering the energy equation, the viscous dissipation cannot be neglected in microchannels, because the ratio $Ec/Re^2 = \nu^2/(c_p \Delta T\, d_h^2)$ does not incorporate the velocity and is proportional to the inverse square of the hydraulic diameter.

With the continuity equation Eq. 3.16, the momentum balance Eq. 3.17, and the boundary conditions of a stationary, incompressible, and fully-developed, one-directional laminar flow, the Poisson differential equation for the pressure loss in a rectangular channel $\partial p/\partial z$ follows to

$$\frac{\partial p}{\partial z} = \eta \left(\frac{\partial^2 w(x,y)}{\partial x^2} + \frac{\partial^2 w(x,y)}{\partial y^2} \right). \qquad (3.20)$$

An analytical solution of this equation can be found with the Prandtl membrane analogy see [239] Chapter C and [240]. The bending of a thin membrane under pressure or the stress in a bar under torsion can be described with the same differential equation. The describing PDEs of the velocity $\vec{w}$ through a certain cross section area is comparable to the displacement of a thin membrane or the stress function in a bar under torsion. That means for the velocity

$$w(x,y) = \frac{\Delta p}{\eta L} \cdot \phi \qquad (3.21)$$

with $\Delta p/L$ as pressure loss per length and ϕ as the torsion function. A solution for the torsion function on a rectangular cross section can be found with a Fourier series:

$$\phi = \sum_{m,n=1}^{\infty} \frac{16(2r_x)^2 (2r_y)^2}{mn\pi^4 (n^2\, 4r_x^2 + m^2\, 4r_y^2)} \sin\left(\frac{m\pi x}{2r_x}\right) \sin\left(\frac{n\pi y}{2r_y}\right), \qquad (3.22)$$

with $2r_x$ and $2r_y$ as the width and depth of the channel, see Figure 3.1. The sum of the first 4 elements is sufficient for rectangular channels with an aspect ratio of approx. $a = 1$. Higher order elements are negligible, but must be considered for flat channels.

The basic equations are valid for laminar flow with a typical Re number below the critical Re number at the transition from laminar to turbulent flow. The critical Re numbers for internal flows are in the range of $Re_c = O(10^3)$, for example,

- circular pipe flow: $\mathrm{Re}_c \approx 2\,300$ ($l = d$, diameter)
- rectangular channel flow: $\mathrm{Re}_c \approx 2\,300$ ($l = d_h$: hydraulic diameter $d_h = 4A/l_p = 2bh/\,(b+h)$)
- plane channel flow: $\mathrm{Re}_c \approx 2\,000$ ($l = h$, distance of the walls)
- plane Couette flow: $\mathrm{Re}_c \approx 1\,800$ ($l = b$, distance of the walls)

There is still some discussion whether the critical Re number for microchannels is lower than that for conventional geometries. The order-of-magnitude is generally accepted, however, surface roughness and entrance effects should definitely be taken into consideration. Careful measurements indicate the same transition regimes and conditions in microchannels as in conventional duct flow [241]

The governing equations can be simplified for long channel geometries with established mean values for the velocity and for constant fluid properties. For different locations 1 (at the inlet) and 2 (at the outlet) of the channel, the basic equations can be written as:

- continuity equation from Eq. 3.16

$$\rho w_2 A_2 = \rho w_1 A_1. \tag{3.23}$$

- momentum equation or mechanical energy equation, also called the Bernoulli equation, from Eq. 3.10 with the height coordinate y

$$\frac{w_2^2}{2} + \frac{p_2}{\rho} + g\,y_2 = \frac{w_1^2}{2} + \frac{p_1}{\rho} + g\,y_1 + w_{t12} - \varphi_{12}. \tag{3.24}$$

- thermal energy from Eq. 3.12

$$e_2 = e_1 + q_{12} + \varphi_{12}. \tag{3.25}$$

The Bernoulli equation, Eq. 3.24, is complemented by the technical energy w_{t12} and the dissipation. It represents the momentum balance and the mechanical energy balance. Three process parameters appear in the equations:

- w_{t12}: specific technical work, positive for added work (pump, actuator, etc.) and negative for produced work (turbine, opened valve, etc.) between location 1 and 2,
- q_{12}: specific heat, positive for added heat (heating, light or electromagnetic waves, etc.) and negative for cooled fluid (heat exchanger, heat loss to ambient, etc.) between location 1 and 2,
- φ_{12}: specific viscous dissipation, simultaneously decreasing the mechanical energy and increasing the inner energy between location 1 and 2. The viscous dissipation is directly linked to the pressure loss Δp.

The technical work is not considered here in detail, for heat transfer, please refer to Chapter 4. The pressure loss Δp_{12} in a slender channel with inlet 1 and outlet 2 is determined from the Bernoulli equation.

$$\Delta p_{12} = \rho \cdot \varphi_{12} = p_1 - p_2 + \frac{\rho}{2}\left(w_1^2 - w_2^2\right) + g\left(y_1 - y_2\right). \tag{3.26}$$

For a constant cross section and negligible gravitation forces, the viscous dissipation in a channel element correlates with the pressure loss:

$$\varphi_{12} = \frac{p_1 - p_2}{\rho}. \tag{3.27}$$

In macroscopic flow situations, the pressure loss is often approximated by the sum of individual pressure loss parts consisting of fittings, bends, valves or straight pipes with the length l_i.

$$\Delta p = \sum_i \left(\lambda_i \frac{l_i}{d_{h,i}} + \zeta_i \right) \cdot \frac{\rho}{2} w_{ref,i}^2 \tag{3.28}$$

The reference velocity $w_{ref,i}$ must be determined for each channel element i. The channel friction factor λ of the straight pipe is determined by the flow regime (laminar – turbulent) and the cross section. The pressure loss coefficient ζ_i is affected by flow internals, such as curves, bends, fittings, and other channel joints.

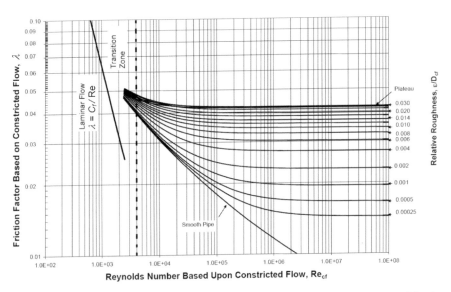

Fig. 3.2. Flow friction factors for straight channels with a rough surface according to Moody or Nikuradse, see also [242] with relative roughness ε. The original diagrams of Nikuradse and Moody are based on experimental values for pressure loss in pipes with sand paper or sand grains glued to the wall (artificial surface roughness).

For fully developed turbulent flow in ducts, the friction factor λ is constant and depends mainly on the relative wall roughness k/d_h, see the modified Moody diagram in Figure 3.2 according to Taylor et al. [242]. A more detailed treatment of the friction factors is given by Gersten [243], with correlations for turbulent flow in rough channels. For laminar flow, which is predominant in microchannels, the coefficient is inversely proportional to the flow velocity, i.e. the Re number.

$$\lambda = \frac{C_f}{Re} \tag{3.29}$$

The wall roughness in this case can be subtracted from the hydraulic diameter of the channel and leads to a narrowing of the channel [242]. The channel friction factor coefficients C_f for various cross sections are given in Table 3.1 for fully developed laminar flow. For further data, see also Shah and London [212], where the laminar flow situation is described in detail.

Table 3.1. Channel friction factor coefficients C_f in fully developed flow through straight microchannels with different cross-sections.

Cross section, char. length	$C_f = \lambda \cdot Re$	$w_{max}/\bar{w}$
circle, D	64	2.000
square, h	56.92	2.0962
rectangular, $h;b$ aspect ratio $\alpha_A = h/b$ see also Eq. 3.67	$96\left[1 - 1.3553\alpha_A + 1.9467\alpha_A^2 - 1.7012\alpha_A^3 + 0.9564\alpha_A^4 - 0.2537\alpha_A^5\right]$	–
slab, $\alpha \to 0$	96	1.500
hexagon	60	–
60° trapezoid $h/b =$ 4.00	55.66	2.181
2.00	55.22	2.162
1.00	56.60	2.119
0.50	62.77	1.969
0.25	72.20	1.766
KOH trapezoid $h/b = 1.00$	56.15	2.137

For known channel friction factor λ in straight channel flow, the pressure loss in the entire channel can be determined. The influence of Re number, velocity, hydraulic diameter, and mass flow rate on the pressure loss is studied in Section 1.4.2. For other channel elements, the pressure loss coefficients ζ_i are often unknown for laminar flow. Using coefficients from tables in literature is not appropriate, because they are valid only for turbulent flow. Hence, Eq. 3.28 should be used very carefully for laminar flow in microchannels.

3.1.4 Compressible flow

The density is assumed to be variable in the above equations, which can be associated with the term *compressibility* and compressible flow. To describe the role of a variable density in microflows, the total differential of the density $\rho(T,p)$ is analyzed here in more detail.

$$d\rho = \left(\frac{\partial \rho}{\partial T}\right)dT + \left(\frac{\partial \rho}{\partial p}\right)dp \tag{3.30}$$

Using reference values $\rho_r = \rho_r(T_r, p_r)$ as well as temperature and pressure differences, dimensionless parameters are introduced:

$$\rho^* = \rho/\rho_r; \quad T^* = (T - T_r)/T_r; \quad p^* = (p - p_r)/p_r, \tag{3.31}$$

leading to the dimensionless differential equation for the density

$$d\rho^* = K_{\rho T}dT^* + K_{\rho p}dp^*. \tag{3.32}$$

The coefficients

$$K_{\rho T} = \left[\frac{T}{\rho}\frac{\partial \rho}{\partial T}\right]_r \quad \text{and} \quad K_{\rho p} = \left[\frac{p}{\rho}\frac{\partial \rho}{\partial p}\right]_r \tag{3.33}$$

are properties of the fluid and indicate the relationship between the density change and temperature or pressure change. As a result, density variations can not be neglected $(d\rho^* \neq 0)$ for Eq. 3.32 under certain conditions:

- when the fluid changes its density due to a temperature change $(K_{\rho T} \neq 0$; perfect gas: $K_{\rho T} = -1)$ then $dT^* \neq 0$ results in $d\rho^* \neq 0$, e.g. for intensive heat transfer with a gas.
- when the fluid changes its density due to a pressure change $(K_{\rho p} \neq 0$; perfect gas: $K_{\rho p} = 1)$ then $dp^* \neq 0$ results in $d\rho^* \neq 0$, e.g. for high pressure losses of a flowing gas.

Within a compressible fluid, the density is a function of pressure and temperature, hence, the mechanical and thermal energy are utilized in the energy equation. With Eqs. 3.24 and 3.25 and the specific enthalpy $h = u + p/\rho$, the total energy balance or first law of thermodynamics can be given as

$$h_2 + \frac{w_2^2}{2} + g \cdot y_2 = h_1 + \frac{w_1^2}{2} + g \cdot y_1 + q_{12} + w_{t12} \tag{3.34}$$

The dissipation term φ_{12} from Eq. 3.24 is no longer included due to an internal reorganization of mechanical energy into thermal energy, and the enthalpy remains constant.

The propagation speed of a pressure disturbance is equal to the speed of sound, which is a main parameter for compressible flow. To derive the correlation for the speed of sound of a perfect gas, a one-dimensional, small pressure wave is considered. In absence of external and viscous forces, the steady Navier-Stokes and continuity equation reduce to

$$w\frac{\partial w}{\partial z} = -\frac{1}{\rho}\frac{\partial p}{\partial \rho}\frac{\partial \rho}{\partial z} \quad \text{and} \quad \rho\frac{\partial w}{\partial z} + w\frac{\partial \rho}{\partial z} = 0, \tag{3.35}$$

respectively. Combining both equations and assuming isentropic conditions ($s =$ const., reversible adiabatic process), the speed of sound is given by

$$c^2 = \frac{\partial p}{\partial \rho}\bigg|_s. \tag{3.36}$$

For a perfect gas and isentropic change of state $pv^\kappa = \text{const.}$, the correlation $\partial p / \partial \rho_s = \kappa(pv)$ holds. With this correlation and the perfect gas equation of state, the general correlation for the speed of sound c can be obtained [244]

$$c = \sqrt{\kappa RT}. \tag{3.37}$$

The isentropic exponent κ varies little for a perfect gas, hence, for a given temperature, the speed of sound c varies only with the individual gas constant R and the isentropic exponent κ. For a given gas, the speed of sound is proportional to $\sqrt{T}$ and almost independent from the pressure, which is important for gas flow heat exchange.

To estimate the influence of compressibility, the speed of sound $c = \sqrt{\kappa RT}$ for a perfect gas is compared with the mean flow velocity of a system. This dimensionless number is also known as the Mach number.

$$\text{Ma} = \frac{w}{c} = \frac{w}{\sqrt{\kappa RT}} \tag{3.38}$$

In a channel with constant cross section $A_1 = A_2$, the continuity equation Eq. 3.23 can be written as

$$\text{Ma}_2 = \text{Ma}_1 \frac{p_1}{p_2}\sqrt{\frac{T_2}{T_1}}. \tag{3.39}$$

For small Ma_1 number at the channel inlet, the Ma number along the channel increases for high pressure losses ($p_1 \gg p_2$). A temperature increase has less influence due to the square root and the measurement in K. For $\text{Ma}_2 > 0.3$, the flow can be reconsidered as compressible, and is called subsonic for $\text{Ma}_2 < 0.9$. In gas dynamics, the most common compressible flow situations are:

- inviscid compressible flow in ducts with varying cross section without heat transfer (isentropic flow),
- inviscid compressible flow in conduits with heat transfer (Rayleigh flow),
- viscous compressible pipe or channel flow with adiabatic walls (Fanno flow).

The theoretical description can be found in advanced textbooks [244, 245, 246]. In microchannel flow, viscous effects are dominant, which is only relevant to the last case. In very long ducts with viscous dissipation, the Ma number can increase up to Ma_2 close to unity, which is called the viscosity-choked flow situation [244]. Flow regimes with $0.9 < \text{Ma}_2 < 1.3$ are called sonic flow. Higher velocities with $\text{Ma}_2 > 1.3$ for supersonic flow can only achieved with the help of special channel arrangements, such as the convergent-divergent Laval nozzle. Iancu and Müller [247] show in a numerical study the influence of shock wave compression on the efficiency of microscale wave rotors. They show with an analytical model that the rotor can reach an efficiency of 70 - 80 % for temperature of approx. 1 000 K.

In cases when the fluid inertia becomes important, the flow governing equations can be non-dimensionalized with $\rho w^2\big|_r$ and p^* can be written as

$$p^* = C \cdot \mathrm{Ma}^2 p^+ \quad \text{with} \quad p^+ = \frac{p - p_r}{(\rho w^2)_r}, \tag{3.40}$$

with the Mach number $\mathrm{Ma} = w_r/c_r$, the velocity of sound c_r at the reference point r, and $C = \rho_r c_r/p_r$. This is the isentropic exponent $\kappa = C$ for a perfect gas. For the appropriate dimensionless variables, density changes arise for $\mathrm{Ma} \neq 0$ according to Eq. 3.40. Above $\mathrm{Ma} = 0.3$, the density and pressure changes can no longer be neglected, e.g. the compressibility factor or relative pressure difference $\Delta p/(1/2\rho w^2)$ for air is approx. 1.023 for $\mathrm{Ma} = 0.3$ [245]. On the other hand, the density may also alter as a result of high temperature changes.

In rarefied gas flow, the mean free path Λ of the molecules also plays an important role for compressibility effects. The dynamic viscosity $\eta = 0.499 \rho \bar{c}_M \Lambda$ for perfect gases acc. to Eq. 2.55 from Sommerfeld [148] and the mean molecular velocity $\bar{c}_M$ acc. to Eq. 2.52 are introduced here to determine the Re number

$$\mathrm{Re} = \frac{\rho w L}{\eta} = \frac{1}{0.499} \frac{w}{\bar{c}_M} \frac{L}{\Lambda}. \tag{3.41}$$

The mean molecular velocity $\bar{c}_M = \sqrt{8/\pi \, RT}$ from Eq. 2.8 and the Maxwell velocity distribution is then introduced with the speed of sound c and the Knudsen number $\mathrm{Kn} = L/\Lambda$ into Eq. 3.41, see Eckert [248] Section 43,

$$\mathrm{Re} = \frac{1}{0.499} \sqrt{\frac{\pi \kappa}{8}} \frac{\mathrm{Ma}}{\mathrm{Kn}} = 1.256 \sqrt{\kappa} \frac{\mathrm{Ma}}{\mathrm{Kn}}. \tag{3.42}$$

Rearranging this equation for the Knudsen number Kn or the Mach number Ma gives

$$\mathrm{Kn} = 1.256 \sqrt{\kappa} \, \frac{\mathrm{Ma}}{\mathrm{Re}} \quad \text{and} \quad \mathrm{Ma} = \frac{0.796}{\sqrt{\kappa}} \, \mathrm{Re} \, \mathrm{Kn}. \tag{3.43}$$

The Mach number Ma and its related compressibility can be expressed with the Re number and the Kn number, which is displayed in Figure 3.3. The diagram was primarily developed for supersonic flow of rarefied gases around re-entry bodies. The main result for microchannel flow is the beginning rarefaction effect for low Re numbers and correspondingly low Ma numbers, for example, the Kn number is approx. 0.1 for $\mathrm{Re} = 1$ and $\mathrm{Ma} = 0.3$. For more information, see also [248].

Furthermore, the influence of the compressibility can be estimated. In a continuum flow ($\mathrm{Kn} < 0.001$) of air with $\kappa = 1.4$, the Re number must be higher than 446 to produce noticeable compressibility effects with $\mathrm{Ma} > 0.3$. This Re number value is simply a guideline and should be checked for each particular application. The speed of sound in rarefied gases with $\mathrm{Kn} \geq 0.001$ should be individually evaluated for each specific case.

3.1.5 Viscous heating and entropy generation in channel flow

Due to the large surface-to-volume ratio and predominantly laminar flow regimes, the pressure loss of fluid flow in microstructured devices can be high compared to

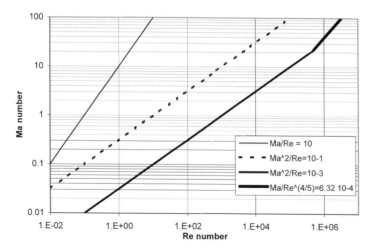

Fig. 3.3. Relationship between the Ma number and the Re number for different Kn numbers in rarefied gas flow (molecular flow Ma/Re> 10; transition flow $\mathrm{Ma}^2/\mathrm{Re} > 0.1$; rarefied gas flow $\mathrm{Ma}^2/\mathrm{Re} > 0.001$), adopted from Eckert [248].

turbulent flow in conventional equipment. Consequently, viscous dissipation effects must be considered in the energy balance. The generation of internal energy from mechanical energy, for example, is called viscous heating due to the temperature rise from the pressure loss. This effect is expressed as φ_{12} in Eq. 3.25 and $\mathrm{Ec}\Phi^*/\mathrm{Re}^2$ in Eq. 3.19. The influence of the viscous dissipation is determined by the thermal boundary conditions of the flow, where two limiting conditions can be distinguished:

- a fully developed, laminar flow, together with parallel viscous dissipation, which causes a constant heat flux to the ambient along the axial coordinate z. The temperature profile due to the heat transfer is independent from the axial coordinate and allows a heat flux density. It can be analytically derived for simple geometries, for example, for a circular pipe [238].

$$\dot{q}_w = \frac{8\eta w^2}{d_h} \tag{3.44}$$

The heat transfer can be relatively large for small diameters or high gas velocities.
- an adiabatic wall, which allows only self-heating of the gas flow. The thermal energy increases the bulk temperature of the fluid, which can be estimated by

$$T_{ad} = T_0 + \frac{\eta w^2}{\lambda_f}\left(16\frac{z/d_h}{\mathrm{Pe}} + 1\right) \tag{3.45}$$

with the bulk temperature T_0 at the entrance $z = 0$ and the Péclet number $\mathrm{Pe} = \mathrm{Re}\cdot\mathrm{Pr}$. This equation is valid for small Re numbers neglecting entrance effects.

These boundary conditions may not apply for practical problems with viscous flow in microchannels, however, they can help to estimate the order-of-magnitude for viscous dissipation. This is important for small diameters $d_h \to 0$, for long channels $z/d_h \to \infty$, and for Pe $\to 0$. In a comprehensive study about fluid flow in microchannels, Hetsroni et al. [249] show that viscous dissipation can lead to an oscillatory regime of laminar flow under certain conditions. The relation of hydraulic diameter to channel length, fluid properties, and the Re number are important factors for the impact of viscous dissipation on flow parameters. Oscillatory flow has been observed for high-viscous liquids, such as transformer oil, flowing in microchannels with $d_h < 20$ μm. In water flow (small kinematic viscosity and large heat capacity), oscillatory flow regimes do not occur a any realistic channel diameter.

To predict the efficient use of energy in mechanical and thermal systems, the second law of thermodynamics is very useful, since the loss of energy quality, for example, through pressure loss or temperature level of heat, is directly linked to the entropy production. The efficiency of heat exchangers or the effort for static mixers are directly linked to the entropy production within the device; for more information see also Bejan [175]. There are two methods for determining the overall entropy production in a system with viscous flow and heat transfer:

- the direct method, where the entropy production of heat transfer and viscous dissipation is determined locally and subsequently integrated over the entire flow domain. In CFD-calculations, this can be applied during post-processing, as all information regarding the local entropy production is obtained from the velocity and temperature field.
- the indirect method, which concerns the global entropy balance over the flow domain surface, which yields the overall entropy production. For an arbitrary control volume V with a surface S, the entropy balance is written as

$$\underbrace{\int_V S_{pro} \, dV}_{\text{production}} = \underbrace{\int_{in} \rho w s \, dA_{in}}_{\text{convection in1}} - \underbrace{\int_{out} \rho w s \, dA_{out}}_{\text{convection out2}} - \underbrace{\int_S \frac{\vec{q}}{T} \, dA}_{\text{heat flux}} \qquad (3.46)$$

with the specific entropy s and the velocity w perpendicular to the surface/cross section. The overall entropy production on the left side is determined from the convective terms and the heat transfer over the entire control surface.

A comparison of the entropy production within different processes gives hints for the energy efficiency of the process. A minimal entropy production indicates an efficient process, but is not a decisive criterion. The entropy production rate is an effective measure, which can indicate high dissipation processes. More information on this broad topic are given by Bejan in his textbook on entropy generation minimization [175].

In the following, the direct method for entropy production is applied for a fully-developed, laminar pipe flow with constant wall heat flux density $\dot{q}_w$ over a circular cross section. For comparison, Erbay et al. [250] calculated the entropy generation in parallel plate microchannels for inlet and slender channel flow. In a circular pipe, the flow and temperature profiles are given by

$$w = 2\bar{w}\left[1 - \left(\frac{r}{R}\right)^2\right],\tag{3.47}$$

$$T - T_w = \frac{-\dot{q}_w R}{\lambda}\left[\frac{3}{4} - \frac{1}{4}\left(\frac{r}{R}\right)^4 - \left(\frac{r}{R}\right)^2\right],\tag{3.48}$$

with the wall temperature T_w. The flow through a capillary produces two entropy terms from viscous dissipation $\dot{S}_{p,d}$ and heat transfer by convection $\dot{S}_{p,c}$, respectively. The velocity and temperature profile can also be given in the dimensionless form with $r^* = r/R$

$$\dot{S}^*_{p,d} \equiv \frac{\dot{S}_{p,d}}{\lambda T_b^2/\dot{q}_w^2} = 16\underbrace{\frac{\Pr \widehat{Ec}}{\widehat{Nu}^2}}_{\Pi} r^{*2},\tag{3.49}$$

$$\dot{S}^*_{p,c} \equiv \frac{\dot{S}_{p,c}}{\lambda T_b^2/\dot{q}_w^2} = \left(2r^* - r^{*3}\right)^2.\tag{3.50}$$

The dimensionless numbers are defined with the bulk temperature T_b, that also expresses the caloric mean temperature,

$$\widehat{Ec} \equiv \frac{w_m^2}{c_p T_b} = Ec\frac{\Delta T}{T_b};\quad \widehat{Nu} \equiv \frac{\dot{q}_w R}{\lambda T_b} = Nu\frac{\Delta T}{T_b};\quad \Pi = 16\frac{\Pr\widehat{Ec}}{\widehat{Nu}^2}.\tag{3.51}$$

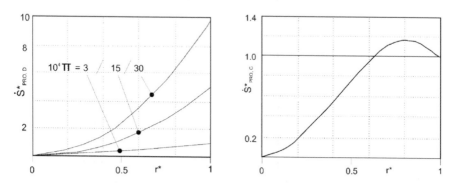

Fig. 3.4. Entropy production rates from viscous dissipation (left) and heat transfer (right) in fully developed pipe flow for three different values of the dimensionless group $\Pi = 16\,\Pr\widehat{Ec}/\widehat{Nu}^2$, see also [45].

The local dimensionless entropy production rates from viscous dissipation and heat transfer in fully-developed pipe flow are shown in Figure 3.4 for three different values of the dimensionless group $\Pi = 16 \, \Pr \widehat{Ec}/\widehat{Nu}^2$, see Eq. 3.49. At the center line, both entropy production rates are zero and increase toward the wall. In turbulent channel flow, the entropy production rate is very high close to the wall, due to the high velocity gradient. In laminar flow, the production rate is distributed more evenly over the entire radius as a result of the velocity and temperature profile. With increasing $\widehat{Ec}$ number and accordingly increasing Π, the viscous dissipation begins to dominate the entropy production rate. The entropy production is also mentioned in Section 2.3.3, for more details see [251].

3.1.6 Fluid dynamic entrance length

At the entrance, the flow in a channel is disturbed by wall friction and asymptotically develops along its way through the channel for $z/d_h \rightarrow \infty$. The fluid dynamics of channel flow are called fully-developed, when changes in the velocity profile are lower than a certain value, often set to 1 %. The pressure loss in the entrance region differs considerably from the developed region. For laminar incompressible flow, the length from the channel entrance can be approximated by the empirical correlation

$$\frac{L_{hyd}}{d_h} = \frac{C_1}{1 + C_2 \text{Re}/C_1} + C_2 \text{Re} \tag{3.52}$$

with the following constants

- pipe flow: $C_1 = 1.2$, $C_2 = 0.224$,
- channel flow: $C_1 = 0.89$, $C_2 = 0.164$.

For low Re numbers, the entrance length is in the same order-of-magnitude than the hydraulic diameter. For higher Re numbers, the entrance length reaches further into the channel before the flow is fully developed. These lengths are also characteristic for damping a flow disturbance along the channel flow. According to Baehr and Stephan [89] p. 373, the entrance length of laminar flow in ducts with circular cross section is given by $L_{hyd}/d_h = 0.056 \, \text{Re}$, where the center line velocity deviates less than 1 % from the final velocity. The pressure loss Δp in the channel entrance depends on the geometry and the Re number and is determined by the pressure loss coefficient [25]

$$\zeta = \left(k_1 + \frac{k_2}{\text{Re}} \right) \left[1 - \frac{d}{d_0} \right] \tag{3.53}$$

with the coefficients $k_1 = 2.32 \pm 0.05$, $k_2 = 159 \pm 30$ for circular pipe flow, the tube diameter d, and the reservoir diameter d_0. This equation is valid for $6 < \text{Re} < \text{Re}_{crit}$. For lower Re numbers, $\text{Re} < 0.1$, the viscosity plays an important role, and the pressure loss Δp in the entrance is described for a circular tube by

$$\Delta p = \frac{32}{3(n+1)d} \sqrt{2 \eta_a \lambda_a} \bar{w} \tag{3.54}$$

and for a slit with width b and height h by

$$\Delta p = \frac{8}{(n+1)h}\sqrt{\eta_a \lambda_a}\bar{w} \tag{3.55}$$

with the flow behavior index n, the apparent shear viscosity η_a, and the apparent elongation viscosity λ_a. The fluid behavior index n describes the viscosity behavior with the shear rate. For Newtonian fluids, the shear stress is the product of shear rate and apparent shear viscosity η_a, meaning $n = 1$. The elongation viscosity λ_a couples the elongation rate with the elongation stress. For Troutonian fluids, the elongation viscosity is only a function of the temperature and can be estimated with $\lambda_a = 3 \cdot \eta_a$. With these simplifications, the entrance pressure loss becomes

$$\Delta p = \frac{32}{\sqrt{6}d}\eta_a \bar{w} \tag{3.56}$$

for circular tubes and

$$\Delta p = \frac{12}{\sqrt{3}h}\eta_a \bar{w} \tag{3.57}$$

for slit flow. The elongation viscosity is also important for the pressure loss in channel reductions and nozzle flow. More details on laminar flow are given by Knoeck [252].

In turbulent flow, the entrance length is much shorter and lies in the range of $10 < L_{hyd}/d_h < 60$. The entrance effect increases not only the transport properties of the flow within the channel, it must also be taken into consideration when choosing the location of measurement devices in channels, as flow measurement devices are sensitive to flow disturbances, and the signal may be distorted in the entrance region. As a rule of thumb for turbulent flow, the inlet length before the sensor should be $10 \times d_h$, and the outlet length after the sensor should be $5 \times d_h$, to minimize measurement errors.

3.2 Convective Fluid Dynamics in Microchannels

Leaving the introduction of fluid dynamics in microchannels, this section focuses on the fluid dynamics of convective fluid motion in microchannels. The flow velocities are relatively high and lead to Re numbers from approx. 10 to 1 000. Typical velocities w range from approx. $0.1 < \bar{w} < 10$ m/s for liquids to $1 < \bar{w} < 50$ m/s for gases. In this range, the momentum forces gain importance, vortices appear in curved flow, and transient flow regimes may arise, but in long straight channels, the flow relaminarizes again and leads to straight, undisturbed streamlines. The channel geometry is determined by the micro fabrication process, particularly Deep Reactive Ion Etching process (DRIE, also known as the Bosch process or Advanced Silicon Etching ASE), which produces vertical walls and rectangular cross sections. In most cases, a two-dimensional channel layout is extruded into the third dimension of the silicon substrate, hence, all channels have similar depth due to the given etching time. Fabrication of variable depth is only possible with additional effort. A slight variation of the channel depth (approx. 5 %) results from inhomogeneous etching depth during the DRIE process and must be considered for sufficient device characterization and during experimentation.

3.2.1 Dean flow in 90° curves

The flow in microchannels is generally regarded as straight laminar flow, see Section 3.1.3. This is correct for straight channels with low flow velocity and, therefore, low Re numbers. For higher Re numbers, turbulence occurs, which is linked to a critical Re number. For example, in straight channels, the flow remains laminar with straight streamlines below a Re number of 2 300, although first flow disturbances with wavy streamlines may appear for even lower Re numbers according to Albring [100]. Natrajan and Christensen [253] experimentally investigated the transitional flow in microscale capillaries and found that the mean velocity profiles begin to deviate from the parabolic laminar profile at Re = 1 900. Li and Olsen [254] measured the velocity profiles and shear stress in square microchannels for Re numbers ranging from 200 to almost 4 000. A recent study of Hof et al. [255] indicates, that turbulence and transient flow structures have a finite life time in long, smooth channels at a finite critical Re number. Turbulent disturbances in channel flow decays, and, eventually, the flow will always relaminarize to steady flow.

In steady flow, streamlines are the track of massless particles (Lagrangian particles, see Section 3.1.2) as they pass through the device and also indicate the local direction of the fluid flow. In unsteady flow, streamlines indicate only the local direction of the fluid flow, the pathway of a massless particle is called a streak line and deviates from streamlines due to the temporal development.

Heat and mass transfer in laminar flow regime is controlled by conduction and diffusion, respectively. In a straight pipe or channel, a parabolic flow profile establishes following a sufficient entrance length, see Eqs. 3.22 and 3.47. The flow in curved microchannels with rectangular cross sections is discussed next due to of their predominant role in microfluidic equipment.

Straight laminar flow changes when the fluid flows through curves, bends, or around obstacles. Centrifugal forces in bends push the fluid from the center of the channel, where the bulk fluid flows with high velocity, to the outward side. At this wall, the fluid is forced either upwards or downwards, producing a symmetric, double vortex filling the entire channel, see Figure 3.5, left side. This flow regime in curved channel elements is often called Dean flow, in honor of Dean's first investigations on this topic [256]. The development of the vortex flow is induced by the centrifugal force, which is mostly dependent on the mean flow velocity w and the bend radius r. The viscous wall friction acts against the centrifugal force and, therefore, dampens the vortex flow. The flow in helical microchannels was studied by Schönfeld and Hardt [257] and Jiang et al. [258] for microfluidic applications and mixing purposes. They characterize the existence and growth of various vortex structures with typical dimensionless numbers. Thomson et al. [259] analytically model the flow with series solutions of low Dean and Germano numbers in helical rectangular ducts. Sudarsan and Ugaz [260] describe a spiral micromixer with vortex formation and enhanced mixing characteristics. Kumar et al. [261] and Pathak et al. [262] describe mixing processes in curved tubes and channels, respectively.

Along the outlet channel, the vortex flow is dampened and eventually dissipates, before returning to straight laminar flow. This complex flow regime is often referred to secondary flow structures, under which many three-dimensional flow structures are sub-summarized. According to Herwig et al. [263], the notion of secondary flow hides to much information about the actual flow situation, where modern CFD tools and measurement techniques are able to give a clear picture. Hence, it should be attempted to describe complex three-dimensional flow structures as much as possible in order to obtain more physical information about the process and to evaluate flow-related transport processes, such as mixing or heat transfer.

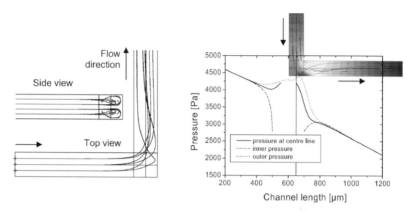

Fig. 3.5. Left: Streamlines and vortices in a 90° bend (L-mixer) with laminar vortex flow, Re = 99, $\bar{w} = 0.85$ m/s; Right: Pressure distribution in the 90° bend at the center plane of the channels, $100 \times 100 \ \mu m^2$, Re = 99.

In the above section, flow phenomena were described in rectangular channels with 90° bends. For various cross sections of conventional pipes, the flow and pressure distribution was theoretically investigated by Adler [264] and experimentally by Nippert [265]. Based on the work of Dean [256] dealing with the laminar flow in curved pipes with low Re numbers, Nippert gives a detailed description of the processes in circular bends larger than 8 mm channel diameter. His investigations deal mainly with turbulent flow regimes. Berger et al. [266] described the flow in curved pipes in a comprehensive review article.

Here, the investigated flow regimes are laminar with vortex formation. No onset of turbulence was observed in the investigated bends. Starting with the pressure distribution in the bend, Figure 3.5 right side, upper part, shows the pressure distribution in the center plane of the channel. The diagram displays the pressure along the inner and outer wall and in the middle of the channel.

In the inlet channel, a uniform pressure distribution in the cross section can be observed. Due to the curvature of the bend the flow is altered into a new direction. At the outer side of the bend, the pressure is increased comparable to the stagnation point of an impinging jet. At the inner side of the bend, the pressure decreases directly at or shortly behind the sharp corner, which often results in recirculation, separation flow, or cavitation. In Figure 3.5 right side, the lines of the inner and center pressure are discontinuous due to their shorter fluid path. Within the bend, the center and the outer pressure increases, but the inner pressure rapidly decreases. Behind the bend, the outer and center pressure drop immediately, and the inner pressure regenerates to equal the other values. Approx. 100 μm behind the bend, a uniform pressure establishes in the cross section. At this point, the vortices are already dampened and straight laminar flow is established again, as shown by the streamlines in Figure 3.5 left side. The pressure loss in the bend results in vortex formation and is the basis for further calculation in mixing theory, see Section 5.4.6. In straight channel sections the pressure decreases linearly as indicated by Eq. 3.21.

3.2.2 Fluid forces in bends

The comparison of two major forces on the fluid flow gives an impression of the importance of geometrical and flow parameters. As already mentioned, the centrifugal force is responsible for vortex development. It is dependent on the mass of the fluid in the bend m, flowing with the angular velocity ω around the 90° bend with the radius $r = d/2$. Due to the complex flow situation, the forces are estimated with mean values for the velocity $\bar{w}$ and the bend radius r. The centrifugal force [267] in a cross-sectional element in the bend can be written as

$$F_C = m\omega^2 r = m\frac{\bar{w}^2}{r}. \tag{3.58}$$

Considering an element with the length l in flow direction, the cross section of the channel with the width $d = 2r$ and the height h, the centrifugal force is given to $F_C = 2\rho h l \bar{w}^2$. The main resistance to the centrifugal force comes from viscous friction at the wall of the element. The force mostly originates from wall shear stress at the U-shaped element surface $(2d)$ and at the outer wall of the channel (h)

$$F_V = \tau A = \eta \left.\frac{\mathrm{d}\,w}{\mathrm{d}r}\right|_W (2d+h)\,l. \tag{3.59}$$

The ratio of the centrifugal force to the viscous force gives an indication of the flow dynamics in the bend and a parameter estimation of flow and geometry. A low ratio F_C/F_V indicates a dominant viscous force and, therefore, a straight laminar flow through the bend. A high ratio indicates vortex development in the bend. The ratio of the two forces gives

$$\frac{F_C}{F_V} = \frac{2\,h}{2d+h} \cdot \frac{\bar{w}^2}{v\,\frac{\mathrm{d}\,w}{\mathrm{d}r}\big|_W}. \tag{3.60}$$

The length l of the element plays no further role. Instead, the aspect ratio of the cross section $\alpha_A = h/d$ must be considered. Furthermore, the velocity gradient at the wall $r = d/2$ can be estimated by a derivation of the velocity field given by Eq. 3.47

$$\left.\frac{\mathrm{d}\,w}{\mathrm{d}r}\right|_{r=d/2} = 2\frac{\bar{w}}{d}, \tag{3.61}$$

which gives a force ratio of:

$$\frac{F_C}{F_V} \approx \frac{2\alpha_A}{2+\alpha_A}\frac{\bar{w}d}{2v} \approx \frac{2\,\alpha_A}{2+\alpha_A}\mathrm{Re}. \qquad (3.62)$$

The force ratio equation shows some similarity to the definition of the Dean number $\mathrm{Dn} = \mathrm{Re}\,(D/R_c)^{0.5} = \mathrm{Re}\,(2d_h/d)^{0.5}$, which describes the development of vortices in curved pipes [268]. The resulting force ratio from Eq. 3.60 and the terms of geometry and flow velocity are displayed in Figure 3.6 over the aspect ratio of a microchannel with constant hydraulic diameter. For flat channels with a low aspect ratio, the first

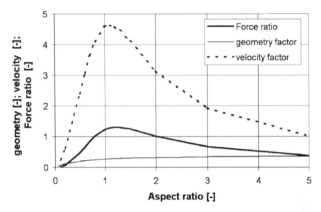

Fig. 3.6. Qualitative display of the ratio of the centrifugal force to the viscous force as well as the geometry and velocity factor dependent on the aspect ratio α_A for a constant Re number.

term of the right side of Eq. 3.60 vanishes and vortex development is suppressed. For high aspect ratios in deep and narrow channels, the first term approaches a value of approx. 2, and the second term of the right side becomes important, being the ratio between the mean velocity and the velocity gradient at the wall. This ratio decreases for narrow channels due to the high shear gradients at the wall and reaches its maximum at aspect ratios of unity, i.e. square cross sections, where the vortex flow is generated most effectively. For a more detailed discussion of the dominant forces in curved pipes see [264, 256]. For mixing applications, the flow situation in a T-mixer is more complicated, because of the "slip flow" at the symmetry plane of the mixing channel, however, the influence parameters are qualitatively similar and treated in the next section.

The combination of curved channel elements have been varied to find an optimal geometry for the channel length between 90° bends, the bend angle itself, or the impact angle in a T-shaped mixer [269]. In Figure 3.7, the flow situation and concentration profile is displayed for an optimized U-shaped or S-shaped 90° bend mixing element, which indicates a swirling flow behind the bends and good convective mixing. The simulations were made for water at 20 °C, a Schmidt number of $\mathrm{Sc} = 3700$ (aqueous Rhodamin B solution), and a mixing ratio of 1:1 with concentration 1 in one half and concentration 0 in the other half of the inlet channel. The light gray tone in the outlet section indicates good mixing of the two components.

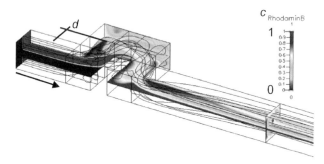

Fig. 3.7. Streamlines and concentration profiles (Sc = 3700 for aqueous Rhodamin B solution) in a combined 90° bend mixer (U-mixer) with channel dimensions of 300×300 μm², offset d, mass flow rate of 250 g/h, (Re = 270.7).

3.2.3 Fluid dynamics in T-junctions with symmetric inlet conditions

In the following section, the fluid dynamics of the mixing process in T-junctions are treated with symmetrical inlet conditions and 1 : 1 mixing of the reactants. Asymmetrical inlet conditions of gas-phase mixing is described in Section 6.4.2, other conditions can be found in [270]. The general, geometrical setup of a T-shaped mixer with rectangular cross section is given in Figure 3.8.

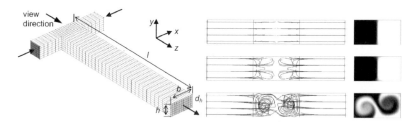

Fig. 3.8. Left: Flow regimes in a rectangular cross section T-shaped micromixer: straight laminar flow, vortex flow, and symmetry breakup with engulfment flow; Middle: concentration profiles of the laminar diffusive and the engulfment mixing; Right: geometrical setup of the T-mixer [271].

The CFD simulations were performed with the package CFD-ACE+ from ESI Group, versions 2002 to 2006, a volume-of-fluid code with additional multi-physics features. For the mixing simulations, the modules *Flow* and *Scalar* are used. A typical model setup of a T-shaped micromixer (T600×300×300 which represents a mixing T with rectangular cross sections, a mixing channel width of 600 μm, two symmetric inlet channels with a width of 300 µm, and 300 µm overall depth) with three-dimensional and transient calculation consist of approx. 387 000 cells with 410 000 grid points. The ordered grid consists of rectangular cuboid elements with a corner length of 5 µm in the channel cross section. In the axial channel direction, the element corner length varies to reduce the element number. The transient simulations were performed with a step width of $1 \cdot 10^{-5}$ s, however, the results were noted for every tenth time step (Re < 500) and every fifth time step for higher Re numbers. The velocities and the diffusive scalar are determined with a 2nd order Limiter method, with a blending value of 0.1. The inertial relaxation of the velocity components and the scalar were set to 0.2. The convergence criteria for all parameters were set to $1 \cdot 10^{-5}$ with a minimum residual of $1 \cdot 10^{-18}$. This was sufficient for the investigated flow situations. A detailed review on the quantification of uncertainty in CFD is given by Roache [272], who states that grid refinement is not always the method of choice to diminish numerical uncertainty. The steady simulations do not converge for Re numbers higher than 240 due to the transient effects. The transient simulations were not sensitive to convergence problems, if the time step was short enough.

The development of the numerical simulation can be observed at the residual development with iteration step, as shown in Figure 3.9. The left diagram shows the residual of the simulation of T-shaped mixer with symmetrical inlet conditions and a Re number of 130 in the mixing channel (600×300 µm^2). The residuals for the velocity components, the pressure, and the species distribution decrease continuously until the convergence criterion is reached.

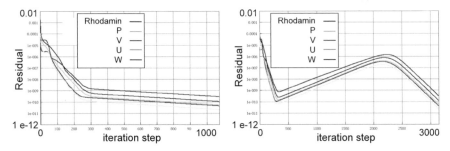

Fig. 3.9. Development of residuals with iteration steps of the simulation of a T$600 \times 300 \times 300$ mixer; left: Re = 130, right: Re = 150.

An interesting phenomenon is the residual development of the velocity components u, v, and w, the pressure, and the concentration, which all show similar behavior. With an increasing iteration step number, the residuals diminish, indicating a vanishing numerical error, which determines the matrices. The typical behavior of the engulfment flow simulation, see next section for flow regime description, was an increasing residual after a certain time, indicating the birth of a new solution, and the swapping of the liquid to the opposite side. Intermediate results of the simulations (iteration step approx. 500) display a symmetrical flow situation (vortex flow) for a Re number of 150 in the mixing channel. Due to small numerical fluctuations or roundoff error, a small portion of one component would pass to the opposite side, initially generating a numerical error. The numerical error of the intermediate solution was mirrored by inclining residual values, see Figure 3.9 right side. Established engulfment flow leads to decreasing residuals and finally to the solution.

The boundary condition at both inlet channels was a constant laminar velocity profile with near parabolic profile approximated by the Fourier series according to Eq. 3.22. For most applications, the Fourier series is truncated after the second or third term. These boundary conditions produce a stiff system for the developing pulsations, which also lead to pressure pulsations. The transient flow in a T600×300×300 mixer was also simulated with constant inlet and outlet pressure, where the mean pressure value was determined from the velocity boundary condition applied at both inlet channels. The inlet pressure was adjusted to the new combination to gain the same Re number as with the constant velocity profile at the inlet. The mean velocity was approx. 0.5 % lower, however, the pulsation frequency was remarkably lower, as the flow and concentration fields have been mostly similar to the constant inlet velocity profile. Aside from the flow and concentration field, the pressure loss is an important value for mixing characterization.

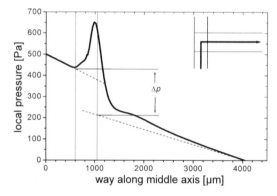

Fig. 3.10. Typical pressure loss in a T-mixer. The dashed lines describes the analytical pressure loss from wall friction.

In Figure 3.10, typical pressure loss along the mixing channel is displayed. In the straight inlet channel and mixing channel, the pressure loss decreases linearly with the channel length, which originates from wall friction. This pressure loss does not affect the mixing process. At the junction of the inlet channels and the mixing channel, a strong increase in pressure loss is observed, which results from the bending of the streamlines and the creation of vortices. This pressure loss must be taken into account when calculating the energy provided for mixing. From simulations, the point, at which this pressure loss dissipates, can also be determined. Simulations have shown that in the observed range of Re numbers, Δl is approximately 3-5 times the width of the mixing channel b_M.

The pressure loss in the entire device of a T-mixer is displayed in Figure 3.11 with piecewise polynomial regression fits. For Re < 10, the linear term is dominant, however, for higher Re numbers, the square term gains importance. At Re = 200, the linear and the square term contribute approximately 50 % to the pressure loss, respectively. Hence, for a first estimation of the pressure loss in a device, the laminar and turbulent contribution should be determined for a certain velocity (or Re number). Then, both contributions can be fitted by geometrical summation of the squares. Yue et al. [42] investigated the pressure loss in T-mixers with single and two-phase flow and determined friction factors with mean deviations within ±20 %. They stated that systematic studies are still required to accurately predict the frictional performance in microchannels.

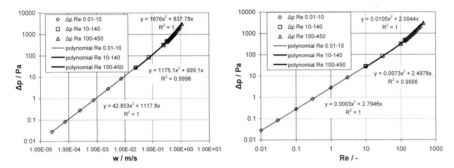

Fig. 3.11. Pressure loss in a T-mixer, here T600×300×300, for different velocities and Re numbers in the mixing channel. The correlations are piecewise polynomial regression fits with the coefficient of determination.

An overview and classification of the investigated flow regimes in symmetrical T-shaped micromixers, in this case the results for T600×300×300, is displayed in Figure 3.12 for 0.01 < Re < 1000 in the mixing channel. The left side shows typical flow patterns with streamlines and the iso-surface for the mean concentration of 0.5, representing the completely mixed state. The iso-surface also indicates the separation interface between the components and illustrates the mixing situation. On the right side of Figure 3.12, the development of the mixing quality α_m is displayed over the Re number; for 0.01 < Re < 100 on a logarithmic scale and for 100 < Re < 1000 on a linear scale for better illustration of the different flow regimes. The mixing quality is the standardized concentration field variance σ_c, see also Section 5.1.2,

$$\alpha_m = 1 - \sqrt{\sigma_c^2 / \sigma_{c,max}^2}, \tag{3.63}$$

and is often used to characterize the state of mixing, see Section 5.3. The mixing quality is given here to visualize the flow patterns and pulsating flow regimes, see next section. The concentration field is determined for a Schmidt number Sc = 3700, representing the diffusion of a large molecule (Rhodamin B) in water. Due to the coarse numerical grid, the numerical diffusion plays an important role in determination of concentration fields. For Re > 130, they only provide a rough indication and a tendency. The concentration profiles reflect the actual physical situation quite well, as experimental investigations reveal [273].

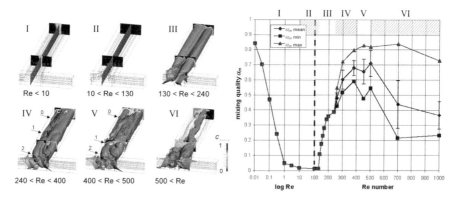

Fig. 3.12. Overview of the flow regimes and Re number ranges in laminar T-junction flow with 1:1 mixing ratio, Sc = 3700, T600×300×300. The mixing quality α_m (right side) is determined at a constant length $l = 5 \cdot d_h$ of the mixing channel, here $l = 2000$ μm. An enlarged image is given in the Appendix with Figure A.19.

To give a short overview of the flow regimes from straight laminar to transient, chaotic flow, the following categorization is introduced for short flow and mixing regime description. This classification is also displayed in Figure 3.12 with typical flow situations and the development of the mixing quality relative to the Re number in the mixing channel.

- Re < 10, Dn < 10, regime I
  straight laminar flow, steady velocity profile with straight streamlines, diffusion dominates the mass transfer;
- 10 < Re < 130 (approx.), Dn > 10, regime II
  symmetric vortex flow, Dean flow with characteristic Dn number;
- 130 < Re < 240 (approx.), regime III
  engulfment flow, breakup of the symmetry, fluid swaps to the opposite side;
- 240 < Re < 400 (approx.), regime IV
  regular, *periodic pulsating flow*, reproducible vortex break down;
- 400 < Re < 500 (approx.), regime V
  quasi-periodic pulsating flow, vortex breakdown, broad range of mixing quality α_m;
- 500 < Re (approx.), regime VI
  chaotic pulsating flow, vortex breakdown, decreased mixing quality and single flushes of fluid swapping to the opposite side.

This list gives no upper limit for high Re numbers, which will lead to turbulent flow. The transition to turbulent flow lies in the range of $1000 < Re < 2300$, but is not investigated in this work. From experimental observations of reactive particle precipitation with high Re numbers, see Section 6.5, it is known that fluid flow may also reach into the opposite inlet channel and can block the inlet with precipitated particles, therefore, flow regimes with Re numbers higher than 1000 are avoided in T-shaped micromixers. High Re numbers are also unsuitable due to high flow velocities producing intolerably high pressure loss. Conversely, as shown in Figure 3.12, a T-shaped micromixer with typical channel dimensions from $50 < d_h < 600$ µm exhibits the most suitable mixing characteristics in the range of $200 < Re < 700$ due to the moderate flow velocities, ranging from $1 < \bar{w} < 50$ m/s for gases and $0.1 < \bar{w} < 10$ m/s for liquids.

3.2.4 Flow regimes in T-shaped micromixers

As demonstrated in the previous section, it is convenient to classify the investigated Re number in the range from 0.01 to 1000 into six domains. The following section provides a more detailed description of the flow regimes from straight laminar to transient, chaotic flow.

Re < 10, Dn < 10, regime I: straight laminar flow

Straight laminar flow with a steady velocity profile governs the transport processes in microchannels, see Figure 3.13. Even in bends and curves, the streamlines follow the geometrical curvature, and an entrance effect within the mixing channel has not been observed. Viscous forces and wall friction control the fluid motion, and diffusion dominates mixing and mass transfer. This flow regime is often applied in lab-on-a-chip devices and is also referred to as creeping flow. Mixing of gaseous components is feasible in this regime due to the low Sc number, in the order-of-magnitude of one. Some good examples of gas-phase mixing in T-shaped micromixers are given by Gobby et al. [274].

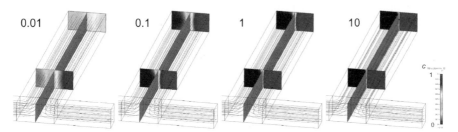

Fig. 3.13. Regime I, straight laminar flow in T600×300×300 mixer; Re numbers from 0.01 to 10; Sc = 3700. See also Figure A.13.

The left image in Figure 3.13 shows the well mixed situation due to diffusion, where the fluid remains in the mixing channel long enough ($Re = 0.01$, $L = 4$ mm gives a residence time of $t_P = 250$ s) to allow sufficient diffusive mass transfer. The diffusion length within $t_P = 250$ s is approx. 500 μm, resulting in a high mixing quality, see Figure 3.12 right side.

10 < Re < 130 (approx.), Dn > 10, regime II: vortex or Dean flow

Here, straight laminar flow is disturbed by the centrifugal forces created by an increasing Re number, and symmetric vortex flow establishes in the mixing channel. The flow regime is comparable with Dean flow and shows characteristic Dn number.

$$\mathrm{Dn} = \mathrm{Re} \left(\frac{d_h/2}{b/2} \right)^{1/2} = 2\,\mathrm{Re} \left(\frac{\alpha_A}{1 + \alpha_A} \right)^{1/2} \tag{3.64}$$

In this case, the T-mixer with rectangular cross sections gives a slightly different characterization as the 90°bend shown in Section 3.2.2.

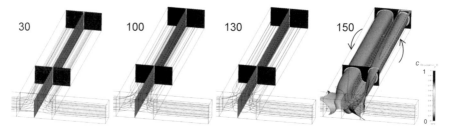

Fig. 3.14. Regime II, curved laminar flow in T600×300×300 mixer, Dean vortices for $30 < \mathrm{Re} < 130$ (approx.) and asymmetrical flow for $\mathrm{Re} = 150$ (engulfment flow regime); Re numbers from 30 to 150; $\mathrm{Sc} = 3700$. See also Figure A.14.

The flow in a T-shaped micromixer displays comparable behavior, with the formation of vortices, as seen in Figure 3.8, left side. With increasing Re number, the flow starts to form secondary vortices at the T-junction near the beginning of the mixing channel. Unfortunately, the component interface lies directly on the symmetry plane of the two components, hence, the mixing of the two components is not influenced by the vortex formation, as can be seen in the development of the mixing quality over the Re number, Figure 3.12 right side.

130 < Re < 240 (approx.), regime III: engulfment flow

As the Re number further increases, the symmetry of the flow structure breaks down and fluid begins to swap from one side to the other side. This effect can be seen in the middle of Figure 3.8, where the concentration distribution in the mixing channel indicates the two co-rotating vortices. From the viewpoint looking onto the mixing channel, a horseshoe-like double vortex is formed at the beginning of the mixing channel and rolls up with mutual wrapping in of both components. This leads to small lamellae, short diffusion lengths, and high mixing quality, which is shown in Figure 3.12 left side. Similar to mixing in 90° bends for higher Re numbers, the mixing process is scale-invariant, depending only on the Re number and the aspect ratio. The inlet Re number, which represents the incoming momentum, is the main parameter of symmetry break-up and flow instability.

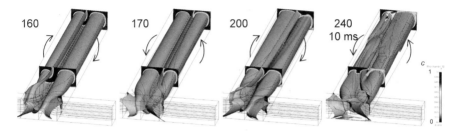

Fig. 3.15. Regime III, asymmetrical, laminar flow in T600×300×300 mixer, engulfment flow; Re numbers from 160 to 200 and Re = 240 at $t = 10$ ms due to periodic oscillating flow; Sc = 3700. See also Figure A.15.

The CFD images in Figure 3.15 show the flow structure at different Re numbers from 160 to 240. The double-vortex can have two different rotational directions, as indicated by the arrows. Considering the flow regime at Re = 160, the inlet flow from the right side is pressed to the bottom of the mixing channel, resulting in a clockwise rotation of the double-vortex, viewed from behind. The situation is different for Re = 170, where the inlet flow from the right side is pushed to the top of the mixing channel, yielding a counter-clockwise rotation of the double-vortex. Both results are valid for the engulfment flow regime indicating the flow bifurcation with higher Re numbers. Furthermore, the vortex rotation is clockwise at Re = 200 and counter-clockwise at Re = 240, see Figure 3.15.

Critical Re number for engulfment flow: The critical Reynolds number Re_{crit}, where engulfment flow occurs, is a very interesting issue, as it allows an estimation of the minimum flow required in convective micromixers to achieve an appropriate convective mixing process. As described in Section 3.2.1 for vortex or Dean flow, one criterion for the occurrence of engulfment flow is the ratio of the centrifugal force to the viscous fluid forces F_c/F_v, see [26]. Considering a T-shaped mixer with the aspect ratio of the mixing channel $\alpha_A = h/b_M$, where h is the height of the channels, the centrifugal force F_c acting on a certain mass $2\,\Delta m$ in the inlet channels is estimated by

$$F_c = 2 \cdot \Delta m \cdot \frac{\bar{w}^2}{R} = 4\rho \cdot h \cdot \Delta l \cdot \bar{w}^2 \tag{3.65}$$

with the radius $R = b_{IC}/2$ set to half of the inlet channel width. Further, the viscous force F_V working on the same mass inside the mixing channel can be calculated by assuming laminar flow and using the Hagen-Poiseuille law, Eq. 1.14 for rectangular channels:

$$F_v = \frac{C_f \cdot \Delta l \cdot \eta \cdot A_M \cdot \bar{w}}{2d_{h,M}^2} = \frac{C_f \cdot \Delta l \cdot \eta \cdot (b_M + h)^2 \cdot \bar{w}}{8b_M \cdot h} \qquad (3.66)$$

where η is the dynamic viscosity of the fluid, A_M is the cross section, and $d_{h,M}$ the hydraulic diameter of the mixing channel. The friction factor C_f can be analytically determined, and is approximated, dependent on the aspect ratio α_A, to

$$C_f(\alpha_A) = 96.383 - \frac{36.5}{0.911 + (0.272 - \log_{10}(\alpha_A))^{3.3}}. \qquad (3.67)$$

This equation holds for $0 < \alpha_A \leq 1$, being symmetric to $1/\alpha_A$, which is the ordinate in Figure 3.16. The correlation given in Table 3.1 for pressure loss in channels with rectangular cross sections gives comparable values with a maximum variation less than 0.6 %.

The ratio of centrifugal to viscous force, Eq. 3.65 divided by Eq. 3.66, produces

$$\frac{F_c}{F_v} = \frac{16d_{h,M} \cdot \bar{w} \cdot h}{v \cdot C_f\,(b_M + h)} = \frac{16}{C_f} \cdot \frac{\alpha_A}{1 + \alpha_A} \cdot \mathrm{Re} = \mathrm{const.} \cdot \mathrm{Re} \qquad (3.68)$$

where the constant factor only depends on the geometry. The Re number is determined in the mixing channel. For a defined geometry, the ratio of centrifugal to viscous forces is directly proportional to the Re number, as already shown by Dean flow in bends, see Section 3.2.2.

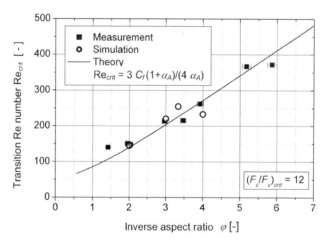

Fig. 3.16. Critical Reynolds number Re_{crit} over the inverse aspect ratio $\varphi = 1/\alpha_A$ at which transition from vortex to engulfment flow occurs, [127]. Comparison of values from measurement and simulation with theoretical results from Eq. 3.69.

To confirm Eq. 3.68, the critical Re number Re_{crit} has been measured using colored fluids and simulated by CFD for a large number of geometries with different aspect ratio, see Figure 3.16. According to Eq. 3.68, the critical Reynolds number Re_{crit} is calculated to

$$\text{Re}_{crit} = \frac{F_c}{F_v} \cdot \frac{C_f(\alpha_A)}{16} \cdot \frac{1+\alpha_A}{\alpha_A} \tag{3.69}$$

The results from experimentation and simulation indicate that engulfment flow occurs at a fixed ratio of F_c to F_v, see Figure 3.16. In the case of symmetrical 1:1 mixing, a critical ratio of $(F_c/F_v)_{crit} = 12$ is determined.

In mixing channels with a very high or very low aspect ratio, the engulfment flow is retarded and the critical Re number is higher. Flat channels have a large surface-to-volume ratio, leading to a stronger dampening of the vortices inside the mixing channel. The values of the critical Re number from experimentation and simulation are close together and can be predicted by an empirical correlation.

Experimental verification of engulfment flow: To validate the results from the numerical simulations, the symmetrical mixing of two fluids and the existence of flow structures must be observed. The high optical contrast of pH indicators was used for this purpose. The 1:1 neutralization reaction of an alkaline solution of di-sodium hydrogen phosphate (pH = 8) with de-ionized water and Bromothymol Blue (pH = 7, green color) resulted in a mixture with a pH value of 7.5, indicated by the blue coloration (dark gray), see Figure 3.17, middle. The concentration of the reactants is very low (mmol/l) and has no great influence on the fluid properties.

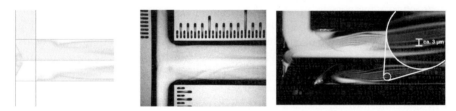

Fig. 3.17. Qualitative comparison between simulation (left) and experiment (middle) for a micromixer with square cross sections (T600×300×300), a mass flow of 260 g/h and Re number of approx. 180; right: LIF image of the concentration field at a depth of 50 µm in a T200×100×100 micromixer with Re = 207, scale indicates a length of 3 µm, see [275] (courtesy of University of Bremen, IUV).

For low flow velocities, a clear segregation line is observed between the two components. The mass transfer is controlled by diffusion in the small interface area. In engulfment flow, the symmetry breaks down and parts of each fluid swap to the opposite side, as described previously. The fluid parts expand and separate, resulting in the filament structure shown in Figure 3.17. Comparison with simulation results displays similar behavior, however, the measured flow structures are much smaller in the range of few micrometer. These structures and their corresponding concentration fields are not resolved in the simulations due to the coarse numerical grid. Laser Induced Fluorescence (LIF) measurements of Schlüter et al. [275] indicate filament widths smaller than 3 µm in the core of the vortex, see Figure 3.17 right side.

The LIF measurements in microchannels together with a confocal laser microscope allow to determine the three-dimensional concentration profile in a T-shaped

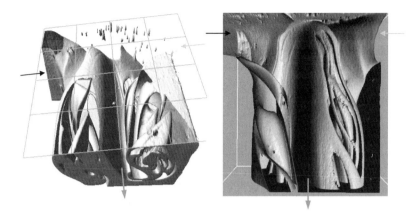

Fig. 3.18. Three dimensional concentration field, micromixer geometry $400 \times 100 \times 285$ µm at the beginning of the mixing channel with Re $= 160$, [273] (courtesy of University of Bremen, IUV).

micromixer with steady flow regime. Figure 3.18 from the group of Prof. Räbiger and Dr. Schlüter at Bremen University, shows the concentration profile and the interface between two components in the T-junction of a micromixer with a cross section of 800×300 µm. The complex flow field and the thin fluid lamellae are clearly visible, caused by the generation of engulfing vortices and the resultant roll-up of the fluids, however, the fluid lamellae are not homogeneously distributed. These flow structures decrease the concentration lamellae width and enhance the mixing enormously. The work group of Dr. Schlüter examined the flow field with particle image velocimetry PIV and confirmed the complex flow structure at the mixing channel entrance. With further increasing volume flow and Re number, the flow regime begins to show transient effects. Unfortunately, accurate LIF and PIV measurement are only feasible in steadystate flow.

Vortex structure in engulfment flow: A closer look at the flow structures of the engulfment flow gives a deeper insight into the convective mixing process in a T-shaped micromixer. The flow field at the entrance of the mixing channel can be divided into three major areas, see also Figure 3.19: stagnation point flow at the backside of the mixing channel, small vortex pair with strong rotation and fast engulfment, and large vortex region with slow rotation. The stagnation region and the small vortex pair are characterized by strong elongation and folding of the fluid elements, therefore, they are the key regions for mixing enhancement. Besides the top view of the streamlines in a T-shaped micromixer, Figure 3.19 shows the rear side view and the mixer from the left and right side. The streamlines indicate the portion of the entrance flow, which transfers to the opposite side of the mixing channel. This transfer area remains in between the streamlines and includes the middle of the inlet flow with the highest velocity. The fluid with lower velocity stays on the inlet side and forms a large vortex. The transfer fluid coming from the left side moves to the

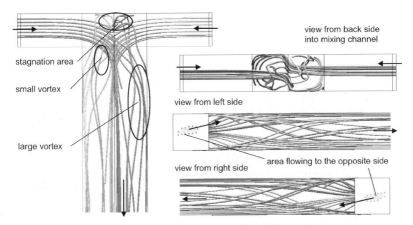

Fig. 3.19. Engulfment flow in a T-shaped micromixer (T200×100×100) with a mass flow of 73 g/h and a corresponding Re number of approx. 135; the streamlines indicate the region of the entrance flow, which moves to the opposite side of the mixing channel.

top of the mixing channel and also joins the large vortex. Only the fluid located at the sharp, outer part of the transfer area streams next to the stagnation area at the rear side of the mixing channel and forms the small vortex, which is displayed in Figure 3.20. Here, extremely rapid mixing occurs due to the fine vortex structure with lamellae width below 1 μm, see Figure 3.17 right side.

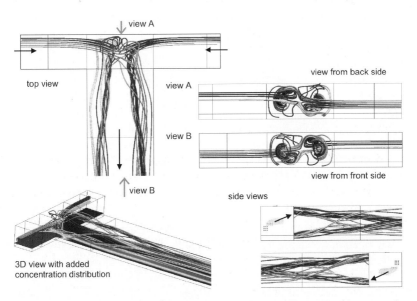

Fig. 3.20. Engulfment flow in a T-shaped micromixer (T200×100×100) with a mass flow of 73 g/h and a corresponding Re number of approx. 135; streamlines for the small vortex pair indicate the area with high mixing intensity.

240 < Re < 400 (approx.), regime IV: periodic pulsating flow

At a Re number of approx. 240, the flow in a T-shaped micromixer exhibits transient behavior, a periodic pulsating flow occurs, which has not yet been reported in literature. From numerical simulations, a reproducible vortex break down and periodic pulsation of the concentration iso-surface can be observed, as a result of one inlet swapping to the other side and drawing back. The high shear rate between the two components at the contact point of the T-junction induces Kelvin-Helmholtz instability of shear flow, leading to unstable flow conditions and periodic fluctuations in flow profiles and mixing quality. This instability is known from turbulent flow, see Herrmann [276], and describes the roll-up of a shear layer forming a vortex sheet without surface tension. The frequency is approx. 470 Hz for Re = 300 in a T600×300×300 micromixer with constant inlet velocity profile, shown in Figure 3.21. The frequency from numerical simulations is not only dependent on the Re number or the geometry, but also on the inlet boundary conditions and the numerical grid size. Preliminary simulations with constant inlet pressure yield a lower fluctuating frequency; for a Re number of 300 in a T600×300×300 micromixer, the frequency drops to approx. 405 Hz with a Strouhal number

$$Sr = \frac{f \cdot d_h}{\bar{w}} = \frac{f \cdot d_h^2}{v \cdot Re} \qquad (3.70)$$

of 0.217, see Figure 3.22. The vortex breakdown in the CFD images is highlighted by arrows and flows downstream into the mixing channel, as shown by the time frames from the CFD simulation.

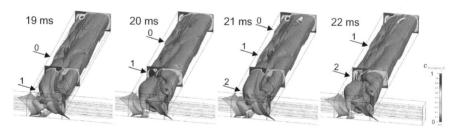

Fig. 3.21. Regime IV, periodic pulsating laminar flow with approx. 470 Hz in T600×300×300 mixer; Re = 300, Sc = 3700, observation times are 19, 20, 21, and 22 ms, see Figure 3.22 and A.16.

The fluctuations are clearly visible in the transient behavior of the mixing quality at a fixed location in the mixing channel, here at 2 000 μm, which is five times the hydraulic diameter. In Figure 3.22 left side, the temporal development of the mixing quality is displayed at the first cut in the mixing channel at $z = 2\,000$ μm. This cut is close to the cuts displayed in every CFD image as a cross section at $z = 1\,600$ μm. The mixing quality curves in Figure 3.22 clearly indicate the regular fluctuations for Re = 240 and 300, but also the quasi-periodic flow situation for Re = 500 and the chaotic behavior for Re = 1 000. The time axis in Figure 3.22, left side, is identical with the temporal data given in Figure 3.21 with transient CFD images. The wide variation of mixing quality in this range of Re numbers is also illustrated in the diagram of Figure 3.12 and analyzed in Chapter 5. The progression of the mixing quality at Re = 240, starting at Re = 200, is slightly dampened, for which a steady numerical solution also exists.

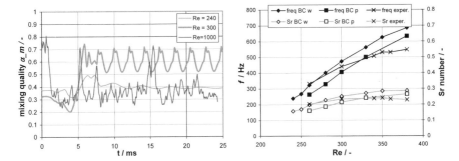

Fig. 3.22. Pulsating flow in a T-shaped micromixer; Left: Fluctuating mixing quality at a mixing channel length of 2 mm of a T600 × 300 × 300 micromixer over time for different Re numbers in the mixing channel. Right: Frequency and Strouhal number Sr of the fluctuation inside the mixing channel over the Re number. The error in determining the frequency from numerical simulations is less than 2.6 % from the actual value.

In Figure 3.22 right side, the frequency and dimensionless Strouhal number Sr are displayed over the Re number in the mixing channel. The periodic pulsations have been observed in the Re number range from 240 up to approx. 400. The fluctuations for Re = 500 as shown in Figures 3.22, left side, and 3.26 are treated as quasi-periodic due to the fine fluctuations in between the large mixing quality reductions.

Simulations of T-shaped micromixers with different hydraulic diameters (T200× 100×100, d_h = 133 μm, Re = 300, f = 4 230 Hz, Sr = 0.25) indicate the similarity of the pulsating flow in relation to the Sr number. Further investigations, numerical simulations and experimental measurements are necessary to clarify and validate the initial results of the pulsating flow regimes. For comparison, the fluctuating flow behind a cylinder, or the vortex street in wake flow behind a wire, has a characteristic Sr number of approx. 0.2 [277], which is in the range of the values observed in the T-shaped micromixer, see Figure 3.23. A microreactor with pulsating wake flow behind posts or cylinders in the mixing channel was investigated by Deshmukh and Vlachos [278] with 3D CFD simulation to enhance mixing for a catalytic surface reaction. Their simulated Sr numbers from 0.13 to 0.16 in the range of 60 < Re < 100 correlate quite well with data from other literature.

Due to the high frequency of flow pulsations with approx. 470 Hz at Re = 300 in the T-mixer, experimental investigations require a good temporal resolution by digital camera or light source. In Figure 3.24, mixing of Uranin fluorescence dye solution with water illustrates the dynamic process within the micromixer. Further experimental tests using improved optical measurement with controllable stroboscope

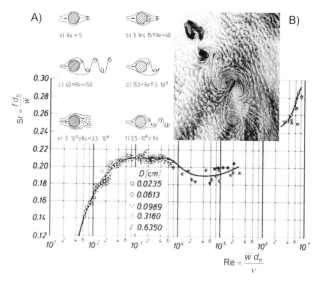

Fig. 3.23. Periodic flow in the wake of a cylinder; A) flow regimes depending on the Re number according to Baehr and Stephan [89] p. 342; B) von Kármán vortex street behind an island according to NASA, SeaWiFS Project; Diagram: Strouhal number Sr depending on the Re number of the separated flow around a cylinder according to Tritton [279, 280].

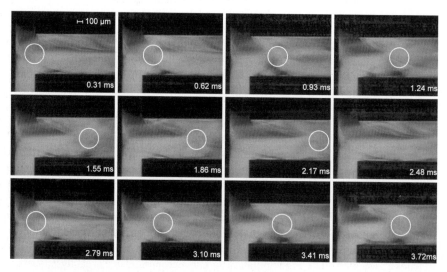

Fig. 3.24. Experimental flow visualization with fluorescence color Uranin in a T-mixer (T600 × 300 × 300) at Re = 300 in the mixing channel, main velocity of approx. 0.75 m/s. Typical flow structures are indicated with a circle. Digital camera (Olympus C5050, VGA 640×480 pixel) with 15 frames/s and stroboscopic light with 85 Hz indicate a mixing fluctuation frequency of approx. 480 Hz, which is in the range of simulation data.

or high-speed camera are necessary to yield more conclusive data. The situation

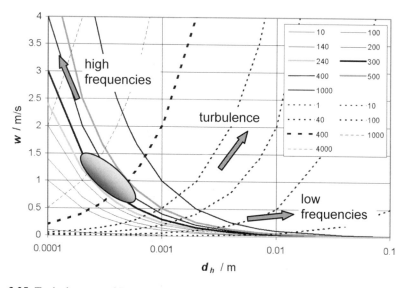

Fig. 3.25. Typical ranges of Re numbers (full line) and frequency (dotted line) depending on the hydraulic diameter d_h and the mean velocity $\bar{w}$ for Sr $= 0.2$ and aqueous systems.

changes with increasing channel diameter, as in typical standard laboratory applications using mixing devices with $3 < d_h < 20$ mm or conventional process engineering equipment with $25 < d_h < 200$ mm. The typical velocities in this equipment lie within the range as mentioned above, however, the Re number is at least one or two orders of magnitude higher, leading to turbulent flow conditions and other mixing regimes. The cross section is very often circular in conventional equipment with pipes, leading to different flow situations. The flow in T-shaped mixers may show similar characteristics, and the momentum of a single fluid element may play an important role in describing the mixing process, however, this is not aim of this work.

Figure 3.25 gives an impression of the small range of hydraulic diameter and mean velocity, where periodic fluctuations may occur in T-shaped devices with rectangular cross section and 1:1 mixing. The upper and lower Re numbers for periodic fluctuations in T-mixers are indicated by bold grey curves which give a quite wide range for microchannels. The area is narrowing for conventional equipment with hydraulic diameters around 1 cm. There, the typical frequency of periodic fluctuations is around 1 Hz. Due to typical flow velocities in conventional equipment, i.e. $\bar{w}$ is 1 to 50 m/s for gases and 0.1 to 10 m/s for liquids, the here described flow regime is not in the typical range. For very small devices with typical dimensions of 100 μm and smaller, the flow velocities must be in the range of more than 2 m/s which is very high and untypical for lab-on-chip devices, hence, periodic fluctuations can be regarded as typical flow regime for microchannels.

400 < Re < 500 (approx.), regime V: quasi-periodic pulsating flow

With increasing Re number, the frequency of the flow pulsations increases as well, see Figure 3.22. The regular pulsation is disturbed by higher-order pulsations, which can be seen in the transient mixing quality at Re = 500 in Figure 3.22. Each plateau with high mixing quality has four small peaks, until the mixing quality drops down, where also one small peak is visible. This quasi-periodic pulsating flow can be characterized by the large mixing quality reduction, whose frequency is lower than the proceeding ones. The mixing quality reduction is related to vortex breakdown, as can be seen in Figure 3.26, where each passing of a vortex disturbance through the 1 600 μm cut indicates a further reduction of the mixing quality in Figure 3.22. The variation of the mixing quality is greater, the mean mixing quality reaches a maximum. Figure 3.26 also shows the birth and progression of a vortex breakdown (1)

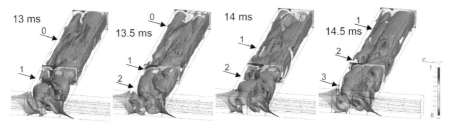

Fig. 3.26. Regime V, quasi-periodic pulsating, laminar flow in T600×300×300 mixer; Re = 500, Sc = 3 700, observation times are t = 13 to 14.5 ms, see Figure 3.22 and A.17.

through the mixing channel. The swirling flow on the backside of the mixing channel indicates the pulsating flow regime.

500 < Re (approx.), regime VI: chaotic pulsating flow

For Re > 500, the flow generates irregular structures, the mean mixing quality decreases, and a wavy interface often builds up in the middle of the mixing channel, see Figure 3.27. The regularity of the pulsation vanishes, chaotic motion of the components takes over, however, the fluid components are more segregated now. Occasionally, fluid transfers to the opposite side and causes a high peak of mixing quality, such as the peak for Re = 1 000 at 15 ms in Figure 3.22.

Figure 3.27 displays the birth and development of a swirling peak at Re = 1 000 in the mixing channel. The swirl delivers a mixing quality peak at z = 1600 μm after 15 ms, also shown in Figure 3.22. Between these swirling peaks, a wavy segregation layer forms and separates both components near the middle of the mixing channel. Only a small portion of the fluid reaches the opposite side and is well mixed, therefore, the mean mixing quality drops below 0.4. The numerical diffusion must be considered, however, the trend is clearly visible.

In this section, the detailed description of the flow regimes in symmetric T-shaped micromixers with 1:1 mixing in a Re number range from creeping flow of 0.01 to

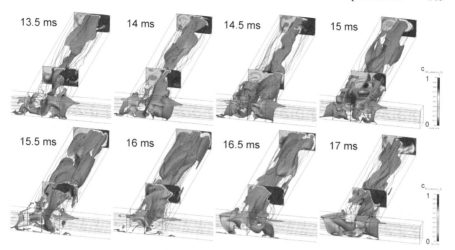

Fig. 3.27. Regime VI, chaotic pulsating, quasi-laminar flow and quasi-vortex flow in T600×300×300 mixer; Re = 1000, Sc = 3700, observation times are t = 13.5 to 15 ms in the top row, t = 15.5 to 17 ms in the bottom row. See also Figure A.18.

1000 with chaotic flow structures. This gives a primary insight, still, further investigations in numerical simulation, experimental observations, and analytical modeling are necessary to develop a clearer image, however, the data presented is sufficient to assist the design process of convective micromixers and microreactors, which will be discussed in Chapter 5 and 6.

3.3 Multiphase Flow

In microstructures, the surface-to-volume ratio increases and surface effects gain importance. Factors such as surface tension or wetting become more dominant compared to volume forces like gravity, pressure, or momentum, therefore, multiphase flow in microchannels with many interfaces acts in a different manner and shows other characteristics than in macrosized devices. Günther and Jensen [281] recently gave a comprehensive review on multiphase microfluidics. They showed typical geometries with relevant forces in specific length and velocity ranges. Highly important are the flow regimes and interfacial shapes of gas-liquid and liquid-liquid flows. Cheng and Mewes [282] gave in another review an overview of two-phase flow with heat transfer. In the next two sections, an overview of some special flow and transport characteristics is given with relevance to micro process engineering. Particulate flow is also important in microchannels, however, this is treated in Sections 4.3.3, 4.3.4, 6.4 and 6.5.

3.3.1 Gas-liquid flow patterns

Adiabatic, two-phase gas-liquid flow in tubes with hydraulic diameters in the order of 1 mm have been extensively studied in the last decade, see [283]. Two-phase flows in microchannels are characterized by low gravitation forces, the disappearance of stratified flow patterns, and a shift in some of the flow-pattern limits. However, two-phase flow patterns in conventional scale and microchannels appear to be morphologically similar [284]. The liquid-ring flow pattern reported by several studies [285] to be unique to microchannels is observed at low gas flow rates and is characterized by axi-symmetrical distribution around the channel inner wall. Generally, four different flow regimes have been observed in microchannels: stratified, bubbly, annular, and intermittent flow patterns, here, for example, in a 976 μm microchannel with the boundaries of gas and liquid flow rates, shown in Figure 3.28.

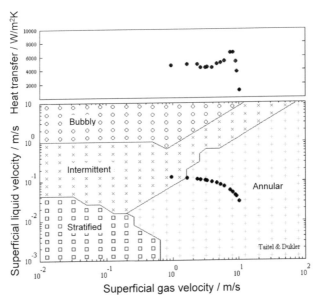

Fig. 3.28. Two-phase flow pattern map (Taitel-Dukler flow regime map, see Bar-Cohen and Rahim [286]) for refrigerant microgap channel flow (Yang and Fujita [287], refrigerant R113, $G = 200$ kg/m^{2}s, $\dot{q} = 50$ kW/m^2, $D_h = 976$ μm single channel).

Intermittent flow patterns of bubbly and churn flow have not been observed in microchannels, probably due to the dominating surface effects. Similar flow patterns have been found in boiling flow, however, more investigations are necessary to clarify the complete picture. A key role is played by the bubble nucleation process, which contradicts the annular flow regime in its early stages. To understand this contradiction, it is important to examine the way, in which flow patterns are defined. This is not a purely semantic topic, since the definition encompasses a set of attributes that are used to identify pressure loss and heat transfer mechanisms, at least in conventional scale practice. An appropriate definition is still a topic of discussion within the heat transfer community, the most interested and active group in this field.

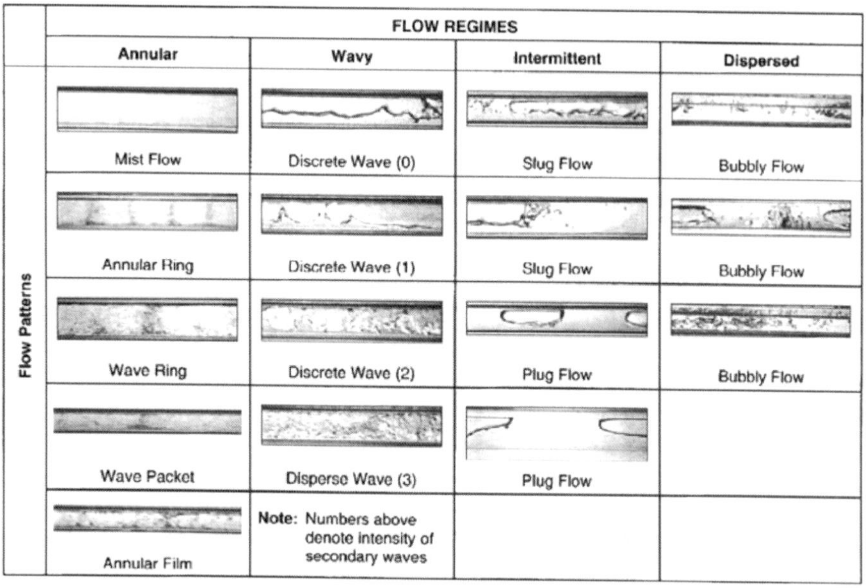

FLOW REGIMES				
	Annular	Wavy	Intermittent	Dispersed
Flow Patterns	Mist Flow	Discrete Wave (0)	Slug Flow	Bubbly Flow
	Annular Ring	Discrete Wave (1)	Slug Flow	Bubbly Flow
	Wave Ring	Discrete Wave (2)	Plug Flow	Bubbly Flow
	Wave Packet	Disperse Wave (3)	Plug Flow	
	Annular Film	Note: Numbers above denote intensity of secondary waves		

Fig. 3.29. General description of two-phase flow patterns in condensation in horizontal tubes, from [288].

This leads us to two-phase flow heat transfer of condensation and evaporation. Boiling and evaporation is only shortly mentioned, leading then to condensation flow patterns. These topics have similarities and differences with those patterns observed in conventional size channels and pipes. In a study of water condensation in an 82.8 μm trapezoidal channel, Wu et al. [289] observed different flow patterns progressing along the channel: full droplet flow, droplet/annular/injection/slug-bubbly, annular/injection/slug-bubbly or full slug-bubbly flow with decreasing mass flow rate. Coleman and Garimella [288] present a comprehensive study on condensation experiments in circular and non-circular tubes with hydraulic diameter between 0.4 and 4.91 mm. They identified four different flow regimes, see Figure 3.29: intermittent, wavy, annular, and dispersed.

Bar-Cohen and Rahim [286] gave a comprehensive overview of the two-phase flow regimes of refrigerants in microchannels. They defined four regimes with stratified, annular, intermittent and bubbly flow dependent on the superficial gas and liquid velocity in the channel shown with the Taitel and Dukler flow regime mapping methodology, see Figure 3.28. More than 90 % of the investigated experimental data point fell in the annular flow regime where also the critical heat flux CHF can be found. The CHF occurs at void fractions from 0.7 to 0.9 in the microchannels.

Detailed measurements and analytical correlation of the CHF in microchannels are given by Qu and Mudawar [290]. The influence of interfacial phenomena on an evaporating meniscus was investigated by Wayner [291] who found a major influence of the Marangoni and capillary forces as well as the disjoining pressure. To achieve a consistent continuum description of the heat flux, the author introduced a slip flow velocity of the meniscus. The processes are found to be very complex, and simple modeling can only provide insight to confirm the general phenomena in evaporation. The situation changes, if a surfactant influences the contact line dynamics [292], where the contact angle increases during condensation and decreases during evaporation .

The size of the intermittent regime increases with decreasing hydraulic diameter, due to larger surface tension forces compared to gravitational forces, which enhances the development of plug and slug flow. With decreasing hydraulic diameter, the wavy flow regime diminishes and is replaced by annular flow. Tube shape has less influence on flow patterns than hydraulic diameter, however, liquid is retained in sharp corners. The research group of Garimella set up a transition criterion for the vapor content (void fraction or quality) x from intermittent to other flow regimes [293].

$$x \leq \frac{a}{G+b} \quad \text{and} \quad \begin{aligned} a &= 69.57 + 22.60 \exp(0.259 d_h) \\ b &= -59.99 + 178.8 \exp(0.383 d_h) \end{aligned} \tag{3.71}$$

with the total mass flux G in kg/m^2 s and the hydraulic diameter d_h in mm. For lower quality x the flow is intermittent, for higher x the flow is either wavy or annular.

3.3.2 Two-phase pressure loss

Two-phase flow in microchannels produce a characteristic pressure drop, which is often an important criterion for the design of heat exchangers and mass transfer equipment. The pressure drop decreases the saturation temperature during evaporation and condensation and also influences the temperature profile. The pressure drop consists of friction, surface tension, acceleration, and hydrostatic components, while the latter can often be neglected. To calculate the pressure drop, two models have been proposed: the homogeneous model with mean flow properties, such as single-phase flow and the "separated flow" model with an artificially segregated flow. The two streams represent the gas and the liquid flow. A commonly used model originates from Lockhart and Martinelli, which employs two friction multipliers, Φ_G^2 or Φ_L^2

$$\left(\frac{\Delta p_f}{\Delta z}\right)_{2P} = \Phi_G^2 \left(\frac{\Delta p_f}{\Delta z}\right)_G \quad \text{or} \quad \left(\frac{\Delta p_f}{\Delta z}\right)_{2P} = \Phi_L^2 \left(\frac{\Delta p_f}{\Delta z}\right)_L . \tag{3.72}$$

with the two-phase frictional pressure drop $(\Delta p_f/\Delta z)_{2P}$ and the pressure drop of the gas $(\Delta p_f/\Delta z)_G$ or liquid $(\Delta p_f/\Delta z)_L$ when they are flowing alone in the channel. The frictional multipliers are correlated in terms of the Lockhart-Martinelli parameter X.

$$X = \frac{(\Delta p_f / \Delta z)_L}{(\Delta p_f / \Delta z)_G} \qquad (3.73)$$

To complete the equations, experimental data are needed for the correlation $\Phi_L^2 = f(X)$.

The pressure drop during condensation in a microchannel has been investigated experimentally and analytically. The research group of Garimella performed pressure drop experiments with R143a in circular and non-circular channels with hydraulic diameter ranging from 0.5 to 5 mm. They observed three different flow regimes: mist/annular/disperse, intermittent/discrete wave, and an overlap zone. For the regime of intermittent/discrete wavy flow, the developed model assumes, among other things: vapor flowing in long solitary bubbles separated from the wall and other bubbles with liquid, flowing faster than the liquid slug ahead and with constant speed/frequency. For annular/mist/disperse flow, a liquid film was assumed to uniformly cover the wall. The pressure drop of two-phase flow in T-junctions have been investigated by Yue et al. [42]. They developed parameters for the Lockhart-Martinelli method to predict the pressure loss of N_2/water mixtures in smooth channels.

3.3.3 Contacting and phase separation

In an own feasibility study, a special silicon chip was developed with channels for aerosol - water contact in a spray nozzle and gas-liquid separation in an impactor. The entire silicon chip is shown in Figure 3.30 with DRIE processed channels and structures covered with a Pyrex lid. The aerosol enters the chip from the left side parallel to a water stream to be contacted. In a spray nozzle or in a Venturi nozzle, see Figure 3.31, left side, the water is mixed with the aerosol-gas stream. The following channel expansions with posts and bends enforce contact between liquid and aerosol to wash out the nanoparticles from the gas flow, see Figure 3.31, middle.

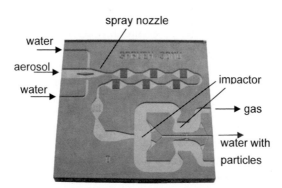

Fig. 3.30. Separator chip (20×20 mm^2 area) from silicon with contactor (spray nozzle), mixing contactors and two-stage impactor for phase separation.

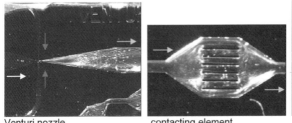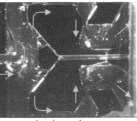

Venturi nozzle   contacting element   separator, impactor

Fig. 3.31. Details of a separator chip with gas-liquid two-phase flow in the separator chip; Left: Venturi nozzle with gas flow entering from the left side, water entering from top and bottom into the nozzle. A two-phase flow exits the nozzle and is lead through the mixing elements. Middle: Mixing element with gas-liquid two-phase flow from left to right. Right: Two-stage impactor for separation of gas-liquid flow.

The T-shaped impactor, which almost fills the lower half of the chip (20×20 mm^2 footprint), should separate the liquid from the gas stream by momentum and capillary forces. The gas-liquid mixture is sprayed onto a perforated wall consisting of 30 μm diameter pillars with a distance of 30 μm, visible as a line in Figure 3.30. A second impactor, following the first one, treats only half of the mass flow rate, therefore, the second impactor is shaped like a 90° bend. Here, two pillar lines are used to separate the liquid from the gas flow by capillary forces. The liquid is collected in a channel network along the center line of the two-stage impactor.

The forced convection two-phase flow in microchannels is difficult to describe, hence, optical investigations of the gas-liquid flow are presented initially. The Venturi nozzle provides for good mixing and contact between the components, see Figure 3.31 left. The rapid flow velocity can be seen directly after the nozzle, as well as the formation of annular-slug flow in the straight wide channel. Each mixing element with channel widening and contact grid leads to good mixing and contact between both phases. The small channels between the grid pillars show some characteristics of bubbly, slug, and annular flow, see Figure 3.31 middle. The two-stage impactor with the gas-liquid flow is shown in Figure 3.31 right. The liquid is collected in the flow dead zones behind the corner of the jet flow and forms a rotating, adhering droplet. Some liquid is entrained and flows to the perforated side. Due to the capillary force, the liquid is sucked into the pillar row, however, due to the high velocity and impact pressure, gas also enters the pillar row and the collecting channel. The bubble formation in the collecting channel is clearly visible and leads to an incomplete gas-liquid separation. Different flow rates and channel velocities exhibit a similar behavior of the two-phase flow separation.

In the future, each contacting and separation element must be investigated separately to effectively and independently control the flow parameters, as well as designs to control the pressure loss and improve fluid distribution in the elements. First experiments with a contact device achieve proper particle transport from gas to liquid. Unfortunately, an accurate mass balance was not performed due to low mass flow rates.

3.3.4 Immiscible liquids

Besides the gas-liquid two-phase flow, the mixture of two immiscible liquids, an emulsion, also creates a two-phase flow with typical surface effects. Additional to the interfacial forces between the two liquid components, the wetting behavior and contact angle between the single liquid and the solid surface are important for the flow regimes and pressure loss of the channel flow. An emulsion consists of small, spherical droplets of one or more immiscible liquids in the continuous phase surrounding the droplets. These droplets typically have a mean diameter $\bar{d}_P$ in the range of 0.1 to 100 µm. Emulsions are kinetically stable against coalescence, when surfactants are added, which cover the interface between both phases, or when the droplet diameter is smaller than the critical Kelvin diameter from classical nucleation theory, see Friedlander [294]. Emulsions are categorized into polydisperse and monodisperse, as shown in Figure 3.32.

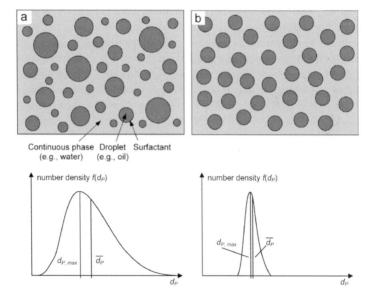

Fig. 3.32. Illustration of a polydisperse emulsion (a) and a monodisperse emulsion (b) with the typical size distribution.

A monodisperse emulsion consists only of uniformly sized droplets with a typical variation coefficient ($\sigma/\bar{d}_P$) of less than 0.05. Monodisperse emulsions are important due to their uniform physical and physico-chemical properties or reduced Ostwald ripening, where the droplet diameter temporarily changes to reach thermodynamic equilibrium. They have attracted much interest in applications like chemistry, pharmaceuticals, cosmetics, foods, electronics, or optics. The production of monodisperse emulsions in conventional equipment requires a high effort in mechanical energy ($10^5 - 10^9$ J/m^3) and certain separation steps. A typical high-pressure homogenizer, for example, has an emulsification efficiency, the ratio of the energy for droplet formation to the supplied energy, of less than 0.001 see Schuchmann and Danner [295].

Microstructured devices have been proposed and investigated to produce mono-disperse emulsions with a narrow size distribution and low energy consumption [28]. Typical devices are Interdigital micromixers, T-shaped micromixers, flow-focusing geometries or tapered capillaries, in which the shear force from the continuous-phase flow is the driving force for droplet generation. In microchannel arrangements and arrays, spontaneous transformation by interfacial tension is the driving force and action of droplet generation during the channel-sized formation of the droplets.

The inherent surface affinity of microchannel wall materials, such as hydrophilic or hydrophobic, plays an important role in the two-phase flow behavior. Plastics and polymers often exhibit hydrophobic characteristics and attract non-polar liquids, while metals and oxides often are hydrophilic, which attracts polar liquids. An overview of commonly used materials and related fabrication techniques is given in Table 3.2 according to Kobayashi and Nakajima [28].

Table 3.2. Typical materials of microchannels for emulsification.

Materials	Fabrication techniques	Inherent surface affinity
Surface-oxidized silicon	Anisotropic wet etching	hydrophilic
	Chemical dry etching	hydrophilic
Quartz glass	Mechanical cutting	hydrophilic
Pyrex glass	Isotropic wet etching	hydrophilic
Photo-structurable glass (Foturan^{TM})	Chemical dry etching	hydrophilic
Silicon nitride	Chemical dry etching	hydrophilic
Poly(dimethylsiloxane) (PDMS)	Soft lithography	hydrophobic
Poly(methylmethacrylate) (PMMA)	Mechanical cutting	hydrophobic
	Injection molding	hydrophobic
Polyurethane	Soft lithography	hydrophobic
Stainless steel	Mechanical cutting	hydrophilic
Nickel	LIGA process	hydrophobic

An effective strategy for successful generation of droplets is to select the to-be-dispersed liquid phase with a surface affinity different from that of the channel. Hence, hydrophobic channels should be used to generate water-in-oil (W/O) emulsions and, vice-versa, hydrophilic channels for oil-in-water (O/W) emulsions. The surface charge is another important factor influencing the two-phase flow of emulsions. Surface-oxidized silicon, for example, has a negative surface charge, when in contact with water [296]. The surface charge is also determined by the ionic strength of the surrounding solution. Noy et al. [297] describe the role of surface charges and pKs-values of surfaces with different functional groups for chemical force microscopy. The use of anionic surfactants, which are repulsed from the negatively charged channel wall, successfully assists the generation of monodisperse (O/W) emulsions. Thus, an appropriate combination of channel material, the two liquid phases, and the surfactant leads to a successful generation of monodispersed emulsions in microchannels.

The force balance between two immiscible liquid phases is given by the Young-Laplace equation

$$\Delta p = \frac{4\sigma_{ab}\cos\theta}{d} \tag{3.74}$$

with the interfacial tension σ_{ab}, the contact angle θ, and the channel diameter d. This equation is strictly valid for circular cross sections, but can assist an order-of-magnitude estimation in rectangular or other cross sections. If the pressure difference applied between the two phases is exceeding the Young-Laplace pressure, the to-be-dispersed phase moves forward in the microchannel. The acting forces can be evaluated from dimensionless numbers, given in Table 3.3. For an order-of-magnitude estimation the following boundary conditions are assumed: hydraulic diameter $d_h = O(10^{-5}$ m), to-be-dispersed phase properties, such as density $\rho = O(10^3$ kg/m^3), viscosity $\eta = O(10^{-2}$ Pa s), velocity $w = O(10^{-3}$ m/s), an interfacial tension $\sigma_{ab} = O(10^{-2}$ N/m), and the gravitational acceleration $g = O(10$ m/s^2).

Table 3.3. Dimensionless numbers for emulsion processes in microchannels and typical order-of-magnitude values.

Definition	Force ratio	Order-of-magnitude
$Re = \frac{\rho w d}{\eta}$	inertial force / viscous force	10^{-3}
$Bo = \frac{\rho g d^2}{\sigma_{ab}}$	gravitational force / interfacial tension force	10^{-4}
$We = \frac{\rho w^2 d}{\sigma_{ab}}$	inertial force / interfacial tension force	10^{-6}
$Ca = \frac{\eta w}{\sigma_{ab}}$	viscous force / interfacial tension force	10^{-3}

The order-of-magnitude estimation of these dimensionless numbers indicates the dominating interfacial tension force (Bond number Bo, Weber number We, and Capillary number Ca) and the laminar flow regime (Re number) inside microchannels. With regard to actual microstructured devices, the present emulsification techniques does not have enough throughput capacities for industrial applications, hence, scale up of the emulsification devices is an important task for the near future. An example of such a device is given in Section 1.5.2. Integration of droplet generation, handling, reaction with the droplets or at the interface, and subsequent droplet analysis should also be addressed by future research and engineering work.

4

Heat Transfer and Micro Heat Exchangers

This chapter aims to introduce heat transfer by conduction and convection in microstructures. Beginning with heat conduction and convective heat transfer of laminar flow in straight microchannels, the transfer in curved and bent microchannels will then be treated in detail. Optimized heat transfer networks are built from simple elements to cool high performance dissipative devices. The flow and temperature distribution over a distinct area is important for appropriate device performance. Finally, design issues for microstructured heat exchangers will be discussed, such as fouling of microchannels or thermal conductivity of the walls between microchannels.

4.1 Heat Transfer Fundamentals

4.1.1 The energy balance

Energy can take various forms in process technology, however, like the mass, the energy itself is conserved according to the energy balance of a control volume.

$$\sum \dot{E}_{in} = \sum \dot{E}_{out} - \sum \dot{E}_{loss} \tag{4.1}$$

The various forms of energy and their summation are described by the first law of thermodynamics with a dissipation term Φ

$$dE_{sys} = dU + dQ + dW + \Phi \tag{4.2}$$

for closed systems and with the temporal derivative of the energy for open systems

$$d\dot{E}_{sys} = d\dot{H} + d\dot{Q} + dP_t + \dot{\Phi}. \tag{4.3}$$

including the technical power P_t. The above equation is comparable with the Bernoulli equation for fluid dynamics, see Section 3.1.2. The dissipation Φ takes into account that the energy conversion from one form into another is accompanied with natural losses Φ. These losses are characterized by an entropy change during a process according to the second law of thermodynamics. For a differential element describing the heat transfer situation, the balanced value X from Eq. 2.26 can be set for heat convection to $X = \rho \, w_i \, e_i$ in the z-direction and for heat conduction to $X = \dot{q} = \dot{Q}/A$ in the y-direction. This leads to the energy equation for a one-dimensional differential element

$$\frac{\partial \, (\rho \, w_i)}{\partial t} \, e = -\frac{\partial \, (\dot{q} \, A)}{\partial z}. \tag{4.4}$$

For a process device with mass flow rate $\dot{m}$, heat flux $\dot{q}$ over the boundary, technical work W_t, or mechanical power P (for example, mechanical work from an active mixer), and chemical reaction, the energy equation is written as

$$\frac{\partial \, (m \, e)}{\partial t} = (\dot{m} \, e)_{\text{in}} - (\dot{m} \, e)_{\text{out}} + \left(p \, \dot{V}\right)_{\text{in}} - \left(p \, \dot{V}\right)_{\text{out}} - p\frac{\partial V}{\partial t} + \dot{Q} + \dot{E}_q + P. \tag{4.5}$$

In short, the temporal energy change in a system consists of the energy flowing in and out, the volume flow in and out, the volume change inside the system, the heat flux over the boundary by convection, conduction, or radiation, the energy produced inside the system (for example from chemical reactions or flow dissipation). On the right side, the technical work or mechanical power P is mentioned, which is brought into or produced by the system. This relatively complex equation can be simplified with the help of assumptions and simplifications in relation to the actual system.

The energy of the fluid can be expressed according to the first law of thermodynamics with kinetic and potential energy

$$m \, e = \rho \, A \left(u + w^2/2 - g \, y\right), \tag{4.6}$$

considering a straight channel in z-direction with a constant cross section in x and y-direction, without chemical reactions and technical work consumed or produced. The direction and sign of gravitational force depends on the geometrical setup and must be adjusted to suit the actual problem. With the first Fourier law $\dot{q} = -\lambda \, \partial \, T/\partial x$ for conductive heat transfer perpendicular to the channel axis, the energy equation can be written in the following form

$$\frac{\partial}{\partial t}\left(\rho \left(u + \frac{w^2}{2}\right)\right) = -\frac{\partial}{\partial z}\left(\rho \, w \left(u + \frac{w^2}{2}\right)\right) + \rho \, g \, y -$$
$$-\frac{\partial}{\partial z}(p \, w) - \frac{\partial}{\partial z}\left(\lambda \, \frac{\partial T}{\partial x}\right) - \frac{\partial}{\partial z}(\tau \, w) + \dot{w}_t, \tag{4.7}$$

with the wall shear stress τ and the specific technical power $\dot{w}_t$. This equation is the so-called Bernoulli equation for the energy balance in channel flow, see also Eq. 3.24. This complex equation (not taking into account other energy transfer processes, such as radiation, or chemical reactions) can be simplified by adjusting to suit each process.

For open systems and flow processes, the inner energy is replaced by the enthalpy $h = u + p/\rho$. For flow processes, the total enthalpy $h_t = u + p/\rho + w^2/2 - g \, y$ is applied together with the kinetic and potential energy part [76] p. 66. With the energy dissipation coming from shear stress and velocity gradient $\varepsilon = \partial \, (\tau \, w)/\partial z$ (only in the z-direction), the enthalpy form of the energy equation can be written as

$$\rho \frac{\mathrm{d}\,h}{\mathrm{d}t} = \frac{\mathrm{d}\,p}{\mathrm{d}t} + \varepsilon - \nabla\dot{q}. \tag{4.8}$$

With the caloric equation of state, the correlation between the inner energy u or enthalpy h, and the temperature ($\mathrm{d}\,u = c_v\,\mathrm{d}\,T$ and $\mathrm{d}\,h = c_p\,\mathrm{d}\,T$) the energy equation can be rewritten as

$$\rho\,c_p \frac{\mathrm{d}\,T}{\mathrm{d}t} = \frac{\mathrm{d}\,p}{\mathrm{d}t} + \varepsilon - \mathrm{div}\,(\lambda\,\mathrm{grad}\,T). \tag{4.9}$$

Solving this equation gives the temperature distribution for the actual process. Besides conduction and convection, heat may also be transported by radiation, which is proportional to T_4 and the emission surface properties. Radiation heat transfer is not treated here; interested readers are referred to [89] Chapter 5. A chemical reaction within the system influences not only the species equation, also the reaction enthalpy must be considered in the energy balance due to the apparent heat consumption or release, see also Eq. 2.38.

$$\Delta H_R = H_p - H_r \tag{4.10}$$

The net heat balance is calculated from the enthalpy of the reactants H_r and the products H_p, see also [173] p. 371. The entropy generation and balance was already described in Section 3.1.5 together with viscous dissipation in channel flow. Real processes are always accompanied by friction and other losses, which are treated in thermodynamics of irreversible processes. Entropy fluxes are also important for coupled irreversible processes, as covered in Chapter 7.

4.1.2 Heat conduction in small systems

Heat transfer is widely diversified and has been covered by many textbooks, journals, and conferences. A comprehensive picture is given by [89, 298] as well as in [299, 300]. Heat transfer in devices and structures on the micro- and nanoscale is also a broad field and is treated extensively in the books of Karniadakis et al. [204], updated in 2006 [301],Nguyen and Wereley [203], and Volz [172].

The first Fourier's law of heat transfer describes the correlation between the steady heat flow and the driving temperature difference, here for one-dimensional heat conduction:

$$\dot{Q} = -\lambda\,A\,(r)\,\frac{\mathrm{d}\,T}{\mathrm{d}r} \tag{4.11}$$

The integration over the coordinate r leads for constant heat conductivity λ to

$$T_1 - T_2 = \dot{Q}\,\frac{1}{\lambda}\int_{r_1}^{r_2}\frac{1}{A\,(r)}\,\mathrm{d}\,r \tag{4.12}$$

The integrated area divided by the thermal conductivity is often called the thermal resistance R_{th} in analogy to electrical resistance. For geometries, such as a plate, cylinder, or sphere, the thermal resistance R_{th} in Eq. 4.12 is determined with the

Table 4.1. Thermal resistance of the simple geometrical elements plate, cylinder, and sphere.

plate or slab	cylindrical element	spherical element
$A = \text{const.}$	$A = 2r\pi L$	$A = 4r^2\pi$
$R_{th} = \frac{1}{\lambda}\frac{r_2 - r_1}{A}$	$R_{th} = \frac{1}{\lambda}\frac{\ln(r_2/r_1)}{2\pi L}$	$R_{th} = \frac{1}{\lambda}\frac{(1/r_1 - 1/r_2)}{4\pi}$

following expressions in Table 4.1. In analogy to the electrical resistance, the thermal resistances of composed systems are arranged in series by simple addition and, in parallel, by the inverse of the sum of the inverse values.

The heat conductivity in microsystems is influenced by the microstructure of the material. Grain boundaries and crystal lattices form additional resistances to the heat transfer, see Section 2.3.6. Carbon nanotubes with asymmetrical, thin metal film on the outside, efficiently conduct thermal energy depending on the film size and structure, see Chang et al. [193]. In regular crystals, the heat transfer coefficient is dependent on the crystal orientation, and the Fourier equation of heat transfer must be expanded to the tensor notation

$$\dot{q} = -\lambda\frac{dT}{dx} \quad \rightarrow \quad \dot{\vec{q}} = -\Lambda \,\text{grad}\, T \tag{4.13}$$

with the heat conductivity tensor

$$\Lambda = \begin{pmatrix} \lambda_{11} & \lambda_{12} & \lambda_{13} \\ \lambda_{21} & \lambda_{22} & \lambda_{23} \\ \lambda_{31} & \lambda_{23} & \lambda_{33} \end{pmatrix}. \tag{4.14}$$

The solution of the three-dimensional heat conduction is often only possible with numerical methods. CFD and FEM programs allow the treatment of complex heat transfer problems.

The time dependent second Fourier law is derived from a differential element with the balance of the heat capacity and the heat conduction

$$\rho\, A(r)\, c_p \frac{\partial T}{\partial t} = \frac{\partial}{\partial r}\left(\lambda A(r)\frac{\partial T}{\partial r}\right). \tag{4.15}$$

Constant material properties, such as heat capacity c_p and heat conductivity λ, and mathematical simplification for different geometrical bodies ($n = 0$ for a plate, $n = 1$ for a cylinder, and $n = 2$ for a sphere) lead to

$$\frac{\partial T}{\partial t} = a\left(\frac{\partial^2 T}{\partial r^2} + \frac{n}{r}\frac{\partial T}{\partial r}\right), \tag{4.16}$$

with the temperature conductivity or heat diffusivity $a = \lambda/\rho\, c_p$. The transient temperature development in a semi-infinite body, displayed in Figure 4.1, is given in one-dimensional form by

$$\frac{\partial T}{\partial t} = a\frac{\partial^2 T}{\partial x^2}; \quad t \geq 0, \ x \geq 0. \tag{4.17}$$

The introduction of the dimensionless temperature θ leads to

$$\frac{\partial \theta}{\partial t} = a\frac{\partial^2 \theta}{\partial x^2} \quad \text{with} \quad \theta = \frac{T - T_0}{T_W - T_0}, \tag{4.18}$$

describing the temperature development in a solid body with defined wall temperature T_W.

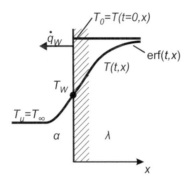

Fig. 4.1. Temperature distribution in a semi-infinite body during cooling.

The solution of Eq. 4.18 is dependent on the boundary condition at the wall, see also Section 5.2. With constant wall temperature $T_W = $ const., the dimensionless temperature is determined by the error function

$$\frac{T - T_0}{T_W - T_0} = \mathrm{erf}(x^*) \quad \text{with} \quad x^* = \frac{x}{2\sqrt{at}} \tag{4.19}$$

and the Gaussian error function

$$\mathrm{erf}(x) = \int_0^x e^{-\xi^2}\,d\xi. \tag{4.20}$$

The wall heat flux can thus be calculated from Eq. 4.19. For a constant heat transfer coefficient α at the wall, the solution of Eq. 4.18 is derived with the help of two dimensionless numbers, the Fourier number

$$\mathrm{Fo} = \frac{at}{x^2} \tag{4.21}$$

and the Biot number

$$\mathrm{Bi} = \frac{\alpha x}{\lambda_s}. \tag{4.22}$$

The Bi number is similar to the Nußelt number Nu from convective heat transfer, except for the heat conductivity of the solid body. The Fourier number is also defined for mass transfer problems $Fo' = D_{mt}/x^2$ with the molecular diffusion coefficient D_m, see Section 6.1.2 with Eq. 6.32. The temperature development during cooling of the body is given by

$$\theta_C = \frac{T - T_\infty}{T_0 - T_\infty} = \mathrm{erf}(x^*) - e^{\mathrm{Fo\,Bi}^2 + \mathrm{Bi}}\,\mathrm{erfc}\left(\sqrt{\mathrm{Fo}}\,\mathrm{Bi} + x^*\right) \qquad (4.23)$$

and for heating

$$\theta_H = 1 - \theta_C. \qquad (4.24)$$

The characteristic time for heating or cooling is drawn from the combination of the Fo and Bi numbers and depends not on the length (semi-infinite body). A miniaturization will not influence the temperature development and the heat flux for a semi-infinite body.

$$t_C = a\,t\left(\frac{\alpha}{\lambda}\right)^2 = \mathrm{Fo\,Bi}^2 \qquad (4.25)$$

The unsteady heat transfer in small systems is rapid and can often be approximated with the fast asymptotic behavior of a steady state solution. For small bodies (high Fo and low Bi number), the temperature distribution can be approximated by asymptotic solutions. The temperature inside a small body is only a function of time

$$\theta = \frac{T - T_\infty}{T_0 - T_\infty} = \exp\left(-\frac{\alpha\,A\,t}{m\,c_p}\right) \quad \text{for } \mathrm{Fo} > 0.3 \text{ and } \mathrm{Bi} < 0.2, \qquad (4.26)$$

or more exact for the three basic geometries (plate $n = 0$, cylinder $n = 1$, sphere $n = 2$):

$$\theta = \exp\left(-(n+1)\,\mathrm{Bi\,Fo}\right). \qquad (4.27)$$

The characteristic time in Eq. 4.26 is given by

$$t_C = \frac{m\,c_p}{\alpha\,A}. \qquad (4.28)$$

With the temperature diffusivity of the solid $a = \lambda/\rho c_p$ and the characteristic length $l \propto A^{0.5}$, this equation can be rewritten

$$t_C = \frac{l^2}{a\,\mathrm{Bi}}. \qquad (4.29)$$

with the Bi number. Similar problems and their solutions for mass transfer are given in Section 6.2, other solutions for more complex geometries are given in [76] Chapter 1.4.

4.1.3 Convective heat transfer in microchannels

The total heat transfer in microstructured devices consists of the heat conduction through the walls and the convective heat transfer from the wall into the fluid in microchannels. For high temperature differences ΔT, radiation heat transfer also must be considered. A comprehensive introduction to convective heat transfer in microchannels is given in Zohar's textbook [302]. Single-phase flow heat transfer, boiling, and condensation in microchannels is a broad area, not exhaustively covered here, however, Kandlikar et al. [303] review results of actual research and applications in their book.

Besides phase or temperature change of the fluid in a microchannel, the pressure loss plays an important role in heat transfer. For convective flow in channels, the pressure loss is given by Eq. 3.28. The friction factor C_f of fully developed laminar flow in straight channels is proportional to 1/Re, further data for various channel cross sections are given in Section 3.1.3, Table 3.1. The heat flux in convective heat transfer is described by the kinetic relation

$$\dot{Q} = \alpha\, A\, \Delta T \qquad (4.30)$$

see also Table 2.2. The temperature difference ΔT occurs between the bulk flow and channel wall with area A. It is convenient to treat the heat transfer coefficient α in dimensionless form of the Nußelt number Nu. For straight laminar flow, the dimensionless heat transfer coefficient, the Nu number, is constant, depending on the boundary conditions.

$$Nu = \frac{\alpha\, d_h}{\lambda} \qquad (4.31)$$

For constant wall heat flux, the Nu number is $Nu_q = 4.3$, for constant wall temperature the Nu number is $Nu_T = 3.66$. In a wide gap or narrow slit, the Nu number is 7.54 ($\dot{q} = $ const.) and 8.24 ($T = $ const.) for double-sided heat transfer, and 4.86 ($\dot{q} = $ const.) and 5.39 ($T = $ const.) for single-sided heat transfer. For smaller channels, the heat transfer coefficient increases due to the constant Nu number. With decreasing channel dimensions, the transfer area and the mass flow through the channel are also decreased, hence, the transported heat is limited by these boundary conditions. To maintain a high heat transfer coefficient with a high transport rate, an optimum channel dimension must be found [304], which is also oriented at the fabrication process.

At the entrance of a channel or behind channel elements, such as channel junctions, expansions or contractions, the disturbed flow enhances the radial transport in the channel. This results in increased pressure loss as well as increased heat or mass transfer. For a locally resolved description of the heat transfer in entrance flow, the dimensionless channel length X^*, starting at the entrance, is defined according to

$$X^* = \frac{L}{d_h}\, Pe = \frac{L}{d_h}\, Re \cdot Pr \qquad (4.32)$$

with the heat transfer Péclet number Pe defined with the Prandtl number $Pr = v/a$. A similar Pe number is also used for mass transfer, but defined with the Schmidt number $Sc = v/D$, see also Section 5.3. The mean Nu number in the entrance flow Nu_{me} is calculated with the mean Nu number in straight channel flow Nu_m according to the following correlation [89]

$$\mathrm{Nu}_{\mathrm{me}} = \frac{\mathrm{Nu}_m}{\tanh\left(2.432\,\mathrm{Pr}^{1/6}\,X^{*\,1/6}\right)}. \qquad (4.33)$$

This equation is valid for the entire channel length X^* and $\mathrm{Pr} > 0.1$. The analytical values for the entrance flow correlate well with the numerical values of the flow and heat transfer simulations for two different flow rates [26]. Beyond a certain length, the velocity and temperature profile does not alter, and this so-called fully-developed flow is described in Section 3.1.1 and in [188] Chapter 32.

In turbulent flow (for $\mathrm{Re} > \mathrm{Re}_{\mathrm{crit}} = 2\,300$ in channel flow) the pressure loss is proportional to the square mean velocity and the heat transfer can be calculated according to Gnielinski [298] Chapter Ga.

$$\mathrm{Nu} = \frac{\xi/8\,(\mathrm{Re} - 1\,000)\,\mathrm{Pr}}{1 + 12.7\sqrt{\xi/8}\left(\mathrm{Pr}^{2/3} - 1\right)}\left(1 + \left(\frac{d_h}{l}\right)^{2/3}\right)K_{\mathrm{Pr}} \qquad (4.34)$$

with $\xi = (1.8\log_{10}(\mathrm{Re}) - 1.5)^{-2}$ and $K_{\mathrm{Pr}} = (\mathrm{Pr}_{\mathrm{fluid}}/\mathrm{Pr}_{\mathrm{wall}})^{0.11}$. This correlation is valid for $0.5 < \mathrm{Pr} < 2\,000$, $2\,300 < \mathrm{Re} < 5 \cdot 10^6$ and $1 < L/d_h < \infty$. Turbulent flow does not often occur in microchannels, however, the manifolds or inlet and outlet headers may produce turbulent conditions. A review of heat transfer in various kinds of microstructured heat exchangers is given by Palm [305], for heat sinks and electronic cooling in [304] and for circular micro pipes in [306]. The influence of the surface roughness in microchannels was investigated in a numerical study of Koo and Kleinstreuer [307]. The authors reveal that the surface roughness effect on heat transfer is less significant than on momentum transfer. In special cases, such as high Pr number fluid flow, thermal dispersion from surface roughness may greatly influence the heat transfer. Multiphase flow heat transfer is treated briefly in Section 3.3 with boiling, evaporation, and condensation, for film flow see [298] Chapter Md, or [299] Chapter 8.5.

4.1.4 Rarefied gases with slip boundary conditions

The above equations are valid for liquids and gases above the continuum limit (Knudsen number $\mathrm{Kn} < 0.01$). For rarefied gases, the transfer processes at the wall differ due to the low number of gas particles. Basic correlations for rarefied gases are given in Section 2.3.1, here, flow processes of rarefied gases are treated. Experiments with rarefied gas flow through a capillary indicate a linearly decreasing volume flow with decreasing total pressure down to a certain value [308]. For even lower pressures, the volume flow remains constant and, subsequently, may even increase. The reduced flow resistance is produced by slip velocity at the wall, which results from the molecular motion and insufficient momentum transfer between the wall and bulk fluid. In the same way, the molecular motion in the gas influences the energy transfer into the fluid. A temperature jump occurs at the wall, which increases the heat transfer resistance, see Figure 4.2.

For rarefied gas flow $(0.01 < \mathrm{Kn} < 0.1)$, the boundary condition of the gas velocity at the wall is described by

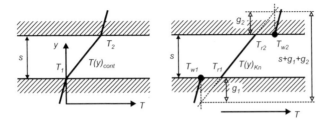

Fig. 4.2. Temperature gradient in a gap; left: linear development for dense gases (Kn < 0.01); right: temperature jump in rarefied gases (Kn > 0.01).

$$w\,(x = 0) = \zeta \left(\frac{\partial w}{\partial x} \right)_{y=0}.$$ (4.35)

The slip length ζ can be calculated with the accommodation coefficient β of the tangential velocity and the mean free path Λ of the molecules. The accommodation coefficient β describes the efficiency of the momentum and energy transfer from molecules to the wall and vice versa.

$$\zeta \approx \frac{2 - \beta}{\beta} \Lambda$$ (4.36)

Experimental data for β can be found in [153]. In [153, 309] and [298] Chapter Mo, methods can be found to determine the friction factor and the resulting volume flow in rarefied gas flow. The temperature jump at the wall is described in a similar way with the temperature jump coefficient g.

$$T\,(x = 0) = T_r = T_W + g \left(\frac{\partial T}{\partial x} \right)_{x=0}$$ (4.37)

The temperature jump coefficient g can be determined via kinetic theory from the thermal accommodation coefficient γ, a material parameter f, and the mean free path Λ.

$$g = \frac{2 - \gamma}{\gamma} \frac{15}{8} f \Lambda$$ (4.38)

The material parameter f is calculated from the linearized Boltzmann transport equation (see Section 2.3.2) and is given by Frohn [309] to

$$f = \frac{16}{15} \frac{\lambda}{\eta\, c_v} \frac{1}{\kappa + 1} = \frac{16}{15} \frac{1}{Pr} \frac{\kappa}{\kappa + 1}.$$ (4.39)

For flow processes, the ratio of the temperature jump coefficient and the characteristic length is important and can be derived from Eq. 4.38.

$$\frac{g}{l_{char}} = \frac{2 - \gamma}{\gamma} \frac{15}{8} f\, Kn$$ (4.40)

For monatomic gases with $\kappa = 5/3$, $\mathrm{Pr} = 2/3$, and complete accommodation $\gamma = 1$, the length ratio reduces to

$$\frac{g}{l_{char}} = \frac{15}{8}\, \mathrm{Kn}. \tag{4.41}$$

The linear temperature distribution in a gap is determined with the jump coefficient g and Eq. 4.37 to

$$\frac{T(y) - T_2}{T_1 - T_2} = \left(1 - \frac{y + g_1}{s + g_1 + g_2}\right). \tag{4.42}$$

The temperature jump coefficient g can be regarded as an additional distance of the gap, see Figure 4.2. For $0.1 < \mathrm{Kn} < 10$, monatomic gases, and complete accommodation, the temperature gradient is only a function of the Kn number.

$$\frac{T(y/s) - T_2}{T_1 - T_2} = \left(1 - \frac{y/s + (15/8)\,\mathrm{Kn}}{1 + (15/4)\,\mathrm{Kn}}\right) \tag{4.43}$$

The derivation of the temperature profile and comparison of the coefficients yields the heat flux for $0.1 < \mathrm{Kn} < 10$, which can be expressed as the ratio to the continuum heat flux.

$$\frac{\dot{q}}{\dot{q}_{cont}} = \frac{1}{1 + (15/4)\,\mathrm{Kn}} \tag{4.44}$$

For $\mathrm{Kn} > 10$ (free molecular flow), the ratio of the heat fluxes in a gap are also dependent on the Kn number.

$$\frac{\dot{q}_{FM}}{\dot{q}_{cont}} = \frac{4}{15\,\mathrm{Kn}} \tag{4.45}$$

The distance between the plates plays no role in this regime; the molecules only hit the walls and not each other. The unsteady heat transfer of rarefied gases can be treated in the same manner, see [298] Chapter Mo.

Assuming a round capillary with the outer radius r_A and constant wall heat flux $\dot{q}$, the heat transfer of rarefied gas flow (for $\mathrm{Kn} < 0.1$) is expressed with the dimensionless heat transfer coefficient, the Nu_q number [298].

$$\mathrm{Nu}_q = \frac{\dot{q}}{(T_W - \bar{T})}\frac{r_A}{\lambda} = \frac{\alpha\, r_A}{\lambda} = 24\left(11 - 6\,\Delta\bar{w} + (\Delta\bar{w})^2 + 24\frac{g}{r_A}\right) \tag{4.46}$$

The dimensionless slip velocity Δw is determined with the slip length from Eq. 4.36 to

$$\Delta w = \frac{w(r_A)}{\bar{w}} = \left(1 + \frac{r_A}{4\,\zeta}\right)^{-1}. \tag{4.47}$$

For non-circular cross sections, the half of the hydraulic diameter d_h can be taken for r_A. The Nu number for constant wall temperature is approx. 5 % higher than for constant wall heat flux. A numerical study with the Monte-Carlo method [310] indicated that the slip flow model correctly represents convective heat transfer between continuum and molecular flow. The influence of the axial heat conduction must be considered, however, the viscous heat dissipation, expansion cooling, and thermal creep can be neglected.

4.1.5 Convective cooling for flow measurement

An interesting application of convective heat transfer is flow velocity sensing within a microchannel or on a surface. A thermal flow sensor consists mainly of the heating element and temperature sensors, which measure the temperature of the heater and the fluid before and after the heater, see Figure 4.3. The streaming fluid is mostly heated by resistance heaters. The temperature can be measured either with electrical resistance sensors or with thermocouples. Thermoelectric temperature and flow measurement has been utilized for more than a century, along with the miniaturization of sensing devices including thermocouples. The thermoelectric effect and related thermodynamics are described in Section 7.2.

The complete sensor arrangement involves three complex domains: fluid flow, heat transfer, and thermoelectric energy and signal conversion. The domain of the mechanical signal to be measured is determined by the flow and velocity profile adjacent to the wall and the heating/sensing device. The flow-induced heat transfer (convection) and the parallel heat conduction in the sensor wall/substrate determine the thermal signal domain, which is measured as temperature. Comparison of two or more temperatures indicates the flow situation and can be transferred into a flow rate from the knowledge of the complete flow situation. The temperature is measured as voltage between the hot and cold ends of the thermocouple and is part of the electrical signal domain [311].

In an early work of van Herwaarden [312], thermoelectric flow sensors are presented, which were adopted in the textbook on silicon sensors from Middelhoek and Audet [54]. The basic measurement principle was not changed, but modified with improved materials, geometry, or device design. The basic sensor setup can be classified into two different techniques or methods. In method A, heating of the fluid imposes a temperature profile in the channel or near the wall (Figure 4.3), which is measured at certain positions by thermocouples. In method B, fluid flow cools the heater, where temperature or electrical power is measured, see Figure 4.4. Besides the integral heater temperature, local temperatures of a heated area can be measured, which allows direction-sensitive velocimetry (Figure 4.5). The temperature downstream of the heater is taken to determine the mean flow velocity in the vicinity of the heater. For commercial flow sensors, the in-line setup according to method A with one heater and two sensors up- and downstream is the preferred model and is utilized in several products from commercial sensor suppliers.

The temperature profile in the heated microchannel without fluid flow is almost symmetrical, and both sensors detect a similar temperature. The fluid flow shifts the temperature profile, so that the downstream sensor measures a higher temperature and the upstream sensor measures the temperature of the incoming fluid. A simulation of the temperature distribution is presented by Ashauer et al. [313]. The heating structure with adjacent thermopiles for measuring the temperature is located on a thin membrane to minimize heat loss through the wall material. The thermocouples are arranged in a pile (series switching) to increase their sensitivity for measuring the temperature difference between the heater and the substrate. A good example of

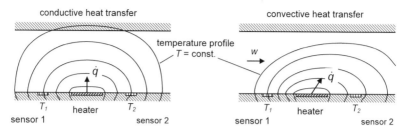

Fig. 4.3. Setup of a thermoelectric flow sensor, method A, in a channel consisting of a heater and two temperature sensors up- and downstream of the heater.

this microstructured device is given by Buchner et al. [314], which is displayed in Figure 4.4.

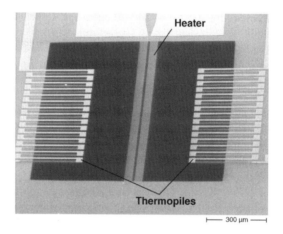

Fig. 4.4. Thermoelectric flow sensor consisting of an electrical heater on a thin membrane with parallel meandering thermoelectric sensors, two thermopiles with 15 thermoelements, method B acc. to Buchner et al. [314], courtesy of Elsevier.

A different scheme with a periodic heater/sensor arrangement was proposed by Al khalfioui et al. [315], to get a higher locally resolved flow measurement, however, the system is more complex and, accordingly, the effort for measuring the physical domain. The flow measurement by a heated membrane with connected thermocouples was proposed by van Oudheusden [316, 311]. The schematic setup is displayed in Figure 4.5. Comparing the four measured temperatures not only determines the flow near the wall, but also the angle-of-attack of the main flow direction. More details on the analysis are given by van Oudheusden [316]. The sensor contacts also serve as mechanical hinges and guarantee good thermal insulation of the heated membrane.

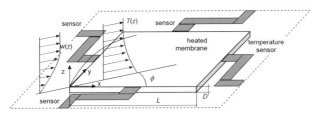

Fig. 4.5. Convective cooling of the sensor device and connected temperature sensors with direct heating of the membrane and temperature measurement; variation of method B for two-dimensional flow measurement acc. to van Oudheusden [316].

The dissipated electrical power in the heater spreads into the flowing fluid and the wall or substrate supporting the heater and sensors. In a steady-state situation, the energy balance of the heater/sensor expresses the correlation between the electrical heating power P on one side and the dissipated heat into the substrate $\dot{Q}_S$ and into the fluid $\dot{Q}_f$ on the other side

$$P = \dot{Q}_S + \dot{Q}_f = G_S\left(T_H - T_S\right) + G_f\left(T_H - T_f\right), \tag{4.48}$$

where T_H, T_S, and T_f are the temperature of the heater, the substrate, and the bulk fluid, respectively. The coefficients G_S and G_f indicate the thermal conductivity of the substrate and the convection by the flowing fluid. Due to the small temperature differences, radiation heat transfer can be neglected in most cases. The flow measurement relies on the variation of G_f with the flow and can be measured, for example, from the temperature difference resulting from the applied electrical power of the heater.

As previously mentioned, the flow rate can be determined either by measuring the temperature of the fluid downstream of the heater (method A) or of the heater itself (method B). In both cases, the fluid temperature upstream of the heater must be known or is set to the ambient temperature. In method A, the fluid temperature is measured downstream of the heater. The wall heat transfer from the heater determines the energy of heating the fluid and increasing its temperature. The location of the temperature sensor influences the measurement and should be not too far downstream from the heater. If the inflowing fluid has the same temperature as the sensor substrate T_S, and the sensor measures the mean temperature of the fluid, the energy dissipation into fluid is determined by

$$\dot{Q}_f = \dot{m}c_p\left(T_S - T_f\right) \tag{4.49}$$

with the mass flow rate $\dot{m}$ and the heat capacity c_p of the fluid. The density of the fluid ρ_f and the channel cross section A_C should be known for accurate sensing. The mean flow velocity $\bar{w}$ is determined with the heater power and the measured temperature difference

$$P = \dot{Q}_S + \rho_f c_p A_C\left(T_S - T_f\right)\bar{w}, \tag{4.50}$$

while the heat dissipation into the substrate $\dot{Q}_S$ must be estimated. The measurement of the fluid temperature downstream is also fairly inaccurate, so the fabricated sensor must always be calibrated. The above equation is the basis to determine the flow velocity and flow rate from the electrical heating power, see Eq. 4.55.

Convective cooling of an electrical heater and measuring its temperature according to method B avoids errors, such as measuring the heated fluid temperature or the unknown heat dissipation into the substrate. For this reason, many sensors are located on a thin membrane or fabricated on polymers or other thermal insulators, which minimizes parasitic heat losses and increases the sensor accuracy. The convective heat transfer into the fluid $\dot{Q}_f$ is determined by the heat transfer coefficient α_W at the wall and the temperature difference.

$$\dot{q}_f = \frac{\dot{Q}_f}{A_H} = \alpha_W \left(T_H - T_f\right) \tag{4.51}$$

The heat transfer from cooling the electrical heater is similar to cooling a plate on one side in surface flow. Here, the influence of the surrounding channel walls is neglected for simplicity. With the help of the boundary layer theory, the wall heat transfer is described by the dimensionless heat transfer coefficient [316], the local Nußelt number Nu:

$$\mathrm{Nu} = \frac{\alpha\, z}{\lambda_f} = 0.332\, \mathrm{Pr}^{1/3}\mathrm{Re}^{1/2} \tag{4.52}$$

with the dimensionless fluid properties (Prandtl number) Pr and dimensionless flow conditions (local Reynolds number) Re. For the entire length of the heater l_H, the overall Nu_L number is given by Welty et al. [317].

$$\mathrm{Nu}_L = \frac{\alpha\, l_H}{\lambda_f} = 0.664\, \mathrm{Pr}^{1/3}\mathrm{Re}^{1/2} \tag{4.53}$$

Both equations are valid for laminar flow with $\mathrm{Re} = \bar{w}\, z/v < 2 \cdot 10^5$. The correlation between flow velocity, dissipation power, and measured temperature difference is given by

$$\frac{\dot{Q}_f}{A_H} = \frac{\mathrm{Nu}\,\lambda_f}{l_H}\left(T_H - T_f\right) = 0.664\, \lambda_f\, \mathrm{Pr}^{1/3}\left(\frac{\bar{w}}{l_H v}\right)^{1/2}\left(T_H - T_f\right) \tag{4.54}$$

where the length coordinate z is substituted with the heater length l_H. Arranging the above equations to correlate the electrical heating power with the flow velocity gives the following equation.

$$P = G_S\left(T_H - T_S\right) + 0.664\, \lambda_f A_H \mathrm{Pr}^{1/3}\left(\frac{\bar{w}}{l_H v}\right)^{1/2}\left(T_H - T_f\right) \tag{4.55}$$

A detailed analysis of the correlation requires many parameters, such as heat conductivity of the substrate, correct fluid properties, or the correct geometry of the sensor. For practical applications, the correlation between electrical power, measured temperature difference, and flow velocity can be reduced to

$$P = \left(C_1 + C_2 w^{1/2}\right)\left(T_H - T_f\right) \tag{4.56}$$

for laminar flow, see also [318]. Both coefficients can be determined by calibration measurements and should not vary with time or flow velocity, however, the fluid properties have to be calibrated to gain accurate measurements.

The local temperature profile of a fluid flowing over a heated plate can be determined with the help of Eq. 4.19 and the heater and inlet temperatures T_H and T_i, respectively, to the following correlation

$$T - T_i = (T_H - T_i)\,\mathrm{erfc}\left(\frac{z}{2\sqrt{a\,t}}\right). \tag{4.57}$$

The length coordinate z and the heating time t of a fluid element can be combined to $\bar{w} = z/t$ resulting in

$$\frac{T - T_i}{T_H - T_i} = \mathrm{erfc}\left(\frac{1}{2}\sqrt{\frac{\bar{w}\,z}{a}}\right). \tag{4.58}$$

Unfortunately, this equation can not be solved directly for the velocity $\bar{w}$ as a function of the local temperature T. However, with appropriate sensor setup, calibration, and data analysis, more information can be extracted from flow measurements.

The application of heat transfer measurements is a wide spread and generally accepted method for determining the flow rate in channels. Despite many applications, no standard sensor arrangement has been introduced similar to the international standard in thermoelectric temperature measurements (IC 584-1 [319]), which would help to apply this robust, but simple measurement principle in many areas and might be a further step for integrated devices.

4.2 Microfluidic Networks for Heat Exchange

Transport processes in micro process engineering are governed by two different mechanisms: the conductive and the convective transfer of a species or energy. The fluid flow in microchannels is often regarded to be laminar with dominant conductive transport. On structured surfaces [320] and in bent and curved flow [26], secondary transversal flow components are introduced into the straight laminar flow, which enhance the transport processes. Thus, two strategies enhance the overall transport in pressure-driven flow passive devices: the implementation of small diameter channels with a short diffusion length and the creation of secondary flow structures perpendicular to the main flow direction.

Aside from high transfer rates, high total flow rates are desirable in many applications of micro devices. The typically low flow rates in single microstructured elements can be enlarged by internal and external numbering-up [321] or equal-up of the desired effects [69, 322]. The internal numbering-up is limited by the available space and the uniform distribution of the fluids. A complete description of the simulation of the inlet and outlet flow manifold for internal numbering-up is given by Tonomura et al. [323, 324]. The uniform distribution of the fluids is very important for the performance of a device with many parallel active elements, either for heat transfer or mass transfer. An excellent example of maldistribution effects in high performance heat exchangers is given by Schlünder [325]. A relative maldistribution of the liquids of approx. 5 % will lead to a corresponding decrease in the heat transfer efficiency.

4.2.1 Status-quo of microfluidic networks for device cooling

Cooling of electronic equipment is often accomplished in long, straight channels, which produces relatively low heat transfer coefficients. Kandlikar [326] states that there is a current research need in single-phase flow cooling for the design of novel microchannel configurations, which provide a high thermal performance with low pressure loss. With a cost effective manufacturing and proper system integration, a heat dissipation rate of up to 10 MW/m^2 (1 kW/cm^2) appears possible. Peles et al. [327] propose advanced heat exchange geometries with posts and fins for cooling purposes. The optimization parameters are the height and distance of the fins to yield high transfer rates. Cheong et al. [328] have measured heat transfer coefficients of over 20 kW/m^{2}K for single-phase liquid water flow in a fin heat exchanger with a unspecified geometry. Heat transfer coefficients in a similar range have been obtained by Overholt et al. [329] with a jet cooling system for electronic equipment.

The setup of different branching levels to spread fluid over an area and collect it again can be managed with the constructal theory of Bejan [330]. The constructal design approach begins with the smallest elements on zero level and connects these with those on the next higher level. This approach is inverse to the fractal description of branched systems, where an element is repeatedly miniaturized until almost infinitely small structures. In nature, systems have a finite smallest size, and , therefore, follow the constructal approach. The optimum size of channel elements and the corresponding area covered depends on the transport velocity of the important quantity, such as the heat flux. The optimum flow distribution and cross sections of the channels on different branching levels are influenced by another biological principle, the so-called Murray's law [331], for example, the branching of blood vessels or plant capillaries. If the sum of the inner radii to the power of three on each branching level is constant, the channel network will need minimal power consumption or exhibit minimal pressure loss for a given flow rate. For channels with circular cross section, the diameter of the highest and largest level element d_n to the power of 3 is equal to the sum of the diameters d_z of the next level elements to the power of 3. In non-circular cross sections, the radius is replaced with the hydraulic diameter.

$$d_n^3 = \sum_i d_{z,i}^3 \tag{4.59}$$

The level with the largest elements has the notation "n" due to n branching levels of the system. The zero level of the system is the smallest level following the notation of the constructal design method. For a symmetrical bifurcation, the above equation gives the following correlation for the diameters on various branching levels z

$$d_n^3 = 2^z \cdot d_z^3 \quad \text{or} \quad d_z/d_n = 2^{-z/3}. \tag{4.60}$$

This correlation serves for an equal wall shear stress in the channels on each branching level z; more details are given by Albring [100] and in two papers of Emerson and coworkers [332, 333]. Examples for microchannel networks are given in Section 4.2.4.

In the following, a fine network of microchannels combined with an appropriate inlet and outlet manifold is proposed to control the temperature distribution over a given area. Based on simulations of simple channel elements, the entire channel system is arranged with the help of Bejan's constructal theory [330], starting with a simple zero level element optimized to remove the heat from a surface. The zero level in constructal design is the smallest level, from which the system is built up, following the bottom-up approach.

The main steps in heat transfer and fluid guidance over the entire system are to spread the fluid over the heated area (inlet manifold), contact the fluid with the heated area (zero level element), heat the fluid to a tolerable temperature, and lift the fluid off the heated area (outlet manifold). The size and shape of zero-level elements are optimized for a high heat transfer with constant heat flux $\dot{q}$ and low pressure loss Δp using the entrance effect, see Kockmann et al. [26]. The flow and temperature profiles in the channel elements are simulated with the commercial CFD-tool CFD-ACE+ to design microchannel structures with high heat transfer coefficients. Single-phase flow is considered and no phase change should occur. Single channel elements are combined as channel networks with characteristics as described.

4.2.2 Single channel element calculation

This section presents design rules and correlations for high performance heat exchangers and microchannel networks. A constant Nußelt number Nu describes the heat transfer in laminar channel flow, see also Section 4.1.3, which takes the following form for constant heat flux $\dot{q}$ and fully developed flow

$$\text{Nu} = \frac{\alpha d_h}{\lambda} = 4.3. \tag{4.61}$$

The heat transfer coefficient α increases with decreasing characteristic hydraulic diameter d_h and constant heat conductivity λ. At the entrance or after a bend, the straight laminar flow is disturbed, an additional pressure loss occurs, and transverse flow components enhance the transport process in the channel, see Section 3.2.2. In a numerical study with low Re numbers ($5 \leq \text{Re} \leq 200$), Rosaguti et al. [334] calculated a heat transfer enhancement of 1.5 times in circular channels compared to straight channel flow. The heat transfer enhancement is described in Eq. 4.33 by the mean Nu_{me} number for both hydrodynamic and thermal entrance flow depending on the dimensionless entrance length $X^+ = x/(d_h \cdot \text{Re} \cdot \text{Pr})$ and the Prandtl number $\text{Pr} = v/a$, see [89]. To determine the optimum channel length x, the above equation is formed to

$$\frac{x}{d_h} = \text{Re} \cdot \left(0.411 \cdot \text{arctanh} \left(\frac{\text{Nu}_m}{\text{Nu}_{me}} \right) \right)^6. \tag{4.62}$$

To yield a heat transfer enhancement of 30 % ($Nu_{me}/Nu_m = 1.3$), the entrance length divided by the hydraulic diameter should be less than 0.005·Re, for a heat transfer enhancement of 10 %, the length should not exceed 0.06·Re. For longer channels, the pressure loss increases without additional benefit to the heat transfer. The simulation of the entrance flow behind a T-junction is shown in Figure 4.6 together with the Nu number over the channel length for different Re numbers.

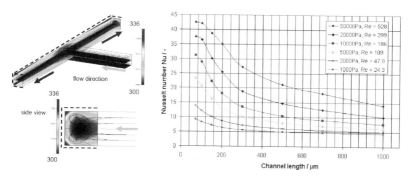

Fig. 4.6. Simulation of the heat transfer in a microchannel with square cross section $d_h = 100$ μm, constant heat flux of 150 W/cm^2, see also Figure A.22; Left: Wall temperature distribution and stream lines, side view; Right: Development of the local Nu number in the curved flow with various pressure losses and Re numbers.

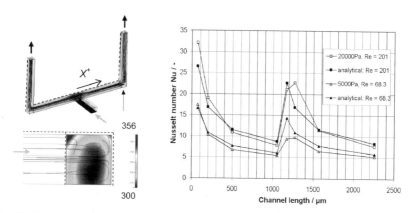

Fig. 4.7. Heat transfer simulation of a fork-shaped microchannel setup with a cross section of 100×100 μm^2, see also Figure A.23; left: Temperature distribution and streamlines; right: Local Nu number over the channel length.

The side view in Figure 4.6 clearly shows the existence of vortices in curved flow, also called Dean flow. This double vortex enhances the heat transfer directly after the bend, but is dampened by viscous forces in the subsequent channel. The Nu number asymptotically approaches the value of laminar flow in straight channels as indicated in Eq. 4.61. Figure 4.7 shows the Nu number in a T-junction and adjacent 90° bends with CFD-simulated and analytical values from Eq. 4.62. Several effects are clearly visible: heat transfer enhancement in curved flow, relaminarization in straight channel flow after curved flow, and a good correlation between numerical and analytical values. Enhanced heat transfer is related to the pressure loss, which is treated in the following for basic channel elements, such as bends and junctions.

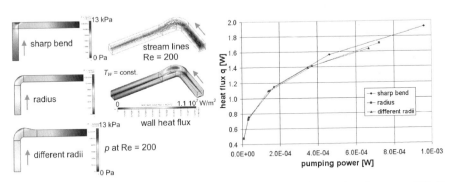

Fig. 4.8. Pressure loss in bends and curves with corresponding heat flux for water, 300 K inlet temperature and 350 K wall temperature, see also Figure A.24. The microchannel with square cross section has a width of 100 µm, an inlet length of 600 µm, and an outlet length of 1 400 µm. The boundary condition at the inlet is a constant laminar velocity profile of water.

The temperature distribution in a channel cross section, as shown in Figure 4.6, indicates a sharp temperature gradient at the wall and a relatively low bulk temperature. To cover a larger area for heat removal and further enhance the heat transfer, the combination of two or more 90° bends has been investigated. The heat transfer simulation with wall temperature distribution, stream lines, and Nu numbers is given in Figure 4.7 for a microchannel setup with forked shape. The numerical Nu numbers are compared with the analytical values from Eq. 4.62. The discrepancies between simulation and analytical values originate from the complex flow in the curved channels. The analytical solution is valid for pipe entrance flow and constant wall temperature, however, the trends correlate well, suggesting, the analytical solution can be used for further design analysis.

In Figure 4.8, the pressure profile, stream lines, and wall heat flux of channel bends with various shapes and rectangular cross section are displayed for a Re number of 200. The diagram indicates the possible heat transfer from the wall to the fluid for given pressure loss or pumping power P necessary to drive the fluid.

$$P = \frac{\dot{m}}{\rho} \Delta p \qquad (4.63)$$

The high pressure loss in a sharp 90° bend also leads to an increased heat flux, especially for high flow rates demanding a high pumping power. Pressure loss savings in channels with circular curved bends ($r = d_h$) also yield a lower heat flux, hence, the curves in the diagram are close related indicating similar performance for each element. Sharp bends generate the highest heat flux for a given pressure loss at high flow rates.

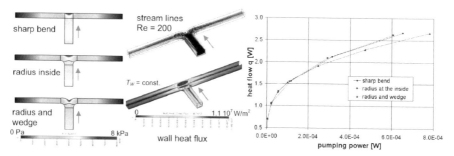

Fig. 4.9. Pressure loss in T-junctions with sharp and round corners, 300 K inlet temperature and 350 K wall temperature, see also Figure A.25. The inlet channel is 170 µm wide, the outlet channel is 100 µm wide, and all channels are 100 µm deep.

The situation is similar for channel junctions or T-junctions, as shown in Figure 4.9, where the pressure profile, stream lines, and wall heat flux are displayed for various junction shapes with rectangular cross section and a Re number of 200. The heat flux is slightly higher in a sharp bend or inside radius bend element for given pumping power, for example, with 0.5 mW pumping power, a heat flux of 2.4 W is transferred from the element walls to the fluid (water) in sharp bend and inner radius bend, whereas only 2.3 W is transferred in the junction with a wedge.

4.2.3 Combined channel elements

Low pressure loss and a low mean driving temperature difference ΔT are essential for an optimum operation and suitable heat exchanger performance. Both effects lead to appropriate geometrical optimization of dendritic channel networks with minimal entropy generation [90], presented under the concept of Bejan's "constructal theory" [330, 335] An example of optimum spacing between the microchannels for heat transfer is given by Favre-Marinet et al. [336]. An argument from constructal theory is the correct and appropriate proportion of heat transfer characteristics for the involved components, in this case, the microchannel and the surrounding material. For optimized device heat transfer, the order-of-magnitude of the transfer capacity of internal channels must be similar to that of the heat conductivity of the surrounding material.

The length of the channel and the area covered (channel plus heat entrained area) of the zero level element also depends on the ratio of the heat transfer coefficient to the heat conductivity of the solid material, see Bejan [173]. The temperature analysis of channel flow and surrounding material is displayed in Figure 4.10 for the channel flow and temperature distribution of the heated area. With increasing Re number,

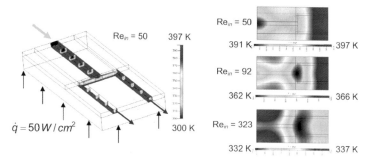

Fig. 4.10. Temperature distribution in a fluidic element with fork-shaped microchannels and surrounding material for different Re numbers in the outlet, see also Figure A.26.

the channel heat transfer also increases, whilst shifting the minimum temperature from the inlet to the junction. For higher Re numbers (Re = 323), the highest temperature is quite close to the inlet channel. With the temperature field, the variance of the temperature distribution also changes with the Re number, as displayed in Figure 4.11 right side. For Re numbers around 100, temperature variance becomes minimal, indicating a homogeneous temperature distribution, and the convective heat transfer is in the range of the heat conduction within the substrate, here silicon with $\lambda \approx 150$ W/mK.

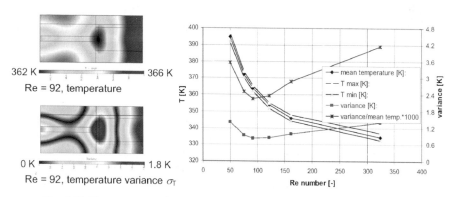

Fig. 4.11. Temperature variance in a microfluidic element, see also Figure A.27.

An optimized temperature distribution with minimal temperature variance in Figure 4.11 also depends on the Re number and the channel geometry. The optimum channel length and covered area are dependent on further constraints of a device, such as heat source, allowable footprint, or substrate material, which are subject to future optimization of channel networks for heat transfer.

4.2.4 Heat exchanger channel network

According to Bejan's constructal theory [330], a network is set up from the lowest level, which is the smallest heat exchange channel in this case, see Figure 4.12. The previous results of the heat transfer in curved microchannels are used to set up a network for enhanced heat transfer. To describe the construction process simply, the heat transfer channels are characterized by the main dimensions and the covered area. Figure 4.12 shows the combination of the branched elements from zero level to the desired covered area. The channel dimensions and the covered area are determined by the actual situation.

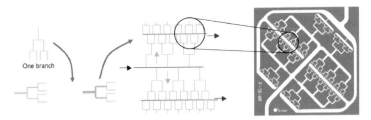

Fig. 4.12. Setup of a two level heat exchanger network from branched channels (left to right), blue indicating the inlet and red the outlet channels.

The channel cross sections are designed according to Murray's law [331], which describes the relation between the channel cross sections in different branched and connected levels as a general rule. First similarity laws for human and animal blood circuit system were reported by Hess in 1915, see Wezler and Sinn [337]. The optimization of a channel network to a minimal pressure loss shows that the cube of the diameters of a parent channel should equal the sum of the cubes of the daughter channel diameters, see Eq. 4.59.

$$\dot{V} \propto C \cdot w \cdot d^2 \text{ and } d_n^3 = \sum_i d_{z,i}^3 \qquad (4.64)$$

This law can be derived from laminar flow in a branched circular tube system, but is also present in biological systems, such as plants [338, 339, 340] and mammals [341]. The application of this biomimetic rule leads to a channel network with low wall shear stress in the channels, which might be one reason for the biological occurrence [331]. Murray's law applied to cooling systems results in structures similar to Figure 4.12 and leads to branched systems as displayed in Figure 4.13. The pressure loss in a network can be calculated with lumped element modeling and with the help of electronic circuit layout routines, see [342].

To maintain a homogeneous fluid flow and temperature distribution, the emphasis must be focused on proper fluid distribution. In Figure 4.13, left side, a rectangular distribution channel system is shown similar to those proposed by Bejan [330]. The middle sketch shows a flow distribution system with variable angles, which is also optimized for short distance connections between the various parts of the heated area.

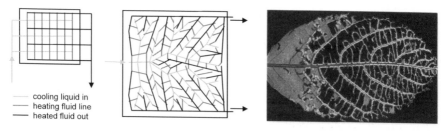

Fig. 4.13. Channel network for supplying a distinct area; left: Rectangular network setup with three levels of heat transfer channels; middle: Dendritic channel network on the heated chip; right: Comparison with a leaf structure (Nachtigall and Blüchel [343]).

The shape of the channel system then resembles dendritic, branched network, which is often found in nature, in a leaf structure or a lung.

Two fluidic network structures recently build are displayed in Figure 4.14. The mask design of a uni-dimensional network allows fluid distribution and area cooling on one layer. The chip footprint of 40×40 mm^2 is covered with four rows of dendritic structures with a minimal channel width of 100 µm. Using DRIE fabrication, the channels have a uniform depth of 300 µm. Four branching levels are realized with wedge-shape junctions for reasonably low pressure loss.

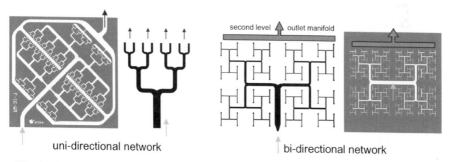

Fig. 4.14. Principle setup of two-dimensional, constructal network with biomimetic origin.

Heat transfer channels are parallelized and arranged hierarchically based on the "constructal" theory of Bejan [330]. The heat transfer in the microchannels is numerically simulated and compared with analytical correlations for entrance flow conditions in optimized structures concerning low pressure loss and high transfer rates. The channel cross sections and the ratio between the different hierarchic branching levels are determined with the help of Murray's law [331] to minimize pressure loss and yield a uniform flow distribution in the network.

The bi-directional network with six or seven branching levels also covers a chip footprint of 40×40 mm^2, but requires a second level for the outlet manifold. Typical Re numbers in the cooling devices range from 50 to 500, producing vortices at bends and curves for enhanced heat transfer rates. Utilizing these design rules for the proper flow distribution in parallel microstructured transfer elements, the benefits of miniaturization can successfully be transferred to other applications with high flow rate demands, see Section 5.7.

4.3 Micro Heat Exchanger Devices

The basic configuration of micro heat exchangers differs little from conventional equipment: heat is transferred from one fluid to the other (for two-flow heat exchanger), with a defined driving temperature difference ΔT between the channels. The balance equations for cooling and heating, as well as the transfer correlation, are valid independently from the length scale addressed here. Next, some special issues of micro heat exchangers will be discussed based on general correlations for plate heat exchangers.

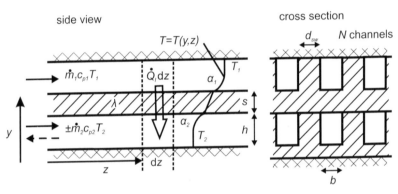

Fig. 4.15. Setup of a differential heat exchanger channel element and cross sectional view.

Energy conservation for cooling and heating of both streams, entering the heat exchanger, is described for a differential element and the entire apparatus. The inlet temperatures are denoted with T_{11} and T_{21}, the outlet temperatures with T_{12} and T_{22} for fluid 1 and fluid 2, respectively. For fluids, which are cooled ($T_{11} > T_{12}$):

$$d\dot{Q}_1 = -\dot{m}_1\, c_{p1}\, dT_1; \quad \dot{Q}_1 = -\dot{m}_1\, c_{p1}\, (T_{11} - T_{12}). \tag{4.65}$$

For fluids, which are heated ($T_{21} > T_{22}$):

$$d\dot{Q}_2 = \dot{m}_2\, c_{p2}\, dT_2; \quad \dot{Q}_2 = \dot{m}_2\, c_{p2}\, (T_{22} - T_{21}). \tag{4.66}$$

If heat loss to the environment is avoided by good thermal insulation or high internal heat transfer, both heat amounts are equal and must be transferred across the channel wall. The heat transfer correlation is described by the kinetic equation or rate equation for heat transfer between the two streams

$$d\dot{Q}_l = k\, b\, (T_1 - T_2)\, dz = k\, A\, dT_{tot} \tag{4.67}$$

with the overall heat transfer coefficient k and the channel width b. The heat transfer area A is drawn from the integration of the channel width over the channel length of all channels N.

$$A = N \int_0^L b\, dz \tag{4.68}$$

The temperature development along the channel length in a single-pass heat exchanger with co-current (left side) and counter-current flow (right side) is shown in Figure 4.16.

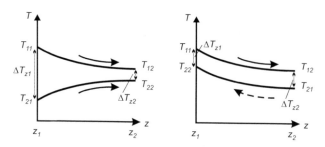

Fig. 4.16. Schematic temperature development along the channel length, left: co-current flow; right: counter-current flow.

In micro heat exchangers, the heat transfer area is difficult to determine due to the small channels and relatively thick walls. For conventional compact heat exchangers with rectangular cross section channels, Harris et al. [344] proposed a "wall efficiency" comparable to the fin efficiency η_{sw} to include the heat conduction in side walls (thickness d_{sw}) of channels. The channel length l, channel width d, and channel height h multiplied with the wall efficiency η_{sw} determine the effective heat exchanger area A_{eff}.

$$A_{eff} = (b + \eta_{sw}\, h) \cdot l \tag{4.69}$$

with the fin efficiency of the side wall from the inner heat transfer coefficient α_i and the heat conductivity λ_w of the wall material.

$$\eta_{sw} = \frac{h}{2} \left(\frac{2\alpha_i}{d_{sw}} \lambda_w \right)^{-0.5} \tanh \left(\left(\frac{2\alpha_i}{d_{sw}} \lambda_w \right)^{-0.5} \frac{h}{2} \right) \tag{4.70}$$

The overall heat transfer coefficient k is the sum of the single heat transfer resistances, given here for a plate

$$k = \left(\frac{1}{\alpha_1} + \frac{s}{\lambda} + \frac{1}{\alpha_2} \right)^{-1}, \tag{4.71}$$

and for arbitrary cross sections with a reference area A_k

$$\frac{k}{A_k} = \left(\frac{1}{\alpha_1 \cdot A_1} + \frac{s}{\lambda \cdot A_w} + \frac{1}{\alpha_2 \cdot A_2} \right)^{-1}. \tag{4.72}$$

Typical overall heat transfer coefficients k of microstructured heat exchangers rang from 2.6 kW/m^2 K for gas/liquid flow to 26 kW/m^2 K for liquid/liquid flow [69]. Conventional plate heat exchangers exhibit very good heat transfer characteristics, ranging from 0.2 to 2.5 kW/m^2 K, under optimum conditions up to 5 kW/m^2 K for gas/liquid and liquid/liquid flow, respectively [298].

The integration of Eq. 4.67, combined with Eqs. 4.65 and 4.66, over the entire heat exchanger device or the channel length leads to the mean logarithmic temperature difference, with the temperature differences at point z_1 and z_2 (e.g. inlet and outlet of the heat exchanger)

$$\dot{Q}_{ges} = k \, A \, \Delta T_{\log} = k \, A \, \frac{\Delta T_{z1} - \Delta T_{z2}}{\ln|\Delta T_{z1}/\Delta T_{z2}|}. \tag{4.73}$$

The mean logarithmic temperature difference is independent of the flow direction inside the single-pass heat exchanger with co-current or counter-current flow. For more complex flow situations inside the heat exchanger, the mean temperature difference is determined for the individual parts see [298].

To determine the mean temperature difference from Eq. 4.73, the outlet temperatures of both streams have to be known and must be calculated iteratively by outlet temperature estimation. For fluids with notably temperature-varying properties or complex flow regimes in the heat exchanger passages, the temperature profiles need to be calculated numerically along the channel length.

4.3.1 Number-of-thermal-units (NTU) concept

With constant overall heat transfer coefficient and fluid properties, the differential heat transfer equations can be solved to determine the outlet temperatures [325]. The heat transfer in a differential channel element with a driving temperature difference $(T_1 - T-2)$ leads to the cooling of mass flow 1 acc. to Eq. 4.65 or heating of mass flow 2 acc. to Eqs. 4.66. The temperature change of mass flow 1 is given by

$$\frac{d\,T_1}{d\,z} = -\frac{b\,k}{\dot{m}_1\,c_{p1}} (T_1 - T_2) \tag{4.74}$$

and for mass flow 2 by

$$\frac{d\,T_2}{d\,z} = \pm \frac{b\,k}{\dot{m}_2\,c_{p2}} (T_1 - T_2) \begin{pmatrix} + : & \text{co} - \text{current} \\ - : & \text{counter} - \text{current} \end{pmatrix}. \tag{4.75}$$

The addition of both equations leads to the differential equation for the driving temperature difference.

$$\frac{d\,(T_1 - T_2)}{T_1 - T_2} = -k\,b\,dz \left(\frac{1}{\dot{m}_1\,c_{p1}} \pm \frac{1}{\dot{m}_2\,c_{p2}} \right) \begin{pmatrix} + : & \text{co} - \text{current} \\ - : & \text{counter} - \text{current} \end{pmatrix} \tag{4.76}$$

To solve this ordinary differential equation, the dimensionless temperatures θ_i and the dimensionless capacity ratio N_i of the heat transfer capacity of the entire device and each flow heat capacity $\dot{m}c_p$ are introduced.

$$\theta_1 = \frac{T_{11} - T_{12}}{T_1 - T_2}, \quad \theta_2 = \frac{T_{22} - T_{21}}{T_1 - T_2}, \quad N_i = \frac{k\,A}{(\dot{m}\,c_p)_i} = \frac{\Delta T_i}{\Delta T_{\log}} \tag{4.77}$$

The dimensionless temperatures also indicate the heat exchanger efficiency, which denote the actual temperature change of one flow compared to the maximal possible temperature change. The capacity ratio N_i has been named number of transfer units (NTU) by Chilton [325] and gives the ratio of a process quantity ($\dot{m}c_p$) to the performance of the device (kA). The number of transfer units N_i has different physical meanings:

- The capacity ratio N_i also expresses the temperature change of the fluid compared to the mean temperature difference $\Delta T_{\log}$, see Eq. 4.77,
- With the residence time in the device $t_P = L_C/\bar{w}$ from Eq. 5.28, the transfer units can be expressed as

$$N_i = \frac{k\,A}{(m\,c_p)_i}\, t_{P,i}. \tag{4.78}$$

A short residence time, which often occurs in micro heat exchangers, produces minimal temperature change in the fluid and a correspondingly high mean temperature difference. The high heat transfer coefficient and a small fluid volume in the device allow a short residence time for heating or cooling.

- With Eq. 4.28, which denotes the characteristic time of unsteady heat transfer or the relaxation time t_{rel}, the number of transfer units N_i can be regarded as the ratio of two characteristic times, where the local heat transfer coefficient α is replaced by the total heat transfer coefficient k.

$$N_i = \frac{t_P}{t_{rel,i}} = \frac{\alpha_i A_i\, t_P}{(mc_p)_i} \tag{4.79}$$

For further calculations, the ratio of the heat capacity fluxes is often used to characterize heat transfer equipment.

$$C_1 = \frac{(\dot{m}\,c_p)_1}{(\dot{m}\,c_p)_2} = \frac{N_2}{N_1}; \quad C_2 = \frac{N_1}{N_2} = 1/C_1 \tag{4.80}$$

To develop the dimensionless temperature profile along the dimensionless channel length $z^* = z/L_C$ (heat exchanger length), Eq. 4.75 gives the energy balance of a channel element (fluid 2) in counter-current flow.

$$N_2\,(T_2 - T_1) + \frac{d\,T_2}{dz^*} = 0 \tag{4.81}$$

The algebraic transformation and integration over the heat exchanger length leads to the correlation for the outlet temperatures, here for counter-current flow. The correlations for other flow situations can be derived in a similar way and are given in [298].

$$\theta_i = -\frac{1 - \exp\left((C_i - 1)N_i\right)}{1 - C_i \exp\left((C_i - 1)N_i\right)}; \quad C_i \neq 1 \qquad (4.82)$$

This equation cannot be used for equal heat capacity fluxes on both sides ($C_i = 1$), which is given by

$$\theta = \frac{N}{N + 1} \qquad (4.83)$$

The local temperatures over the heat exchanger length can be derived from the energy balance and are given here for counter-current flow.

$$\theta_1(z^*) = \frac{T_1(z^*) - T'_1}{T_2 - T'_1} = -\frac{N_1}{N_2 - N_1}\left(1 - \exp\left(-(N_2 - N_1)z^*\right)\right) \qquad (4.84)$$

$$\theta_2(z^*) = \frac{T_2(z^*) - T_2}{T'_1 - T_2} = \frac{N_2}{N_2 - N_1}\left(1 - \exp\left(-(N_2 - N_1)z^*\right)\right) \qquad (4.85)$$

The co-current flow is similar and can be found in other literature [298, 299, 325]. With this set of equations, the temperatures and heat fluxes can be determined for a given heat exchanger. Conversely, the heat transfer area, channel cross sections, and wall dimensions can be determined for heat exchanger layout. The above equations can also be applied to mass transfer equipment, if the same physical processes are present.

4.3.2 Design issues for exchange equipment

In process engineering, there are two design criteria for heat exchangers: to determine the heat transfer area and number of channels for a given process task (temperatures and flow rates) or to determine the outlet conditions and characteristics of a given heat exchanger (A, k, geometry). Further information can be found in [298, 300].

Whilst microstructured heat exchangers deliver high transfer rates, they also have drawbacks, application limits, and factors to consider during design and operation. For example, the axial heat conduction in the relatively thick walls must be considered. Equal distribution on a large number of channels O(1 000) is a major issue for manifold design, as well as the need to mitigate fouling or blocking of single passages or complete parts.

The axial wall conduction in microstructured heat exchangers with gaseous flow was addressed by Stief et al. [39] in a numerical investigation of micro heat exchangers. The order-of-magnitude estimation gives an optimum heat conductivity for the wall material of approx. 2.5 W/mK, which correlates with glass or oxide ceramics. Aside from this particular example of gas flow heat exchange, further hints were given for microsystem design. This order-of-magnitude estimation is a good application of the correlations for heat exchangers, see also Section 1.4.1. The optimum wall heat conductivity for heat exchangers with water is determined to approx. 11 W/mK, which is in the range of stainless steel [345]. In a numerical study, Weigand and

Gassner [346] investigated the influence of wall conduction for heat transfer in laminar and turbulent flow. The authors found a major influence of axial wall conduction for low Péclet numbers and short channel length.

The temperature profile of the fluids in a counter-current heat exchanger is schematically illustrated in Figure 4.16 right side, and more detailed in Figure 4.17. Three cases are displayed:

- moderate axial wall heat conductivity with high convective heat transfer (broken line with temperature difference at one side ΔT_{z2}, case a), which corresponds to the heat exchanger efficiency),
- moderate axial wall heat conductivity with moderate convective heat transfer (straight line with temperature difference at one side ΔT_{z2}, case b), and
- high axial wall heat conductivity with moderate or even high convective heat transfer (fine dotted line with temperature difference at one side ΔT_{z2}, case c).

It is clearly visible that the first case a) exhibits the best heat exchanger efficiency θ, displayed as the temperature difference at the entrance or outlet of the heat exchanger. The last case, with high axial heat transfer, also exhibits high heat transfer rates at both ends of the heat exchanger, however, in the middle of the device, only a marginal amount of heat is transferred. The temperatures of both fluids are almost identical in the middle of the device with little temperature change. The area in the middle of the heat exchanger provides inefficient heat transfer and diminishes the device performance.

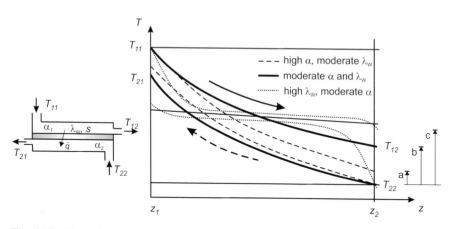

Fig. 4.17. Schematic temperature profiles in a counter-current heat exchanger with different wall material and heat transfer coefficients. The fine-dotted line displays the temperature profile in a microstructured heat exchanger with high heat conductivity in the wall and low overall heat transfer coefficient. The temperature difference at the end of the heat exchanger corresponds to the heat exchanger efficiency, see also Figure 1.4.

To estimate the influence of axial heat conduction in the wall, a heat transfer equation similar to Eq. 4.67 is discussed, where the transported heat depends on three factors.

$$\dot{Q}_i = k_i \cdot A_i \cdot \Delta T_{W\,i} \tag{4.86}$$

Axial heat conduction along the heat exchanger length imposes an additional heat flux Q_z on the wall, which increases the temperature difference for the wall. The original temperature difference of the wall originates from the heat transfer in y-direction. The temperature difference over a small wall element in channel direction is dependent on the local mean temperature difference, which is the driving force for the heat transfer between both flows. The heat transfer in z-direction within a wall element, regarding one channel, is dependent on the following factors:

$$k_z = \frac{\lambda_W}{dz}; \quad A_z = b\,s; \quad \Delta T_{Wz} = \frac{s}{l}\left(\frac{\theta_1 + \theta_2}{2}\right)(T_1 - T_2) \tag{4.87}$$

The same wall element dz transports the heat in radial direction depending on the following factors:

$$k_y = \left(\frac{1}{\alpha_1} + \frac{s}{\lambda} + \frac{1}{\alpha_2}\right)^{-1}; \quad A_y = b\,l\,f_A; \quad \Delta T_{Wy} = \frac{k\,s}{\lambda_W}\Delta T_{\log} \tag{4.88}$$

with the geometrical factor f_A representing the wall cross section compared to the channel cross section $f_A = A_z/A_C$. This ratio is close to one for thin side walls and less for very thick side walls, which enable additional axial heat transfer. The comparison of axial heat fluxes with radial heat flux gives the following dimensionless groups:

$$\frac{\dot{Q}_z}{\dot{Q}_y} \rightarrow \quad a) \ \frac{k_z}{k_y} = 1 + \frac{\lambda_W}{s}\left(\frac{1}{\alpha_1} + \frac{1}{\alpha_2}\right)$$

$$b) \ \frac{A_z}{A_y} = \frac{s}{l\,f_A} \tag{4.89}$$

$$c) \ \frac{\Delta T_{Wz}}{\Delta T_{Wy}} = \frac{\lambda_W}{k\cdot l}\frac{(T_1 - T_2)}{\Delta T_{\log}}\left(\frac{\theta_1 + \theta_2}{2}\right)$$

High values for these three ratios allow large axial heat transfer in a heat exchanger. The ratios indicate the high influence of axial heat conduction, and not only for high wall heat conductivity in a gas flow heat exchanger (low α). A small heat exchanger ($s \approx l$) with thick walls ($f_A \ll 1$) is sensible to the axial heat conduction, as is a heat exchanger with high temperature load ($T_1 - T_2 \gg \Delta T_{\log}$). Eq. (4.89) gives an indication of the relevance of axial heat conduction. Nevertheless, the actual temperature profile is quite complex for different flow regimes, geometrical arrangements, heat capacity flows, and wall heat conductivities. Further information can be found in Stief et al. [39], however, for the actual design of a microstructured heat exchanger, a more detailed analysis with numerical methods needs to be carried out.

Micro heat exchangers consist of a large number of channels to provide a high flow rate. Hence, the channels are arranged in parallel, but also due to fabrication reasons. Foli et al. [347] present a numerical optimization study for micro heat exchangers in automotive applications with minimal pressure drop and maximal heat transfer. With a Pareto analysis, the authors determined optimal cross-sectional dimensions for constant channel volume, and compared various channel cross sections with the aid of CFD calculations.

Generally, the headers or flow manifolds at the entrance and at the outlet are very important for relatively even distribution of the fluids. The effect of fluid maldistribution can be described by the mass flow difference $\Delta \dot{m}$ between the single channels, see Schlünder [325] and Section 6.3.5. In maldistribution situations, channels with low flow rates undergo a high temperature change, while the temperature change is reduced in channels with high flow rates. Behind the passages, the fluid streams mix, resulting in a mean temperature which is lower than the maximum attainable temperature for cooling or heating, therefore, the efficiency of heat exchangers is decreased due to maldistribution in the channels. A rough estimation for the asymptotic efficiency $\theta = 1$ shows the following behavior for highly efficient heat exchangers

$$\max \theta = 1 - \frac{\Delta \dot{m}}{\dot{m}}. \tag{4.90}$$

Thus, the mass flux deviation over the channels directly lowers the efficiency of the heat exchanger. Particularly in micro heat exchangers with the large number of passages, potentially high transfer rates, and equipment loading, a uniform distribution of the fluid streams is very important for equipment performance. High-performance devices are typically very sensitive to load changes, differences in fluid properties, or variation in device geometries due to fouling or corrosion.

For the entire residence time within the equipment, the flow in the manifolds and fluidic connection cannot be neglected in microstructured devices. The dead volumes in the headers must be avoided or minimized by proper design, see also Section 5.6.3 and 6.3.5.

4.3.3 Fouling and blocking of equipment

Fouling and blocking of small passages with geometrical dimensions from 10 μm up to millimeter size is the major problem during testing and operating microstructured devices, see [13] Chapter 1. Although the problem is omnipresent in microsystems technology, fouling still attracts too little attention in the technical-scientific investigation of microsystems. Hessel et al. [22] Chapter 1 and 2, present two fouling mitigation techniques in more detail. A separation layer of a flow component is placed between the wall and the particulate or reacting flow, which hinders the migration and attachment of solids to the wall. The impingement of two liquid jets minimizes a reacting flow contact with the wall (Impinging Jet Micro Mixer [348]). Disposable devices for single use often circumvent the fouling problem in analytical applications where device cost are not a major issue.

Experimental results from Wengeler et al. [349] indicate two major fouling effects in microchannels of T-shaped micromixers with particulate, laminar flow, see Figure 4.18. Very small particles (< 30 nm with $\rho_P \approx 2.5$ g/cm^3) migrate to the wall by self-diffusion in the carrier fluid, and larger particles (> 150 nm with $\rho_P \approx 2.5$ g/cm^3) accumulate at the wall behind bends, nozzles, or expansions, where a flow velocity component exists perpendicular to the wall. Smaller particles follow the flow and do not come into contact with the wall. To mitigate particulate fouling, inlet flow filtering is recommended where possible, however, this is only reasonable for particles larger than 150 nm, as smaller particles will pass through the equipment.

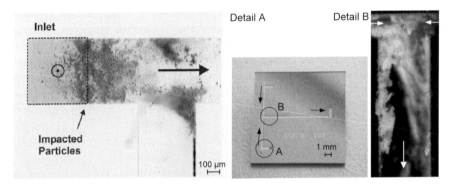

Fig. 4.18. Major areas of particle deposition in a T-shaped micromixer with perpendicular inlet channels, acc. to Wengeler et al. [349].

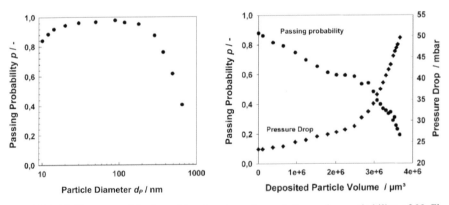

Fig. 4.19. NaCl nanoparticle deposition in micromixers; left: passing probability of NaCl nanoparticles with different diameter; right: deposited nanoparticles and increasing pressure loss indicate the constriction of microchannels by deposited particles, acc. to Wengeler et al. [349].

A rough surface has a high specific surface energy and facilitates the attachment of particles [350, 351], using construction materials consisting of fine powder (metal or ceramic sinter process). Due to the high surface-to-volume ratio, various surface activities have to be considered, for example, nickel Ni as a catalyst for hydrocarbons. Ni atoms at the surface induce the growth of carbon nanotubes and carbon black, which can constrict the channels. Other metals, such as platinum Pt, gold Au, manganese Mn, or copper Cu, may have similar effects, hence, the application of alloys has to be considered very carefully. Material experience from larger equipment may not be valid for microstructured devices due to different surface-to-volume ratios. Other mechanisms of fouling in microstructures, such as corrosion, chemical reactions, polymerization, surface precipitation, or adhesion, are not addressed here. More experimental and operational data needs to be collected in the future to develop a better understanding of the fouling problem.

The best method to prevent fouling in microchannels is through appropriate design of the channel, junctions, expansions, nozzles, and the elimination of dead zones [27]. Larger channel diameters allow high flow velocities, and smooth surfaces from electrochemical polishing also help to mitigate precipitation. High flow velocities in long straight microchannels will prevent particles attaching to the wall. Particles already attached may be washed away by the high shear gradient, particularly, for pulsating flow. For elements with high fouling risk, cleanable or disposable elements should be considered, and special arrangement should be provided for maintenance and cleaning operation. Due to the complex flow and wide variety of different applications, fouling-sensitive design principles and the collection of more experimental data is essential for the successful operation of microstructured devices.

4.3.4 Particle deposition in microchannels

To determine particle deposition in microreactor channels, NaCl aerosol was generated by atomizing a salt solution with a Collision Atomizer and subsequent drying of these droplets. The dry aerosol was then charged in a radioactive Kr^{85} charger and classified by a differential mobility analyzer (DMA; Grimm Model 5.4-900). To obtain mostly uncharged particles and avoid electrostatic deposition, the aerosol was again put through a Kr^{85} charger. The aerosol was mixed 1:1 with filtered nitrogen in the microreactor chips, and the particle concentration at the outlet of the mixing structure was then measured with a condensation particle counter (CPC; Grimm Model 5.403), which is capable of counting particles as small as 7.5 nm. For reference, the particle concentration is measured without the reactor.

The particle loss of a monodisperse NaCl aerosol flowing through the microreactor is measured depending on the particle diameter, see Figure 4.21. The Stokes number St is used as the dimensionless particle diameter responsible for the particle deposition

$$St = \frac{C\,d_P^2\rho_P\bar{w}}{18\eta d_h},$$

(4.91)

with the electrodynamic constant C from the electrostatic measurement principle, see Wengeler et al. [349].

The role of the Stokes number St for particle deposition is described by Heim et al. [352]. The relative particle loss in the investigated microreactors is displayed in Figure 4.21 for different St numbers. Small particles are more likely to pass through the microchannels because they are better able to follow the flow. Larger particles

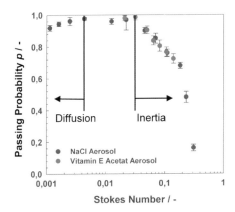

Fig. 4.20. Particle passing probability depending on the Stokes number. Small particles contact the wall due to Brownian motion, large particles attach to the wall due to their inertia [352].

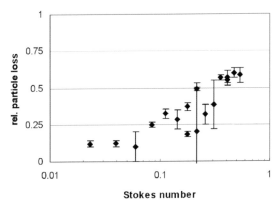

Fig. 4.21. Particle loss in a T-shaped micromixer ($200 \times 300 \ \mu m^2$ inlet, $500 \times 300 \ \mu m^2$ mixing channel) versus the Stokes number as a measure of the normalized particle density [352].

(St > 0.1 corresponding with a particle diameter of approx. 300 nm) are prone to attach to the wall.

Concluding this chapter, microstructured equipment and related transport processes promise the successful application in various fields, where high transfer rates, intelligent and prolific combinations of microstructures in macro devices, and new process routes are needed. High gradients and a high specific surface in devices with various construction materials lead to fast equilibrium state. The characteristic dimensions of microstructured internals are in the range of boundary layers, where high gradients enforce the transfer processes. Cascading of effects and integration of various elements enhance and guide the entire process. This incorporates new possibilities for difficult processes and opens new opportunities for the future in the area of process intensification as an enabling route for new technologies.

5

Diffusion, Mixing, and Mass Transfer Equipment

This chapter along with the succeeding chapter on chemical reactions forms the heart of this work. It is based on fluid dynamics of mixing in Chapter 3 and supplemented by applications in Chapter 6. The chapter gives an introduction of the major mixing principles in microfluidic devices: diffusive, distributive, and convective mixing, with major focus on the latter. Convective micromixers allow high flow rates with low pressure loss and low fouling or blockage risk for fast and effective mixing of different fluids. Typical characteristics of investigated micromixers are translated into design criteria with related fabrication technology. Chaotic advection helps to describe convective mixing regimes, their characteristics and assists the design of efficient micromixers. Finally, high throughput devices are presented to mix 25 kg/h aqueous solution below milliseconds while demanding a pressure loss of less than 1 bar.

5.1 Mixing Processes and Their Characterization

Mixing is a fundamental process, which influences many other transport processes, such as heat transfer, chemical reactions, or separation processes. Figure 5.1 gives an overview of mixing length scales and typical related flow situations, represented by the Reynolds number Re of various mixing and mass transfer processes in nature and engineering. The length scales range from molecular length to astrophysical dimensions over 18 orders-of-magnitude. Even more complex are the Re numbers, stretching over 40 orders of magnitude from creeping flow over laminar flow to highly turbulent flow conditions. Ottino [353] gives a concise review on mathematical concepts to treat this complex area, recently updated by a monograph of Sturman et al. [354].

Three major principles, diffusion, fluid distribution, and convection, determine the mixing process, see Figure 5.2. In microfluidic systems with predominantly laminar flow, diffusion is the prevalent process and geometry determines the time scales. However, for many micro process engineering applications, convection is systema-

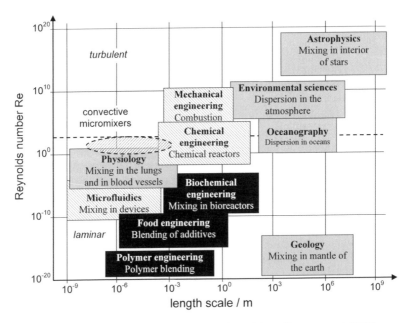

Fig. 5.1. Spectrum of mixing problems, adopted from Sturman et al. [354].

tically used to enhance the transport properties, aside from the reduction of the diffusion length in small equipment or by distributive mixing.

The field of micromixers has made rapid advances during the last two decades. From the first passive multilamination and split-and-recombine micromixers presented in the mid-90s [355] to more complex active mixing devices (e.g. [356]), a large variety of different micromixers have already been presented in literature. Two comprehensive reviews of existing micromixers are given by Nguyen [357] and Hessel et al. [22]. In process technology, mixing is a key unit operation and plays a crucial role in microreactors. The benefits of micro process technology result mainly from the fast transport processes and effective mixing in microstructured devices [3], aside from small internal volumes.

Due to their complex structure, active micromixers are often only used for special purposes, mainly in analytics. Passive micromixers are better suited for chemical production purposes due to their simple setup and robust operation. Recently, a number of micromixers were presented for chemical production, e.g. [34], [120], or [358]. These mixers show good mixing behavior combined with high mass flow rates.

5.1.1 Mixing principles and description

The main mixing principles for microfluidic devices are illustrated in Figure 5.2. Distributive mixing is dependent on repeated folding and stretching of fluid elements and lamellae to decrease the lamella width. By focusing the flow, fluid flow is accelerated

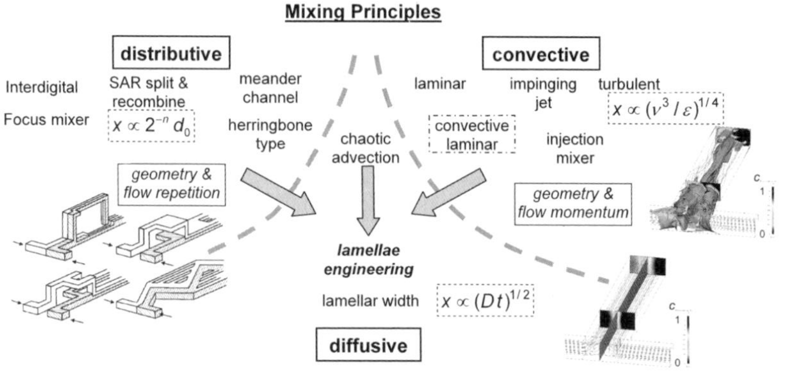

Mixing Principles

Fig. 5.2. Mixing principles for microfluidic devices.

and stretched, and the lamella width is also reduced. Some mixer types and applications are named in Figure 5.2, however, the amount investigated is much larger, see Hessel [22] Chapter 1. The typical length scale for mixing depends on the repetition rate of folding or stretching.

Convective mixing in microstructured devices is dependent on curved and bent flow inducing complex three-dimensional flow patterns and vortices, which increase the radial exchange processes, see Section 3.2.1. The typical length scale is determined by the energy dissipation into vortices, which is displayed for turbulent flow in Figure 5.2. Both distributive and convective mixing lead to fluid lamellae, whereas diffusive mixing results in a homogeneous distribution of species or energy. From this viewpoint, Mae has coined the idea of "lamellae engineering" or micro fluid segment design [359] with the combination of channel or device geometry with fluid flow structures.

As already described in Section 2.3, the diffusive mass transport of the component i within the surrounding component j is described by Fick's first law (one-dimensional form), see Bird et al. [2] Chapter 17.

$$\dot{n}_i = \frac{\partial n_i}{\partial t} = -\rho_m \, A \, D_{ij} \frac{\partial c_i}{\partial x} \tag{5.1}$$

Here, the species flux n_i depends on the molar density ρ_m, the transfer cross section A, the binary diffusion coefficient D_{ij}, and the concentration gradient. Typical values for the binary diffusion coefficient D_{ij} are $10^{-4} - 10^{-5}$ m^2/s for gases, $10^{-8} - 10^{-9}$ m^2/s for self-diffusion in liquids, $10^{-9} - 10^{-11}$ m^2/s for small particles in liquids or large molecules, and $10^{-14} - 10^{-19}$ m^2/s for atoms or molecules in solids. Other important diffusion types, like multi-component diffusion or diffusion in micro-porous systems are described in [65] Chapter 3 or in [2] Chapter 19. Fick's second law describes the temporal behavior of the concentration profile, here in one-dimensional form.

$$\frac{\partial c_i}{\partial t} = D \frac{\partial^2 c_i}{\partial x^2} \quad \text{or} \quad \frac{\partial c_i}{\partial t} = \text{div} \left(D_{ij} \text{grad} c_i \right) \tag{5.2}$$

A more detailed discussion with analytic solutions can be found in [360, 361]. An order-of-magnitude estimation from this equation gives the characteristic mean diffusion length in one direction $\bar{x}^2 = 2\,D_{ij}\,t$, see Eq. 1.1. The factor 2 presents the uniform diffusion from a plane into both perpendicular directions. This equation was derived by Einstein in 1905 with the examination of molecular fluctuations [362]. The typical diffusion length for gases and liquids is shown in Figure 1.2. Thermal fluctuations restrict the sedimentation of small particles in liquids, which can also be described with Eq. 1.1. For example, particles smaller than 1 μm in diameter hardly settle down in water, because their sedimentation velocity is too near to the fluctuation velocity. An additional convective mass flow with mean velocity $\bar{w}$ extends Eq. 5.2 to the following PDE, given in vector notation due to the different directions of diffusion and convection.

$$\frac{\partial c_i}{\partial t} + (\vec{w} \cdot \mathrm{grad})\ c_i = \mathrm{div}\,(D_{ij}\,\mathrm{grad}\,c_i) \tag{5.3}$$

Often, this equation can be reduced to one or two dimensions. With the velocity, a second time scale must be considered along with the diffusive time scale.

5.1.2 Mixing characterization

To compare various mixing states and mixing equipment, some fundamental definitions are helpful. The Danckwerts segregation intensity is defined with the mean square deviation of the concentration profile of the component i in a cross section [363] Chapter 1.

$$I = \sigma^2/\sigma_{\mathrm{max}}^2 \text{ with } \sigma^2 = \int_A (c_i - \bar{c}_i)^2 \mathrm{d}A \quad \text{and} \quad \sigma_{\mathrm{max}}^2 = \bar{c}_i\,(c_{i,\mathrm{max}} - \bar{c}_i) \tag{5.4}$$

The segregation intensity I can be transformed to a value between 0 (completely segregated) and 1 (completely mixed), the so-called mixing quality α_m.

$$\alpha_m = 1 - \sqrt{\sigma^2/\sigma_{\mathrm{max}}^2} = 1 - \sigma/\sigma_{\mathrm{max}} \tag{5.5}$$

The standard deviation can be derived from a concentration distribution or from a temperature profile in a cross section, see Section 6.4.2. Zlokarnik [364] Chapter 8, uses a modified standard deviation

$$\sqrt{S} = \sigma/\bar{c} \tag{5.6}$$

of the concentration field with mean concentration $\bar{c}$, to characterize conventional static mixers, such as Sulzer SMX or Kenics mixer. The pressure loss and mixing intensity in typical static mixers are illustrated in Figure 5.3.

The mixing intensity is measured with color dispersion experiments along the pipe length and decreases linearly with increasing tube length for all in-line mixers. This curve is also called the operating line of a static mixer. The incline of the mixer's operating line depends on the pressure loss within the static mixer $C_f \cdot \mathrm{Re}$, where a high pressure loss causes intensive mixing.

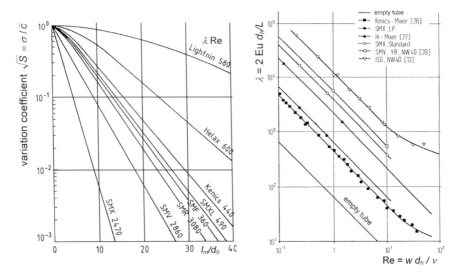

Fig. 5.3. Mixing characteristics and pressure loss within static mixers, according to Zlokarnik [364].

The variation coefficient $\sqrt{S}$ is coupled with the mixing quality α_m

$$\sqrt{S} = \frac{\sigma_{max}}{\bar{c}}(1 - \alpha_m),\tag{5.7}$$

or conversely

$$\alpha_m = 1 - \frac{\bar{c}}{\sigma_{max}}\sqrt{S}.\tag{5.8}$$

The mixing quality α_m equally weights each location on a cross section with same weight. The flow velocity varies over the cross section of a mixing channel, which leads to different local flow rates. If the fluid is collected in a vessel without further mixing, locations with high flow velocity contribute a high volume flow with related mixing quality. To achieve a mixing characterization with better representation of the flow situation, the velocity-weighted mixing quality $\alpha_{\dot{V}}$ in a cross section of the mixing channel is defined as a normalized standard deviation of the concentration field in a given cross section of the mixing zone:

$$\alpha_{\dot{V}} = 1 - \sqrt{\frac{\left(\sigma_{\dot{V}}^2(c)\right)}{\left(\sigma_{\dot{V}\,max}^2(c)\right)}}\tag{5.9}$$

where $\sigma_{\dot{V}}(c)$ is the standard deviation of concentration over the volume flow $\dot{V}$

$$\sigma_{\dot{V}}^2(c) = \frac{1}{A_M \bar{w}}\int_{A_M}\left(c - \frac{\int_{A_M} c\,w\,dA}{\int_{A_M} w\,dA}\right)^2 w\,dA.\tag{5.10}$$

The concentration c is weighted with the corresponding velocity w at the same location with the corresponding cross section A, and the mean velocity $\bar{w}$ with the
entire cross section A_M. Often, concentration values are measured at defined points
or drawn from simulation results on a numerical grid. For discrete values with a total number N of a rectangular, equal-distance mesh, the velocity-weighted standard
deviation is given by

$$\sigma_V^2(c) = \frac{1}{\sum_{i=1}^{N} w_i} \sum_{i=1}^{N} \left(\left(c_i - \frac{\sum_{k=1}^{N} c_k w_k}{\sum_{k=1}^{N} w_k} \right)^2 w_i \right). \tag{5.11}$$

The concentration c_i at a grid point i is weighted with the corresponding velocity
w_i at the grid point i. The velocity-weighted concentration variance gives a more realistic indication of the mean concentration than the geometry-weighted value. The
velocity-weighted mean concentration would be the mean concentration in a large
vessel, where the mixture will flow in and homogenize. The velocity-weighted concentration variance gives the concentration gradients at high velocities more weight,
where the gradients move very fast through the channel. The maximum deviation
$\sigma_{V\max}$ of a completely segregated flow of volume flow $\dot{V}_1$ with concentration c_1 and
volume flow $\dot{V}_2$ with concentration c_2 is calculated according to

$$\sigma_{V\max}^2(c) = \frac{\dot{V}_1 \dot{V}_2}{\left(\dot{V}_1 + \dot{V}_2 \right)^2} (c_1 - c_2)^2 \tag{5.12}$$

for unmixed fluids. The velocity-weighted standard deviation provides a more accurate representation of the actual mixing status than the area-based method.

5.1.3 Potential of diffusive mixing

In order to analyze the level of mixing quality, it is essential to document the grid
scale, on which the measurements are based. Unfortunately, the grid scale of mixing
quality is yet to be discussed extensively in literature, however, a further mixing
characterization, which is still under discussion, was presented by Bothe et al. [365]
and describes the potential of diffusive mixing.

$$\Phi_{L1} = \frac{1}{|A|} \int_A \|\mathrm{grad} f\| \, dA \quad \text{with} \quad f = \frac{c}{c_{max}} \tag{5.13}$$

Here, $\|\mathrm{grad} f\|$ denotes the length of standardized concentration gradient on the cross
section A, also called $L1$ norm of grad f. The parameter Φ_{L1} with the unit [1/m] can
be physically interpreted in various ways: as the specific contact area between two
components, as a numerical value for the driving force of diffusive mass transfer, or
as the mixing potential. The inverse value of Φ_{L1} indicates the mean distance between the two components and, therefore, the length scale of segregation. Schlüter
et al. [275] present an initial application with experimental data. Bothe and Warnecke [366] discuss the meaning of the mixing potential in correlation with other
measures for mixing.

A similar mixing scale was proposed by Chao et al. [367] using the normalized L2 norm of a concentration field

$$\Phi_{L2} = \sqrt{\frac{\int_A |\text{grad } f|^2 \, dA}{|A|}} \tag{5.14}$$

Chao et al. [367] apply the L2 norm of grad f to different stretching and folding mixers, as well as a discrete baker's transformation, and describe the mixing progress in micromixers. On a rectangular grid of a cross section A with node distance Δx in x direction and Δy in y direction, the above equation can be given in discrete form by

$$\Phi_{L2} = \sqrt{\frac{1}{A} \sum_{i=1}^{N_x} \sum_{j=1}^{N_y} \left[(\text{grad } c)_{j,i} \right]^2 \Delta x_i \, \Delta y_j}$$

$$= \sqrt{\frac{1}{A} \left(\sum_{i=1}^{N_x} \sum_{j=1}^{N_y} \left[\left(\frac{\partial c}{\partial x} \Big|_{j,i} \right)^2 + \left(\frac{\partial c}{\partial y} \Big|_{j,i} \right)^2 \right] \Delta x_i \, \Delta y_j \right)} \tag{5.15}$$

with the area $\Delta x_i \, \Delta y_j$ belonging to the grid point i, j. The partial derivatives of the concentration at node i, j are determined with the linear difference to the nodes in the direct vicinity.

$$\frac{\partial c}{\partial x} \Big|_{j,i} = \frac{c_{j,i+1} - c_{j,i-1}}{2\Delta x} \quad \text{and} \quad \frac{\partial c}{\partial y} \Big|_{j,i} = \frac{c_{j+1,i} - c_{j-1,i}}{2\Delta y} \tag{5.16}$$

The gradients at boundary nodes are determined with the difference to the inner side. For mixing of two streams with different temperatures in a T-mixer, the typical curve of the L2 norm of temperature distribution is displayed in Figure 5.4 over the mixing channel length. The temperature distribution is comparable with concentration gradients in mixers, here the Prandtl number Pr determines the diffusive transfer rates instead of the Schmidt number Sc.

Convective mixing at the mixing channel entrance rapidly increases the interface between the two components and leads to high concentration or temperature gradients. The high gradient facilitates high diffusive heat or mass transfer resulting in a diminishing L2 norm along the mixing channel. High gradients at the entrance of the mixing channel develop a rapid diffusive mixing and promote the attainment of complete mixing. When designing mixing devices, high gradients and L2 norms will facilitate rapid mixing and are a good indication of efficient mixing devices.

5.1.4 Stoichiometric mixing and diffusion process

The potential and concentration distribution on a cross section are the starting point for further analysis of mixing process. Stoichiometric and rapid mixing is important for the selectivity of chemical reactions. In most cases, the inlet concentrations c_1 and c_2 are chosen to achieve a stoichiometric ratio of the reactants with complete

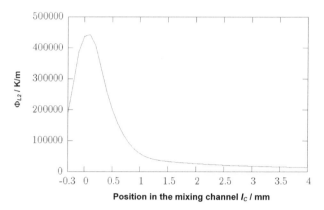

Fig. 5.4. Mean temperature gradient ($L2$ norm) in a T-mixer, Prandtl number Pr ≈ 0.6 for air.

homogenization $\bar{c}$. The non-stoichiometric state is characterized by the concentration Δc deviating from the mean concentration $\bar{c}$.

$$\Delta c = |c_i - \bar{c}| \tag{5.17}$$

In this case with 1:1 ratio mixing and normalized concentrations $c_1 = 1$ and $c_2 = 0$, the mean concentration is $\bar{c} = 0.5$, and concentration deviation Δc is in the range from 0 (stoichiometric mixed) and 0.5 (completely unmixed). The local mixing quality $\alpha_{m,i}$ correlates with the concentration deviation acc. to $\alpha_{m,i} = 1 - \sqrt{\Delta c^2 / \sigma_{max}^2}$.

The local gradient of the concentration distribution determines the potential of the diffusive mixing Φ, see Eq. 5.15. The potential depends on the concentration gradient and on the actual grid size and can reach a maximum value on a grid point of a rectangular grid with $\Delta x = \Delta y$,

$$\Phi_{max} = \frac{|c_1 - c_2|}{2\Delta x} \tag{5.18}$$

for maximum gradient in only one major direction x or y, or, more unlikely for numerical or experimental results, for maximum gradient in both major directions

$$\Phi_{max} = \frac{|c_1 - c_2|}{\sqrt{2}\Delta x}. \tag{5.19}$$

The actual potential of diffusive mixing is divided by the maximum value from Eq. 5.18 to determine the relative potential

$$\Phi_c = \frac{\Phi}{\Phi_{max}}. \tag{5.20}$$

In this case with 1:1 ratio mixing and normalized concentrations, the relative potential of diffusive mixing is in the range from 0 (no concentration gradient with completed mixing) to 1 (maximum gradient with rapid diffusive mixing). Figure 5.5

shows the development of the concentration variation Δc from the mean concentration $\bar{c}$ and the relative potential of diffusive mixing Φ_c at each point of the numerical grid at various locations in the mixing channel for engulfment flow in a T-mixer T600×300×300 at Re = 200. The mean concentration deviation and mean potential are given in gray lines with endpoints. Additionally, an image of the cross section with the relative potential of diffusive mixing Φ_c is given in the upper right corner of each diagram to display the local distribution. The shape of engulfment flow with two co-rotating vortices is clearly visible, indicating also the flow development along the mixing channel.

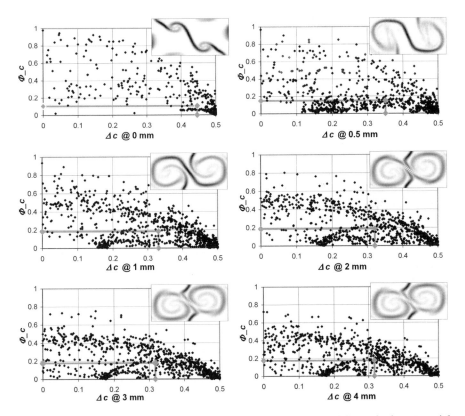

Fig. 5.5. Local description of non-stoichiometric mixed stated Δc and dimensionless potential of diffusive mixing Φ_c (*L2* norm) in a T-mixer (T600 × 300 × 300) at various locations in the mixing channel. The concentration gradient is displayed in the cross sections for engulfment flow with Re = 200 in the mixing channel. Dark areas mean high concentration gradient.

The diagram in Figure 5.5 can be divided in four areas with different meanings for the mixing process. Points in the upper right corner with high relative potential for diffusive mixing Φ_c and high concentration deviation Δc represent locations with poor stoichiometric mixture, but high potential to leave this situation and migrate to the lower left corner of the diagram. Here, with low Φ_c and low Δc, nearly stoichiometric homogenization is achieved, and no potential to leave this area is present. This is the best area and desired for mixing applications.

Points in the upper left corner of the diagram with high Φ_c and low Δc have a nearly stoichiometric state, but also a high potential to migrate away and worsen the situation due to unmixed states in the close vicinity. The worst case for mixing is represented by points in the lower right corner of the diagram with low Φ_c and high Δc. The location in the mixer is far away from being stoichiometric mixed and has low potential to reach this state. At the beginning of the mixing channel (0 mm), many points are located in the lower right area with high deviation from mixed state and low potential for diffusive mixing. Only points near the interface between the components and at the channel walls are located in the middle of the diagram. The mean potential of diffusive mixing Φ is quite low with 0.105, while the concentration deviation Δc is very high with 0.446. The concentration gradient clearly shows the interface between the components and two small vortices.

Further down the mixing channel, the vortices grow and enlarge the interface between the components. At a mixing channel length of 0.5 mm, many points are located at the bottom of the diagram with low Φ_c. The mean concentration deviation reduces to 0.35, while the potential increases to 0.148 at 0.5 mm and reaches a maximum at 2 mm with $\bar{\Phi} = 0.187$, see also Figure 5.4. The points representing different locations in the cross section are moving to the stoichiometric state and forming a "particle swarm" around two major lines. This grouping in the diagram must be investigated in more detail. At 1 mm length, the vortices are covering almost the entire cross section leading to a wide spreading of points. A second interface reaches from the top and bottom wall into the channel and is enclosed by the vortices. More points are located in regions with high potential Φ_c. At 2 mm length with maximum Φ_c, many points are grouped around a line reaching from $\Phi_c = 0.5$ to $\Delta c = 0.5$ with a slight upward bow, representing locations on the interface clearly indicated in the cross section. Only few points are located above this line having the potential to rapidly migrate further down.

With further increasing channel length (3 and 4 mm), the points are slowly moving to the lower left corner of the diagram. Numerical diffusion certainly assists this process and conceals the mixing process, however, these diagrams only give a trend and allow to compare different mixing processes. Further investigations on the basis of these diagrams are needed to get a sophisticated picture of the complex mixing process. As a first example, the mixer configurations in Section 5.4.3 are characterized with the relative potential of diffusive mixing and the concentration deviation at the exit of the device.

5.2 Diffusive Mass Transport and Concentration Distribution in Fluids

To provide an impression of the concentration field development with time, some analytical and numerical solutions are given for the diffusion equations. Starting

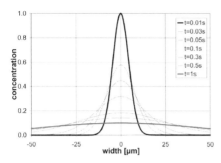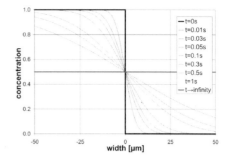

Fig. 5.6. Left: Diffusion from a source peak with $n_0/(\rho_m A) = 1/89206.2$ m, and $D = 10^{-9}$ m^2/s; Right: Diffusion between two semi-infinite bodies, according to Schönbucher [65] Chapter 3, and Jost [360] p. 36.

from a concentration peak c_0, the concentration profile is determined as a one-dimensional distribution of a concentration peak, see [89] Chapter 1.5,

$$c(x,t) = \frac{n_0}{\rho_m A \sqrt{4\pi Dt}} \exp\left(-\frac{x^2}{4Dt}\right) \tag{5.21}$$

with the molar density ρ_m and the total species amount n_0 that diffuses into the surrounding area A. The temporal development of the concentration distribution of a peak is displayed in Figure 5.6 left for a diffusion coefficient $D = 10^{-9}$ m^2/s.

The time-dependent diffusion into a plate is determined by an infinite Fourier series. The exact solution of this series expansion is time-consuming; charts and short cuts for various geometries and boundary conditions are proposed by Geankoplis [66] Chapter 7.

Derived from the source integral, Eq. 5.21, the one-dimensional diffusion between two semi-infinite bodies can be described by

$$c(x,t) = \frac{c_0}{2} \left(1 - \text{erf}\left(\frac{x}{\sqrt{4Dt}}\right)\right) = \frac{c_0}{2} \text{erfc}\left(\frac{x}{\sqrt{4Dt}}\right) \tag{5.22}$$

with the Gaussian error function $\text{erf}(x) = 1/\sqrt{(2\pi)} \int_{-\infty}^{x} \exp(\xi^2/2)d\xi$ and the complementary error function $\text{erfc}(x) = 1 - \text{erf}(x)$. The temporal development of the concentration distribution is displayed in Figure 5.6 for two adjacent semi-infinite bodies with the same diffusion coefficient.

The concentration homogenization between small fluid lamellae is an important diffusion situation in micro process technology. These fluid lamellae are produced in lamination micromixers, as serial lamination in split-and-recombine mixers (SAR), or as parallel lamination in interdigital mixers or multi-lamination mixers, and lead to a short diffusion length. The temporal development of the concentration profile may be determined in a number of ways. In [368], a harmonic series expansion is used to determine the solution of the diffusion equation, Eq. 5.2, for a SAR micromixer. The combination with exponential diffusion determines an integral mixing residual

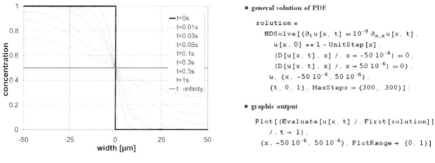

Fig. 5.7. Time development of the concentration profile in symmetrical fluid lamellae with 1:1 mixing; Mathematica® routine to solve and plot the concentration profile numerically.

as a measurement of the mixing quality in the SAR mixer. Convective micromixers induce small lamellae by convective and curved flow and operate at higher Re numbers.

The numerical solution of the diffusion equation, Eq. 5.2, is shown in Figure 5.7 with symmetrical boundary conditions. With this approach, different parameters, properties, and geometrical arrangements can be simulated and compared. An analytical solution with a Fourier series is given by Kluge and Neugebauer [55].

For high mass flux conditions, which may occur in micro processes, correction factors are proposed by Geankoplis [66] Chapter 7. Further interesting examples are given by Fulford and Broadbridge [369] for diffusion of heat and matter in convective flow systems.

5.3 Convective Mass Transport

5.3.1 Analogy between heat and mass transfer

Species transport in convective flow can be described by the following mechanisms: diffusion (Eqs. 5.1 and 5.2), convection in the flow direction (Eq. 5.23), and convective transport in radial direction (Eq. 5.24):

$$\dot{n}_i = -\rho_m \frac{A}{s} D_{ij} \Delta c_i, \qquad (5.23)$$

$$\dot{n}_i = \rho_m A w c_i, \qquad (5.24)$$

respectively. For more complex transport processes or flow situations, the rate equation for mass transfer is expressed with a mass transfer coefficient β, similar to the heat transfer coefficient

$$\dot{n}_i = \beta_i \rho_m A \Delta c_i = \beta_i \rho_m A \left(c_{i,W} - c_{i,\text{bulk}} \right). \qquad (5.25)$$

There are no generally valid solutions for the transport equations, however, solutions exist in special cases for the momentum, heat, and mass transfer, see for example [213]. For a given process, some simplifications will help to find analytical solutions, see also Baerns et al. [135] p. 82. Numerical solutions are suitable for more complex systems, but also work with simplifications.

The analogy between heat and mass transfer with similar fundamental equations gives for similar geometrical and process conditions the same solutions. For laminar flow in a straight channel, the transfer of species and heat from the wall into the bulk flow can be solved analytically [135] p. 82, [89] Chapter 3.8, and is described by dimensionless parameters. The dimensionless heat transfer coefficient, the Nußelt number Nu, is determined for laminar flow in three cases: a) in a pipe with constant heat flux $Nu_q = 4.36$; b) in a pipe with constant wall temperature $Nu_T = 3.6568$; c) over a plate $Nu_z = 0.664\,Re^{1/2}\,Pr^{1/3}$, see also 4.1.3. These equations are valid for thermal and hydrodynamic developed flow. For mass transfer from the wall to the bulk of the fluid, the dimensionless mass transfer coefficient, the Sherwood number Sh, is determined in an analogous way, e.g. for a plate

$$Sh_z = \frac{\beta_{i,z}\,z}{D_{ij}} = 0.664\,Re^{1/2}Sc^{1/3}. \tag{5.26}$$

By elimination of the Re number, the analogy between the heat and mass transfer can be expressed as

$$Sh_z = Nu\left(\frac{Sc}{Pr}\right)^{1/3} = Nu\,Le^{1/3} \tag{5.27}$$

for laminar flow, where the Lewis number $Le = a/D$ describes the ratio between the heat and mass transfer, see Section 2.3.4. The turbulent transport in mixing can be expressed by mixing length theory or engulfment theory, see [124], for example. Mixing from the wall into the bulk of the fluid is important for catalytic wall reactions, adsorption processes at the wall, or precipitation at the wall, and is described in Section 6.2.

5.3.2 Mixing time scales and chemical reactions

Besides the mixing time scale, the flow through a device and the transformation in a device have their own time characteristics. The channel length divided by the mean velocity determines the mean residence time t_P of a fluid in a straight channel:

$$t_P = \frac{l}{\bar{w}} \tag{5.28}$$

With equipment miniaturization, the channels become shorter and the mean residence time t_P decreases. Considering laminar flow with mixing solely through diffusion, the mixing time t_D also decreases with the channel width b or hydraulic diameter d_h

$$t_D \propto \frac{b^2}{D} \propto \frac{d_h^2}{D}. \tag{5.29}$$

The ratio of the mixing time t_D to the residence time t_P indicates the degree of mixing in a channel by diffusion. If the ratio is smaller than one, diffusion has sufficient time to mix completely in the channel.

$$\frac{t_D}{t_P} = \frac{d_h^2}{D}\frac{\bar{w}}{l} = \frac{d_h}{l}\text{ Re} \cdot \text{Sc} = \text{Pe}\frac{d_h}{l} \tag{5.30}$$

The introduction of the Reynolds number Re (with the hydraulic diameter d_h as the characteristic length), the Schmidt number Sc, and their product, the Péclet number Pe, gives the conclusion that high Pe numbers lead to incomplete mixing. For increasing velocities and, therefore, higher Re or Pe numbers, the mixing quality α_m decreases proportionally to 1/Re, see also Figure 5.10. A smaller channel with constant length will increase the mixing quality due to the shorter diffusion length.

Typical diffusion lengths and times are shown in Figure 5.8, lower part, together with characteristic length and time scales of mixing equipment in the upper part. For given length scale, typical diffusion times in gases are well below 10^{-3} s in microstructured devices, see also Figure 1.2. For liquids with lower diffusion coefficient, typical diffusion times are in the range of 10^{-3} s for convective micromixer, however, they are reasonably longer in straight laminar flow. Additionally, typical length scales of conventional and microstructured process equipment are displayed with the related length scales of mixing processes and time scales of flow pulsations described in Section 3.2.4. The length scale of macro-mixing, which is convective mixing in turbulent eddies, is in the same range than the typical geometric structures of micromixers or microreactors. This is one reason, why length scales of meso-mixing in conventional equipment, see Section 5.4.5, are similar to length scales of convective mixing in microstructured equipment. Consequently, meso-mixing in microstructured equipment covers only a short range, until micro-mixing with molecular diffusion starts to dominate. In convective micromixers, small fluidic structures down to a few micrometer are produced leading to rapid mixing, while relatively large channel structures serve for high flow rates and low fouling potential.

In the upper part of Figure 5.8, typical frequencies of flow pulsations, such as vortex shedding behind a cylinder or vortex break down in T-mixers, are given for gas and liquid flow in channels with typical length dimensions. For mixing processes, a frequency range from 1 to 1 000 Hz has a major influence, which is generated in channel elements with typical length scales from 1 to 20 mm and from 100 μm to 7 mm for gas and liquid flow, respectively. In microstructures, gas flow produce high frequencies, while liquid flows have frequencies around 1 000 Hz, which have a large influence on the mixing process, see Section 5.4.

Time and length scales in conventional equipment originate from turbulent transport processes. Standard mixing characterization, such as the model presented by Bałdyga and Bourne, see [124], is based on turbulent transport. Hence, a short excursion in turbulent flow description is beneficial for a better understanding of different time and length scale models. The outline follows the introduction of Foias et al. [234].

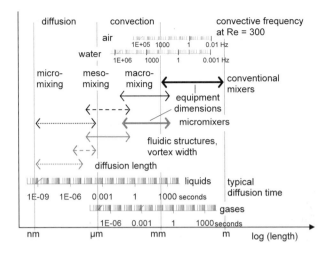

Fig. 5.8. Typical length scales of equipment, fluid structures and molecular processes, indicating the opportunity in microstructures to get closer to the molecular scale.

Turbulent flows show self-similar statistical properties at length scales smaller than the scale at which energy is provided to the flow. Conventionally, turbulent flow is visualized as a cascade of large eddies, large-scale structures of the flow, breaking up successively into ever smaller-sized eddies and fine scale structures. This cascade, or flow of kinetic energy from large to small scales, is thought to occur at length scales, where the viscosity has no major influence. Energy dissipation, the transfer of energy from one length scale to a smaller one, results solely from the presence of the nonlinear (inertial) term $(\vec{w} \cdot \mathrm{div})\vec{w}$ in the Navier-Stokes equations, see Section 3.1.1. The energy dissipation rate ε, the dissipation per mass and time, is constant in space and time for isotropic turbulence. Additionally, the energy dissipation rate is relative on every length scale with dominant inertial fluid motion until the molecular dissipation begins to dominate. That length scale, denoted by l_K as the Kolmogorov length scale, is at the end of the inertial range and the beginning of the viscous dissipation range.

To determine the Kolmogorov length and time scale, l_K and t_K, respectively, dimensional analysis and order-of-magnitude estimation assist in combining the main parameters of energy dissipation ε, viscosity ν, and velocity w. Dimensional analysis, see Section 2.4.1, gives for the energy dissipation

$$\varepsilon = L^2/T^3. \tag{5.31}$$

and for the velocity

$$w = L/T. \tag{5.32}$$

The dissipation length l_K is where the viscous term of the Navier-Stokes equation

$$\nu \,\mathrm{div}(\mathrm{grad}\vec{w}) = \nu \Delta \vec{w} \propto \frac{\nu \vec{w}}{L^2} = \frac{\nu}{L\,T} \tag{5.33}$$

starts to dominate the inertial term

$$(\vec{w} \cdot \text{div})\vec{w} \propto \frac{\vec{w}}{L} = \frac{L}{T^2}, \tag{5.34}$$

and is given by the following inequality

$$v\text{div}(\text{grad } \vec{w}) > (\vec{w} \cdot \text{div}) \, \vec{w} \quad \Rightarrow \quad \frac{v}{LT} > \frac{L}{T^2}. \tag{5.35}$$

Therefore,

$$L^2 < v \, T, \text{ and with } T = \left(\frac{L^2}{\varepsilon}\right)^{1/3} \text{ follows}$$

$$L^{4/3} < \left(\frac{v^3}{\varepsilon}\right)^{1/3} \tag{5.36}$$

and for the Kolmogorov length

$$l_K = \left(\frac{v^3}{\varepsilon}\right)^{1/4}. \tag{5.37}$$

Kolmogorov stated that, in three-dimensional turbulent flows, the eddies with a length sensibly shorter than l_K are of no dynamic consequence. The relevant time scale for dissipative effects is similarly found from dimensional analysis to

$$t_K = \left(\frac{v}{\varepsilon}\right)^{1/2}. \tag{5.38}$$

Further discussion on this topic with statistical treatment of turbulent flow is given by Foias et al. [234]. Mass transfer processes have a typical length and time scale according to Fick's law or the Einstein-Smoluchovski equation, see Section 1.3,

$$l^2 = D \, t. \tag{5.39}$$

Transport processes of momentum and species have the same time scale, t_K and t_B, respectively, however, the length scale is altered by the fluid properties, viscosity v, and diffusion coefficient D. The relevant mass transfer length scale l_B, the so-called Batchelor length scale, is determined with the inertia/dissipation time scale

$$t_B = t_K = \left(\frac{v}{\varepsilon}\right)^{1/2} = \frac{l_B^2}{D}. \tag{5.40}$$

Elimination of the energy dissipation rate

$$\varepsilon = \frac{v^3}{l_K^4} \tag{5.41}$$

gives for the mass transfer length scale in turbulent flow

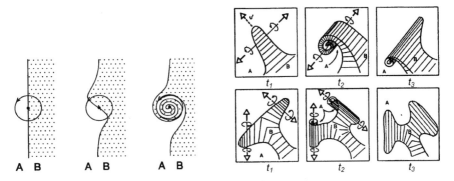

Fig. 5.9. Engulfment of fluid segments; left: formation of lamellar structure through vortex action; right: rolling in of a fluid segment or elongated tongue, adapted from Cybulski et al. [370].

$$l_B = \left(\frac{D}{\nu}\right)^{1/2} l_K. \tag{5.42}$$

With the Schmidt number $\mathrm{Sc} = \nu/D$, the smallest length scale of diffusive mass transfer in turbulent flow relative to the momentum length scale is

$$l_B = \mathrm{Sc}^{-1/2} l_K. \tag{5.43}$$

This length scale is coupled with the Kolmogorov length scale by the inverse square root of the Sc number. The Batchelor micro-scale physically represents a size of the region, within which a molecule moves due to diffusional forces. Outside these regions on larger scales, molecules move due to convective effects, turbulent fluctuations, or laminar striation motion. For three-dimensional turbulent flow, Cybulski et al. [370] propose the correlation $l_B = \mathrm{Sc}^{-1/3} l_K$ between momentum and mass transfer in mixing that leads to a coarser length scale for mass transfer. The actual physical problem determines the typical length scale: in fully developed, three-dimensional, isotropic turbulence, the diffusive mass transfer occurs on length scales proportional to $\mathrm{Sc}^{-1/3} l_K$; in elongated, stretched flow, such as those in convective micromixers (T-mixer with Re = 200 to 1 000), the diffusive mass transfer is dominant on length scales proportional to $\mathrm{Sc}^{-1/2} l_K$. Typical fluid motion is displayed in Figure 5.9 for turbulent flow segments. Further information on turbulent mixing can be found in the comprehensive work of Bałdyga and Bourne [371].

For gases with Sc numbers in the order of O(1), mass transfer occurs on the same length scale as the momentum transfer and convective flow. Liquids with Sc number in the order of O(1 000) or even higher, exhibit very short mass transfer length scales. Numerical grids for mixing simulation must be very fine to correctly resolve the mass transfer length scale. Experimental measurement of concentration profiles in liquid mixing must be on the Batchelor length scale to give the correct physical representation. Large numerical grids or rough experimental measurements give an indication or data, which needs to be handled carefully.

To overcome the difficulties of short length scales in simulation, sub-grid and lumped models have been developed to describe the mixing behavior and characteristics of technical systems. Often, typical characteristic times are taken to compare different mixing processes and chemical reactions with characteristic time $t_R \propto f(k, c)$. For example, the dispersion of a small fluid stream into a large vessel is observed until the concentration variation is lower than 0.01. The corresponding characteristic time t_{99} for blending in a stirred vessel can be described by the following empirical correlation [370]

$$t_{99} = C \cdot \frac{1}{N} \cdot \left(\frac{D_r}{D_s}\right)^a \cdot \left(\frac{H_r}{D_r}\right)^b \tag{5.44}$$

with the stirrer speed and diameter N and D_s, and the vessel height and diameter, H_r and D_r, respectively. The empirical constant C and the exponents a and b must be determined from experimental data.

The disintegration of an injected stream into turbulent pipe flow, also referred to as meso-mixing by turbulence, is described by

$$t_D = \frac{\dot{V}_A}{\bar{w}_{in} D_{\text{eff}}} \tag{5.45}$$

with the volume flow rate $\dot{V}_A$ of the component A and the turbulent, effective diffusivity D_{eff}. The main effort is focused on determining the turbulent diffusivity, which comes from experimental data. The characteristic time constant for meso-mixing by the inertial-convective process of injection nozzle mixing is given by

$$t_{ms} = \frac{3}{2} \left(\frac{5}{\pi}\right)^{2/3} \left(\frac{L_c^{2/3}}{\varepsilon^{1/3}}\right) \tag{5.46}$$

where L_c is the integral scale of concentration fluctuations in the vicinity of the feed pipe. This scale is initially proportional to $L_c \propto \sqrt{\dot{V}_f / w_f}$, with the feed flow rate $\dot{V}_f$ and the feed velocity w_f and describes the disintegration of the injected stream.

The mixing process with a length scale of the boundary between convective and diffusive mixing is called micro-mixing, and is relative to the Batchelor length scale. The time scale of final homogenization of the components is given by the diffusion time within the smallest length scale. Bałdyga and coworkers give the following time scale estimation for micro-mixing

$$t_{Ds} \propto \left(\frac{\nu}{\varepsilon_l}\right)^{1/2} \ln (\text{Sc}) \tag{5.47}$$

or more precisely defined, see Cybulski et al. [370]

$$t_{Ds} \propto \left(\frac{\nu}{\varepsilon_l}\right)^{1/2} \text{arsinh}(0.05 \, \text{Sc}) \tag{5.48}$$

with ε_l as the local rate of energy dissipation into the fluid. Velocity fluctuations on smaller length scales result from local vortices with relatively short lifespans.

The vortex gradually deforms and decays, and finally loses its shape, returning to the state of isotropy. The vortex formation at the interface of two components is called "engulfment", see Figure 5.9 left side. During the vortex lifespan, the size of the laminated structure l increases up to 12 times the Kolmogorov micro-scale. Hence, the characteristic time is given for the engulfment process in turbulent flow by Bałdyga and Bourne [371]

$$t_E = 12 \left(\frac{v}{\varepsilon_l} \right)^{1/2} \operatorname{arsinh}(0.05 \; \text{Sc}), \tag{5.49}$$

which is also the characteristic time of micro-mixing. The local energy dissipation rate ε_l is determined with the help of local velocity fluctuations, for example, in a stirred vessel or in a conventional static in-line mixer. The integral energy dissipation rate ε is the power of the mixing device related to the fluid mass, in which the energy is dissipated. For a stirred vessel, ε is the ratio of the stirrer power to the fluid mass in the vessel. A continuous flow static mixer is driven by the fluid pressure loss Δp, hence, the energy dissipation rate is given by

$$\varepsilon = \frac{\Delta p \, \dot{V}}{m} = \frac{\Delta p \, \bar{w}}{L_C \, \rho} \tag{5.50}$$

with the volume flow rate $\dot{V}$ through the mixer and the mass m within the mixer.

The engulfment process is governed by the growth of turbulent eddies with a certain rate. The engulfment rate constant is given by

$$E = 0.05776 \left(\frac{\varepsilon_l}{v} \right)^{1/2}, \tag{5.51}$$

the inverse of E gives the characteristic engulfment time

$$E^{-1} = t_E = 17.24 \left(\frac{v}{\varepsilon_l} \right)^{1/2}. \tag{5.52}$$

The relative importance of inertial-convective meso-mixing and micro-mixing by engulfment can be evaluated using the parameter M, introduced by Bałdyga et al. see [370]

$$M = E \cdot t_{ms} = \frac{t_{ms}}{t_E}. \tag{5.53}$$

Inertial-convective meso-mixing controls the process if $M \gg 1$, micro-mixing limitations prevail if $M \ll 1$, and both are of similar significance if $M \approx 1$.

To describe chemical reactions in continuous flow systems, the time scale of the chemical reaction must also be considered. The concentration development of an educt or product over time depends on the reaction order m and the kinetic reaction coefficient k. For an mth-order reaction, the reaction rate is determined with the concentration c_j

$$\frac{\mathrm{d}\,c_j}{\mathrm{d}t} \propto r = k\,c_j^m. \tag{5.54}$$

Other reaction types can be found in literature, for example, Perry's Chemical Engineering Handbook [41] Chapter 7. The characteristic time t_R for a reaction with constant volume can be determined with an order-of-magnitude estimation to

$$t_R \propto \frac{1}{k\,c_j^{m-1}} \tag{5.55}$$

Other methods are working with the reaction half time or the characteristic time of 90 % or 95 % of reaction progress. Comparing the time scales and the corresponding dimensionless ratios, the Damköhler number Da, is very important for the design of chemical reactors, hence, for microreactors, see Section 6.3. The design of mass transfer equipment requires close inspection of the local transport processes, fluid distribution, and pressure loss [372]. The entire apparatus can be described by integral parameters of an effective diffusion coefficient D_{eff} and a logarithmic concentration difference $\mathrm{d}c_{\log}$, comparable to heat transfer equipment. The method deriving integral parameters is described in the following section.

5.4 Characteristics of Convective Micromixers

Convective micromixers are those where the mixing process is induced by arising vortices caused by centrifugal forces in bent flow, also known as "secondary vortices", for example, T-shaped mixers [373], L-bends, or S-bends [258]. The advantage of these micromixers is the fast mixing process combined with a relatively high throughput, achieved by using large microchannels with characteristic dimensions of several 100 μm at Re numbers larger than 200.

A large number of micromixers use various internals or wall structures in straight channels (passive mixers) and external forces (active mixers) to induce secondary flow structures. A good overview of the flow principles of chaotic advection and the geometrical arrangement of the internals can be found in [357]. In the context of micro process engineering, chaotic advection is the splitting, stretching, folding, and breaking down of the flow by internal elements at low or intermediate Re numbers. Nguyen [357] distinguishes three different ranges of Re number: Re > 100 as high, 10 < Re < 100 as intermediate, and Re < 10 as low. For high Re numbers, the internals are obstacles on the wall or in the channel, as well as meandering or zig-zag-shaped channels. The pressure loss caused by the internals is one important design issue for determining the geometrical shape.

A mixing device can generally be divided into two parts, the *contacting elements*, such as T- or Y-shaped junctions, and the solely *mixing elements* like straight channels, bends, curves, joints, or internals of the channel. According to fluid dynamics, the mixing mechanism consists of two steps. The initial two lamellae or compartments are split and stretched into a multitude of smaller lamellae by convective vorticity and elongation. At the same time, diffusive mass transfer occurs between the lamellae, which is enhanced by further thinning of lamellae. The time scales of these two mixing steps are very similar, so a clear distinction cannot be made between convective and diffusive mixing and mass transfer. The time scale for mixing of liquids is in the range of milliseconds and below, due to the very small characteristic length of fluid lamellae. Hence, this model is only valid for micromixers, because the timescales for convection and diffusion are the same. More complex contacting elements for micromixers and their arrangement on a silicon chip can be found in Section 5.7 and [374]. The optimized combination of contacting elements, meandering channels, and their arrangement on a single silicon chip was investigated by Kockmann et al. [375]. This section gives a comprehensive description of liquid mixing behavior in passive convective micromixers. The mixing time is derived from the Re number and specific energy dissipation, which is compared to turbulent mixing in conventional batch reactors.

5.4.1 Mixing behavior in 90°bends

Convection in microchannels plays a major role for micromixers with high mass flow rates, see [374]. The convective effects are induced by bent, curved, or differently structured complete channel parts. Additional elements, such as posts or grooves in microchannels, also induce secondary flow structures, however, the main channel remains straight. These secondary flow structures are also called advection flow.

In Figure 5.10, the mixing of two components with an equal fraction (1:1 mixture) is numerically calculated for a channel with square cross section ($100 \times 100 \ \mu m^2$) and a 90° bend. The mixing quality α_m is determined after a constant mixing channel length of 1 000 μm, see [376]. Besides diffusion, the two components mix by convective effects for Re numbers higher than approx. 10, see also Section 3.2. For lower Re numbers, the straight laminar flow allows only diffusive mixing. The mixing quality α_m at constant mixing channel length is proportional to the inverse of the Re number due to the residence time of the fluid, see Eq. 5.30. Smaller channels produce higher mixing quality due to their short diffusion length, see Figure 5.10 left side.

Due to a convective enlargement of the interface between the components, the mixing quality α_m increases with the Re number for Re > 10. The centrifugal force from the curved flow forces faster fluid parts from the middle of the channel to the outer wall of the bend and elongates the component interface. The mushroom-like interface structure is clearly visible in Figure 5.10, right side.

The Re number and the aspect ratio of the channel is important for the mixing quality. The channel width is now included in the Re number. This shows that the convection in microchannels is scale-invariant. The mixing in curved microchannels displays similar behavior, as shown by Jiang et al. [258] and is described in Section 3.2.3.

The mixing quality in the single 90° bend (L-mixer) increases from 0.1 for low Re numbers up to 0.45 for a Re number larger than 270, see Figure 5.11. The combination of two 90° bends (S-mixer) increases the mixer performance, a Re number of approx. 270 results in a mixing quality close to 0.65. The combination of four 90°

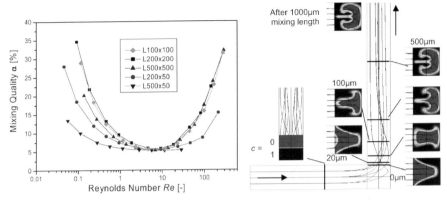

Fig. 5.10. Left: Mixing quality α_m after the 90° bend of five various mixers (L-shaped, channel width×depth) over the Re number; Right: Mixer geometry (cross section 100×100 μm^2) and concentration profiles at various channel locations, Re = 99, w = 0.85 m/s, [26].

bends (U-mixer) raises the mixing quality above 0.7 for high Re numbers. At these flow rates the difference between the S- and the U-mixer decreases and will vanish at higher flow rates. The mixing quality is determined at the channel outlet and can be directly compared, due to the relaminarized flow in the mixing channel with a length of 4 800 µm. The convective mixing process is completed at this point and the effect of diffusive mixing is comparable for all arrangements.

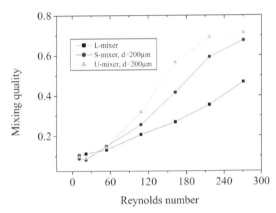

Fig. 5.11. Simulated mixing quality at the outlet of the 90° bend mixer (4 800 µm mixing channel length) with three combinations over the Re number, L-mixer has one 90° bend, S-mixer has two 90° bends with a distance of d = 200 µm, U-mixer has four 90° bends, as shown in Figure 3.7.

5.4.2 Mixing behavior of T-shaped micromixers

T-shaped micromixers are very suitable for fundamental investigation of mixing processes in convective micromixers. One of the first systematic investigation was made by Bökenkamp et al. [377], who showed very fast mixing times in T-shaped mixing devices. Gobby et al. [274] presented a study of gas phase mixing in T-shaped micromixers with low Re numbers, while Wong et al. [378] investigated liquid-phase mixing in a experimental and numerical study. The first, more systematic investigations were presented by Engler et al. [379, 373] and Hoffmann et al. [380], showing different flow regimes depending on the Re number, see also Chapter 3.2.3. For Re < 10, the flow shows straight symmetric streamlines and diffusive mixing at the interface of both components. With increasing Re number, the streamlines indicate four symmetric vortices at the entrance of the mixing channel and diffusive mixing, yet. Above a certain critical Re number, the mixing quality suddenly increases in T-shaped micromixers due to the break-down of flow symmetry. This has been named "engulfment flow" due to the similar flow structure in turbulent mixing, see Figure 5.9. Engulfment flow in experimental investigations of T-shaped micromixers has also been presented by Wong et al. [378].

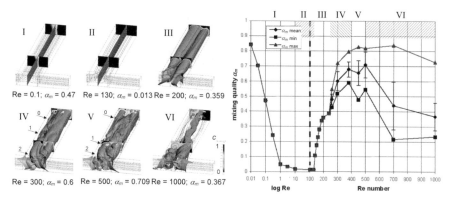

Fig. 5.12. Overview of the flow regimes in laminar T-junction flow with 1:1 mixing ratio. Development of the mixing quality with the Re number in the mixing channel, see also Figure A.19. The mixing quality α_m is determined at a constant length $l = 5 \cdot d_h$ of the mixing channel, here $l = 2\,000$ μm.

In Section 3.2, the flow structures within T-shaped micromixer devices were presented, classified into six regimes with Re numbers below 1 000 in the mixing channel. This classification is also displayed in Figure 5.12 with typical flow situations and the development of the mixing quality with Re number.

- Re < 10, Dn < 10, regime I
  straight laminar flow, steady velocity profile with straight streamlines, diffusion dominates the mass transfer;

- 10 < Re < 130 (approx.), Dn > 10, regime II
  symmetric vortex flow, Dean flow with characteristic Dn number;
- 130 < Re < 240 (approx.), regime III
  engulfment flow, break down of the flow symmetry, fluid swaps to the opposite side;
- 240 < Re < 400 (approx.), regime IV
  regular, *periodic pulsating flow*, reproducible vortex break down;
- 400 < Re < 500 (approx.), regime V
  quasi-periodic pulsating flow, vortex breakdown, wider distribution of mixing quality α_m;
- 500 < Re (approx.), regime VI
  chaotic pulsating flow, vortex breakdown, decrease of mixing quality and single flushes of fluid swapping to the opposite side.

A detailed description of the flow regimes was given in Section 3.2.4. Similar behavior is exhibited by the wake flow behind a cylinder, given in Figure 3.23. Deshmukh and Vlachos [278] employ this transient flow regime for mixing enhancement in a microreactor for catalytic reactions. Flow fluctuations result in varying mixing quality, which has also an influence on the performance of chemical reactions within the T-mixer. An interesting comparison is given in Figure 5.13, where the development of the mixing quality with the Re number in the mixing channel is displayed parallel to the segregation index, the standardized selectivity of the iodide-iodate reaction, see Section 6.3.6. The segregation index is inversely proportional to the mixing quality,

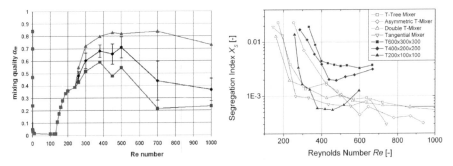

Fig. 5.13. Left: Mixing quality at a length of 2 mm into the mixing channel of a T-mixer T600×300×300 over the Re number in the mixing channel; Right: Experimental data from the iodide-iodate-reaction (Villermaux-Dushman). The segregation index is inverse proportional to the mixing quality α_m, [374] and is explained in Section 6.3.6.

a low value means low iodine concentration and high reaction selectivity. The selectivity is used to characterize the mixing quality of the process. A high selectivity means a fast and homogeneous mixing process. Figure 5.13 clearly indicates the different flow regimes and the pulsating flow on the left side from simulation, which is also confirmed by the selectivity. The selectivity in the T-mixer (T600×300×300) reaches a constant level at Re numbers of approx. Re = 400. In smaller mixers, the segregation reaches a constant level and even increases again with increasing Re number, indicating the influence of the transient flow regimes.

The mixing in T-junctions at sufficiently high Re numbers (> approx. 150, depending on the geometry) is a combination of convection and diffusion effects. The interplay of the vortex generation with the interface between the components can be described by characteristic points, which also indicate the mixing process. In the framework of chaotic advection, Ottino [125] defines two characteristic points, the elliptic point with converging flow and the hyperbolic point with diverging flow, see also Section 5.5. Both can be found in the concentration profile of the T-shaped micromixer. The elliptic point is located in the center of the vortex (2 points in the engulfment flow) and indicates a stable flow with contracting fluid lamellae (mixing by thinning). The hyperbolic point is located in the center of the mixing channel between the two vortices. This point is combined with converging flow and an elongation effect of the fluid lamellae (mixing by stretching). The location of these characteristic points and the location of the interface between the components are very important for the mixing process. Further information about chaotic advection and mixing is given in Section 5.5.

The engulfment flow regime begins with rolling-up of two parallel vortices reaching into the mixing channel. With increasing Re number, multiple smaller vortices appear in the vicinity of the two major vortices. The numerical grid of this simulation results does not allow a resolution of the fine vortex and filament structure. The working group of Prof. Manhart at TU Munich presented concentration profiles for T-mixers T200×100×100 with high resolution due to the combination of CFD simulation with particle tracking methods [381]. In Figure 5.14, concentration profiles

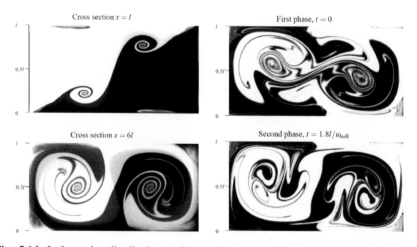

Fig. 5.14. Left: scalar distribution at Re = 186, Sc = 3571 in a T-shaped micromixer T200×100×100 with channel height l; Right: unsteady scalar field at Re = 240, Sc = 3571, $z = 6\,l = 4.5\,d_h$; depicted at two different phases normalized to the residence time in the mixing channel, see also [381].

at Re = 186 and Re = 240 are displayed, indicating the complex flow profile with steady laminar flow. The particle tracking method is combined with a diffusion process of fine particles following a streamline. This method gives an impression for the particle distribution during the mixing process, however, the actual physical process is not correctly reflected. The mutual influence of adjacent streamlines with different concentrations along their way through the device is not covered, hence, the mixing history can not be calculated, which is important to incorporate chemical reactions and their temporal development.

The scalar was sampled on a grid with a cell size of $0.005l \times 0.005l \times 0.005l$, resulting in approx. $8 \cdot 10^4$ grid points on the mixing channel cross section. The number of particles traced was $36 \cdot 10^6$ with an effective particle density of 562.5 particles per cell. For details on the numerical scheme and particle tracking method, see [381]. For periodically fluctuating flow at Re = 240 and Sc = 3571, the scalar field is shown at only one cross section of the mixer at $z = 6 \cdot l = 4.5 \cdot d_{h,M}$. The scalar field is computed at different phases of the flow, which was close to the reference cross section in other simulations $z = 4 \cdot d_{h,M}$ or $5 \cdot d_{h,M}$.

In Figure 5.15, the mixing quality is displayed over the mixing channel length and over a period of 5 ms combined with corresponding CFD simulations of the pulsating flow regime. The cuts with concentrations profile within the mixing chan-

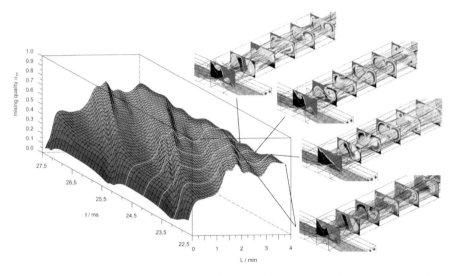

Fig. 5.15. Mixing quality α_m profile in a T-mixer T600×300×300 for Re = 300 in the mixing channel at various times. The cross sections in the mixing channel have a distance of 500 µm.

nel indicate the build-up and collapse of the major vortices reaching into the mixing channel. Between the two fluids swapping to the opposite side, a strong shear layer is built up, which causes a Kelvin-Helmholtz instability and periodic roll-up. Both vortices initially appear at the outer side of the shear layer close to the wall at the beginning of the mixing channel, then migrate to the middle axis, where they collapse. Simultaneously, a new vortex pair is formed in the shear layer close to the wall and moves to the center again. This periodic process leads to a pigtail-like concentration profile within the mixing channel including well-mixed areas but also areas with undisturbed fluid segments. The mixing quality spreads with increasing Re number, however, the mean mixing quality decreases, as shown also in Figure 5.16.

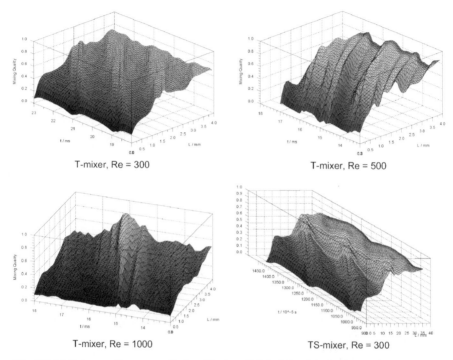

Fig. 5.16. Mixing quality α_m profile in a T-mixer T600×300×300 for Re = 300, 500, 1 000, and for a TS-mixer with Re = 300 in the mixing channel at various times.

Three-dimensional images of the mixing quality over the mixing channel length and a typical period are given in Figure 5.16 for three different Re numbers in a T-shaped micromixer and for the TS-mixer, a T-mixer with two adjacent 90° bends with $R = b_M$. The TS-mixer is displayed with other mixer combinations in Figure 5.17. The relatively even distribution of the mixing quality in the mixing channel for Re = 300 is disturbed with increasing Re number. The mixing quality at Re = 500 shows higher peaks, but also higher gradients, while the minimal value remains almost constant. At Re = 1 000, one high peak is achieved, however, at other times, the mixing quality is relatively low. To improve the mixing quality, combinations of mixing elements have been proposed and realized, see the next section and Section 5.7.

5.4.3 T-shaped micromixers and combinations

The mixing quality in T-shaped micromixers reach values up to 0.8 on a relatively coarse numerical grid, however, no complete mixed state is achieved, which is nearly impossible with numerical simulations [366]. Further mixing elements are added to the contact element to enhance the mixing process, see Figure 5.17. The mixing qua-

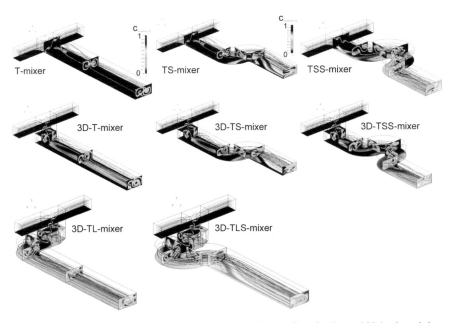

Fig. 5.17. Geometry with streamlines and concentration profiles for Re = 200 in the mixing channel of a T-, TS-, and TSS-mixer, the 3D-T-, 3D-TS-, and 3D-TSS-mixer, as well as the 3D-TL- and 3D-TLS-mixer, see also Figure A.11.

lity of the TS-mixer, given in Figure 5.16 right bottom, is similar to the distribution of the simple T-mixer at the beginning of the mixing channel. With the two bends, the flow is mixed a second time leading to a smooth distribution at the channel outlet. The mean mixing quality is approx. 0.1 higher than the value of the simple T-mixer, representing an enhancement of almost 15 %.

In Figure 5.17, eight further, combined mixers are displayed. They can be classified into three groups: the in-plane mixers of T-, TS-, and TSS-mixer, the 3D-T-, 3D-TS-, and 3D-TSS-mixer, as well as the 3D-TL- and 3D-TLS-mixer. This classification is based on three basic elements, the T-, the L- and the S-mixer, which are combined to various mixer configurations. The 3D-T-mixer is a simple T-shaped junction with adjacent rectangular 90° bend for three-dimensional arrangement of mixing devices. This mixer type is also characterized in Section 5.7. The 3D-TL-mixer is a variation of the 3D-T-mixer where the L-bend turns in the direction of one inlet channel. A second curved bend leads the mixing channel in the conventional direction of the other mixer types. This combination of bends shifts the hyperbolic point after the T-mixer to the channel wall and produces two elliptic points at the component interface, which leads to enhanced mixing. The TS- and the TSS-mixer consist of a simple T-mixer with two or four curved 90° bends for minimum pressure loss compared to rectangular bends, see Section 5.7. Other combinations can similarly be described. The mixing quality over the mixing channel length of eight devices is compared in Figure 5.18 for Re = 200.

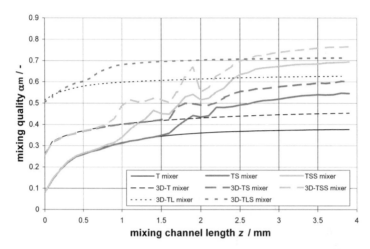

Fig. 5.18. Comparison of the mixing quality along the mixing channel of simple and complex micromixers, see Figure 5.17. The mixing length starts after the L-bend at the beginning of the mixing channel.

The T-mixer displays the base line of mixing quality along the mixing channel, where the impact of other devices are clearly visible. The 3D-T-mixer starts with a higher mixing quality, due to the L-bend, but closely follows the T-mixer. In the TS-mixer, additional vortices start to enhance the mixing process at 800 μm mixing channel length and soon exceed the 3D-T-mixer. The combination of the 3D-T-mixer with four bends produces the best mixing quality. Also for low Re numbers, the combination of a T-mixer as contacting element and further channel elements is the best approach for complete mixing. Transient flow regimes diminish the mixing quality with increasing Re number in simple T-mixers, however, they enhance the mixing quality in combined mixing elements, which have a dampening effect on the peaks and lead to increased mixing quality. The above values for mixing quality are only an indication, due to numerical diffusion, see Section 5.7.1, however, they are beneficial for the appropriate design of efficient microstructured mixing devices.

In Section 5.1.4, the stoichiometric mixing, concentration deviation Δc, and the relative potential of diffusive mixing Φ_c have been introduced, which are now applied to the investigated mixer configurations. For each grid point at the exit of the mixer device, the relative potential Φ_c is displayed over the concentration deviation

Δc, which indicates the distance of the point from the stoichiometric mixture. The diagram for the T-mixer was already shown in Figure 5.5 for the cross section at a length of 4 mm. Many points are grouped around a curved line from $\Phi_c = 0.5$ to $\Delta c = 0.5$. Additionally, the potential of diffusive mixing is displayed in Figure 5.19 over the concentration deviation at the outlet cross section of the combined mixing elements. The points of the 3D-T-mixer are located under a curved line from $\Phi_c = 0.6$ to $\Delta c = 0.5$ and cover almost the whole area. The relative potential Φ_c is almost equal to the T-mixer, but the concentration deviation is about 0.05 lower, and therefore better. At the exit of the 3D-TL-mixer, the points are located under a curved line $\Phi_c = 0.4$ to $\Delta c = 0.45$ and the mean concentration deviation is approx. 0.15. Each additional bend enhances the mixing performance.

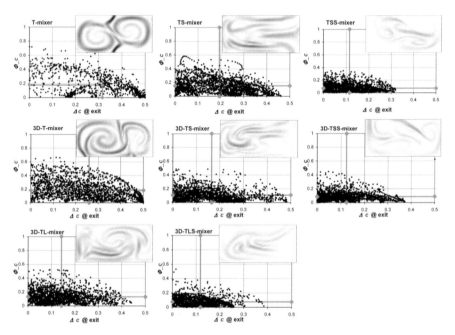

Fig. 5.19. Non-stoichiometric mixing and mixed-states for different mixer configurations at the exit of the mixing device. Mean values are indicated with the gray lines.

With increasing number of mixing elements from the adjacent S-bends, the concentration difference decreases as well as the relative potential for diffusive mixing. The points migrate toward the lower left corner closer to stoichiometric mixture. The TSS-mixer is the best mixer device with compact grouping of all point representing location of the entire cross section, due to some unmixed regions in the 3D-TSS-mixer. This is surprising, because the TSS-mixer has a lower mixing quality α_m at the exit than the 3D-TLS-mixer and the 3D-TSS-mixer, see Figure 5.18.

The effort of the mixing process, the pressure loss Δp at Re $= 200$ is also interesting to compare for the different mixer types. The T-mixer causes a pressure loss of 834 Pa, while the TSS-mixer has a almost double value of Δp with 1613 Pa. The pressure loss of the 3D-T-mixer and the 3D-TSS-mixer has a value of 1041 and 1852 Pa, respectively, while the 3D-TL-mixer and 3D-TLS-mixer have a pressure loss of 1260 and 1637 Pa, respectively. The pressure loss Δp in the different devices is similar to the mixing quality α_m, however, there is a discrepancy to the potential of diffusive mixing and concentration deviation for stoichiometric mixing.

With these results, it is important to characterize a mixer device with different parameters, for example, stoichiometric mixing is better described with the concentration deviation Δc instead of the concentration variance. It will be interesting to investigate and characterize further mixer concepts and devices with the proposed methods.

5.4.4 Mixing times of convective micromixers

The experimental determination and characterization of mixing quality in micromixers is still a subject under investigation, however, colored liquids [368] and some chemical reactions give good results. Fast pH-indicators with color change are used for flow indication [382]. For slow mixing processes, as in split-and-recombine (SAR) mixers, the metal complex reaction of iron rhodanide is used for the flow visualization [368].

For a quantitative characterization, a pH-based color reaction using Bromothymol Blue dye is implemented in two T-shaped micromixers made of silicon with similar, only scaled geometry (one with a mixing channel cross section of $600 \times 300\ \mu m^2$ and the other with a cross section $400 \times 200\ \mu m^2$). Details are given in [127]. Using this dye, the two inlet fluids appear colorless and green, respectively, while the mixed regions appear blue or dark gray. Applying a green color filter, the unmixed regions can be clearly distinguished from the mixed regions.

Optical measurements for high Re numbers are influenced by the shutter time of photographic equipment, meaning the high-frequency flow fluctuations can not be resolved and, hence, remain undetected. The optically measured mixing quality value is in the range of the maximum value from simulation, which indicates a constant mixing length for Re ≥ 350. From parallel chemical reactions, see Section 6.3.6, and from particle precipitation, see Section 6.5, flow fluctuations and decreasing mixing quality, higher segregation and non-stoichiometric mixed regimes are detected with higher Re numbers in the mixing channel.

Images of the mixing channel from above are made over a range of Re numbers from 190 to 1 000, see for example in Figure 5.20. At defined points inside the mixing channel, small slices of the image data are extracted, and the mean green color value is calculated using an image processing system. By plotting the resulting mean value over the position z in the mixing channel, see Figure 5.20 left, the data correlates very well with an exponential first order decay of the form

$$I_m(z) = I_{m,\infty}(1 - e^{-z/l_m}) \tag{5.56}$$

with l_m as the characteristic mixing length and $I_{m,\infty}$ as the asymptotic intensity value. The R_2 correlation values of the regression curves acc. to Eq. 5.56 lie between 0.975

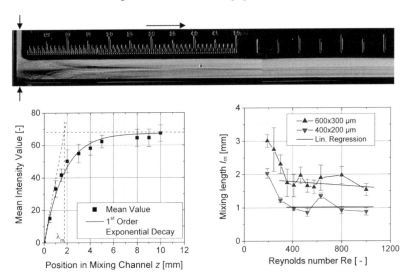

Fig. 5.20. An image processed photograph of the mixing channel of a T-mixer with a cross section of 600×300 µm showing different fluid lamellae. Below: Mean mixing intensity value over the mixing channel length of a T-mixer with a cross section of 600×300 µm; Left: at a Re number of 520, indicating the characteristic mixing length l_m; Right: characteristic mixing length over the Re number for two scaled T-mixers. More information are given by Engler [127]

and 0.995 for all experimental measurements. When evaluating the mixing length for the two investigated T-mixers over the Re number, one sees that the mixing length becomes almost constant for $Re > 350$, see Figure 5.20 right. It can therefore be stated that the mixing length l_m is a characteristic value for the quality of the mixing process of a convective micromixer. For the investigated T-mixers, it holds $l_m \approx 1.7$ mm for the 600×300 µm^2 T-mixer and $l_m \approx 1.0$ mm for the 400×200 µm^2 T-mixer. In regard to scale-up, the factor between these two mixing lengths is supposed to be 1.5. The deviations probably originate from a comparatively lower camera resolution for the smaller mixer. Using the hydraulic diameter of the two T-mixers, a characteristic mixing length of $l_m \approx 4d_h$ can be estimated.

Furthermore, using the mean velocity and normalizing the mixing degree between 0 and 1, the intensity of mixing over time can be expressed using Eq. 5.56:

$$I_m(t) = 1 - e^{-t/t_m} \tag{5.57}$$

where $t_m = l_m/\bar{w}$ is the characteristic mixing time. The asymptotic behavior of the mixing quality along the mixing channel has also been observed in static mixers with color dispersion, see Zlokarnik [364] Chapter 8. The exponential fitting function is the mathematical basis for characteristic parameters of the mixing process, such as time t_m or mixing length l_m. The characteristic mixing time t_m can be further expressed by means of the Re number using these results to

$$t_m = \frac{l_m}{\bar{w}} = \frac{l_m \cdot d_h}{v \cdot Re} =: \frac{d_h^2}{MP \cdot v} \cdot \frac{1}{Re} = const. \cdot Re^{-1} \tag{5.58}$$

with the mixing performance $MP = d_h/l_m$, see Section 5.4.7, as a factor expressing the performance of the mixing process inside the device. The larger the MP value, the shorter the characteristic mixing length is inside a certain mixing channel with respect to its hydraulic diameter, and the better the mixing process becomes. Therefore, the value can be interpreted as something of a quality factor for the specific mixing device. Table 5.1 gives typical mixing performance data for investigated mixer types, described in more detail in Section 5.7.

Table 5.1. Mixing performance for three of the investigated convective micromixers, see Section 5.7 and two commercially available micromixers from IMM [348].

Mixer Type	d_h [mm]	$l_{m,95}$ [mm]	MP [-]
T600×300×300	0.4	4.8	0.083
3D-T Mixer	0.3	1.3	0.23
Tangential Mixer	0.3	1.2	0.25
T-Tree Mixer	0.34	1.6	0.21
Caterpillar SAR (IMM)	1.2	350	0.0034
SuperFocus (IMM)	0.91	350	0.0026

Above a certain Re number, which can be regarded as the point, at which full homogenization is achieved (in this case Re > 350), the mixing time inside a convective micromixer is inversely proportional to the Re number, with a coefficient being dependent on the mixing performance, the mixer geometry, and the involved fluid.

As a simplification, the Convective Lamination Model [127] assumes that the fluid lamellae all have an equal width, which is reduced from the initial width $b_M/2$ to some smaller thickness, but limited to a minimum value due to the dampening of the vortices further down the mixing channel. Furthermore, the width becomes even smaller with higher Re numbers.

The final step in the convective mixing process is accomplished by diffusive mass transfer between the very thin filaments. Applying Einstein's equation (solution of Eq. 5.2) of the mean diffusion path on the final lamella width $d_{L,\infty}$ gives:

$$d_{L,\infty} \approx \sqrt{2 \cdot D \cdot t_m}, \tag{5.59}$$

where $D \approx 10^{-9}$ m²/s is the diffusion coefficient of the dye or solvent in water, and t_m is the mixing time. The influence of convective effects on the mixing time is neglected as a rough estimation. For example, inserting $t_m = 200$ µs for the best convective mixers with aqueous solutions gives a final lamella width of $d_{L,\infty} \approx 0.6$ µm.

This shows that the initial lamella width of $d_{L,0} = b_{IC} = 300$ μm is reduced by a factor of 500 by the convective effects and vortex generation. The convective lamination process is not only effective in the sense that a large number of very small lamellae is created, but also very fast, which is the reason for the extremely fast mixing times achieved in convective micromixers.

5.4.5 Energy dissipation into mixing

Mixing times in the range of 100 μs for aqueous solutions can be achieved by convective micromixers. This is only possible with an enormous specific energy input into the mixing process, which is determined in this section. The specific energy dissipation ε describes the energy input into the micromixer. It consist of three main parts:

1. the energy consumed by fluid friction at the channel walls,
2. the energy consumed by the creation of vortices and interface enlargement,
3. the energy consumed by internal dissipation without mixing, for examples, vortex generation, which does not disturb the interface between the components and only leading to viscous heating of the fluid.

For the mixing process, only the energy consumed by vortex creation and interface enlargement is relevant, because fluid friction and viscous heating do not contribute to the mixing process. Both, mixing and and viscous heating can not be exactly distinguished and are counted together for practical reasons. Thus, only the specific energy dissipation by vortices is used, which holds:

$$\varepsilon = \frac{\Delta p_V \cdot \dot{V}}{m} \tag{5.60}$$

where Δp_V is the pressure loss caused by vortex creation, $\dot{V}$ is the volume flow rate, and m is the mass, into which the energy is dissipated, see also Section 5.3.2.

The basic correlation of the pressure loss in a channel system is described by the simplified Bernoulli equation, see Section 1.4.2 and 3.1.3, consisting of straight channels ($\lambda_c l/d_h$) and channel fittings or components, such as junctions, bends, or cross section variations ζ.

$$\Delta p = \left(\sum \lambda_c \frac{L}{d_h} + \sum \zeta_{vortex} \right) \frac{\rho}{2} \bar{w}^2 \tag{5.61}$$

The friction force originates from the channels walls and is determined by

$$\Delta p = \sum \left(\frac{C_f}{Re} \frac{L}{d_h} \right) \cdot \frac{\rho}{2} \bar{w}^2 \tag{5.62}$$

For laminar flow, the resistance in a straight channel is inversely proportional to the Re number. The pressure loss caused by vortex creation is proportional to the square of the mean velocity for convective micromixers

$$\Delta p_V = \sum (\zeta_{vortex}) \cdot \frac{\rho}{2} \bar{w}^2. \tag{5.63}$$

The mass, into which the mean energy used for the mixing process is dissipated, can be closely estimated with the help of the characteristic mixing length

$$m = \rho \cdot V \approx \rho \cdot A_M \cdot l_m \tag{5.64}$$

with the cross section A_m of the mixing channel. Inserting Eq. 5.64 into 5.60 and defining the Euler number Eu_V for vortex pressure loss

$$Eu_V = \frac{\Delta p_V}{\rho \cdot \bar{w}^2} = \frac{\zeta_{vortex}}{2} = \text{const.} \tag{5.65}$$

gives for the mean specific energy dissipation

$$\bar{\varepsilon} = \frac{\Delta p_V \cdot A_M \cdot \bar{w}}{\rho \cdot A_M \cdot l_m} = \frac{Eu_V \cdot \bar{w}^3}{l_m} = \text{const.} \cdot \bar{w}^3. \tag{5.66}$$

The Euler number can be easily calculated from simulations of convective mixers (details are given in [127]). Results from investigated mixers, see Section 5.7, are summarized in Table 5.2.

Table 5.2. Euler vortex number, mixing length, and mean specific energy dissipation for three of the mixing elements and a T-mixer with comparable cross section [127].

Mixing Element	Eu_V [-]	l_m [mm]	$\bar{\varepsilon}$ @ Re= 1 000 [W/kg]
T-mixer 600×300	1.7	1.6	16 600
3D-T mixer	2.2	0.43	189 500
Tangential mixer	6.9	0.40	638 900
Double-T mixer	1.1	0.45	38 200

The mean specific energy dissipation $\bar{\varepsilon}$ into the mixing process is calculated for these mixers, which are displayed in Figure 5.21. Extremely high values of over 600 000 W/kg are achieved. This is many orders-of-magnitude higher than the mean energy dissipation in conventional batch reactors, where typical values of several 10 W/kg are achieved, see Cybulski et al. [370]. These high values are the main reason for the very short mixing times in convective micromixers and show the potential of these devices.

The correlation between mixing time and mean specific energy dissipation results from combining Eqs. 5.58 and 5.66:

$$t_m = \frac{l_m}{\bar{w}} = l_m \cdot \sqrt[3]{\frac{Eu_V}{l_m \cdot \bar{\varepsilon}}} = \sqrt[3]{\frac{Eu_V \cdot l_m^2}{\bar{\varepsilon}}} = \text{const.} \cdot \sqrt[3]{\frac{d_h^2}{\bar{\varepsilon}}}. \tag{5.67}$$

The proportional constant contains the geometry, pressure loss, and mixing performance of the investigated device. For the mixers shown in this study, it is calculated

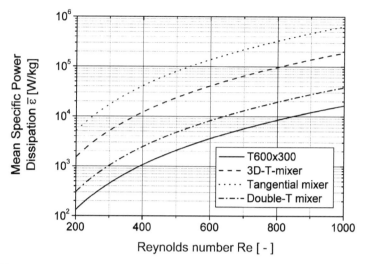

Fig. 5.21. Mean specific energy dissipation over the Re number for three of the optimized micromixers and a T-mixer with comparable cross section acc. to Eq. 5.66, see Engler [127].

from the preceding values (Table 5.2), which lie in the range from 1.6 to 3.2, depending on the specific device. The mixing time is inversely proportional to the third root of the mean specific energy dissipation. With increasing energy dissipation, the mixing times become shorter. Properties of the mixing device, like geometry and mixing performance MP, are included in the constant coefficient.

With these results, a very interesting comparison can be made with conventional mixing theory. The disintegration of a feed stream into a fully turbulent regime is called meso-mixing, which occurs when a fluid volume with characteristic dimensions of several millimeters is decreased in size by turbulent mixing. Following Bourne [124], the mixing time for meso-mixing is expressed with the typical geometrical dimension and the energy dissipation, see Eq. 5.46,

$$t_{m,\text{meso}} \approx 2 \cdot \sqrt[3]{\frac{d_{FS}^2}{\varepsilon}} \qquad (5.68)$$

where d_{FS} is the diameter of the feed stream. Thus, a strong correlation between Eq. 5.67 and Eq. 5.68 is apparent, which is surprising, as convective micromixers operate in the laminar regime, while Eq. 5.68 is valid for turbulent flow. Furthermore, this is a good indication for the high potential of convective mixing, as the specific energy input is similarly transformed into mixing, but the overall value of specific energy dissipation is orders-of-magnitudes higher in convective micromixers.

5.4.6 Mixing effectiveness

The characterization and comparison of different micromixers is a complex task due to their wide range of application. The Re number is one basic parameter for describing flow characteristics. Low Re numbers are typical for analytical applications with small fluid amounts where diffusion is the dominant transport process. Microreactors for high throughput screening or for production purposes operate with higher Re numbers and often use convective effects for mixing enhancement [357].

Another important parameter for characterizing the mixing process is the mixing time. Mixing is a transient process with asymptotic behavior of the mixing quality. For long mixing times, the mixing quality α_m tends toward one, and concentration gradients vanish. This asymptotic behavior can be described by a characteristic time, which is given by the gradient at the start of mixing or at a certain value of the mixing quality, for example, 95 %, see Figure 5.20. A short mixing time results from a high mass transfer rate and gives a high selectivity and yield for complex chemical reactions. This general statement needs to be adjusted to the actual reaction type, where other conditions may prevail.

In process engineering, static in-line mixers are characterized by the pressure loss, which occurs over the channel length with inserts [364], Chapter 8, [383], or Section 5.1.2. For Re > 100, the pressure loss in a channel with mixing elements is described by Eq. 5.61 with the friction factor of a straight channel C_f and the pressure loss coefficient ζ of channel elements, such as bends or T-junctions. The pressure loss coefficient of various in-line pipe or channel elements are combined to give the pressure loss for the entire system. This can be expressed as the dimensionless Euler number Eu, see also Eq. 5.65

$$\text{Eu} = \frac{\Delta p}{\rho \cdot \bar{w}^2} = \frac{1}{2}\left(\zeta + C_f \frac{l}{d_h}\right) \tag{5.69}$$

The pressure loss in the mixing channel is the effort required to achieve effective mixing. The revenue of a mixing device is a short mixing length with complete mixing of a volume flow rate, expressed with the Re number. The ratio of the revenue to the effort of a mixing device is called the mixer effectiveness ME_{II}, which was introduced by Engler et al. [374].

$$\text{ME}_{II} = \frac{\text{revenue}}{\text{effort}} = \frac{\text{Re}}{\text{Eu}} \frac{d_h}{l_m} \tag{5.70}$$

With the mixing length $l_m = \bar{w}\, t_m$ and almost square channel cross sections $A_m \approx d_h^2$, the mixer effectiveness can be written as

$$\text{ME}_{II} = \frac{\dot{m}^2}{\eta\, d_h^2\, \Delta p\, t_m} \tag{5.71}$$

Unfortunately, micromixers from literature are often not characterized in such a detail that they can be compared with the mixer effectiveness. From the effectiveness results of five micromixers, it can clearly be seen that diffusive mixers are less effective than convective mixing devices. The dissipated energy is mainly used for the generation of small fluid lamellae, which shortens the diffusion length. The performance data from three of the investigated micromixers are given in Table 5.3.

They are displayed together with the data of two commercially available micromixers from IMM, a split-and-recombine mixer and a multilamination-focusing micromixer [348]. The mixing performance MP of the investigated micromixers was already displayed in Section 5.4.4, Table 5.1. The mixing time for 95 % mixing $t_{m,95}$ has been chosen for better comparison. Convective mixers have very short mixing times and very low pressure losses, however, the mass flow rates are smaller than those of the commercially available micromixers.

Table 5.3. Performance data of three of the investigated convective micromixers and two commercially available micromixers from IMM [348]. The first three mixer types are combined with three 180° meandering channel bends, see Section 5.7.

Mixer Type	$\dot{m}$ [kg/h]	d_h [µm]	A [mm^2]	$t_{m,95}$ [ms]	Δp_{tot} [bar]	ME$_I$ [-]	ME$_{II}$ [-]
3D-T Mixer	20	300	0.09	0.3	1.2	0.514	9 526.0
Tangential Mixer	20	300	0.09	0.5	0.093	4.0	1 714.7
T-Tree Mixer	20	340	0.12	0.9	3.0	0.061	988.9
Caterpillar SAR (IMM)	90	1 200	1.2	20	17	0.00061	12.8
SuperFocus (IMM)	350	910	2.5	5	10	0.021	2 282.9

The far right column in Table 5.3 shows very high values for the mixer effectiveness of the 3D-T-mixer with ME$_{II}$ ≈ 10 000. The second convective mixer, the Tangential mixer, also has a high ME$_{II}$ value of 1 710. This mixer is not as effective as the 3D-T-mixer due to the higher pressure loss. The caterpillar mixer from IMM shows a much lower ME$_{II}$ value of 12.8, indicating that the mixer works less effectively than the convective mixers due to the high pressure loss of 17 bar. The main application of this mixer is the homogenization of viscous media with repeated, stepwise increase of the interface between the components. The SuperFocus mixer is designed for high flow rates and shows, therefore, a remarkably high mixer effectiveness, while the mixing effectiveness is considerably low. This mixer allows a high mass flow rate, however, the multitude of thin lamellae with a width of 100 µm and larger do not mix rapidly. A close look at these characteristic mixing numbers allows to analyze of mixing devices and performance improve.

In Eq. 5.70, the Re number represents the flow rate through a mixing device. To characterize the mixing process itself, it is sufficient to only consider the pressure loss and the mixing length.

$$\text{ME}_I = \frac{\text{revenue}}{\text{effort}} = \frac{1}{\text{Eu}} \frac{d_h}{l_m} \qquad (5.72)$$

With the above simplifications, the mixing effectiveness can be written as

$$ME_I = \frac{\dot{m}}{d_h \, \Delta p \, t_m}.$$
(5.73)

Figure 5.22, left side, displays the mixing effectiveness of various mixing principles over the Re number in the inlet channel. The mixing effectiveness ME_I consists of primary parameters and combines the requirements given above, being a dimensionless number, which increases with increasing mass flow, decreasing pressure loss, and decreasing mixing time. It should, therefore, be suitable to serve as a tool for comparing different micromixers for use in chemical production processes.

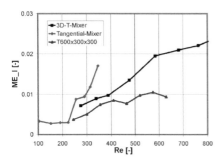 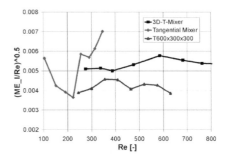

Fig. 5.22. Left: Mixing effectiveness of various mixing principles, see also [128]; Right: dimensionless group to compare mixing times.

With the Re number and the mixing time t_m, the mixing effectiveness can be written as

$$ME_I = Re \frac{\eta}{\Delta p \, t_m}.$$
(5.74)

Introducing the pressure loss coefficients from Eq. 5.61, the mixing effectiveness can be written as

$$ME_I = \frac{2}{\left(\zeta + C_f \frac{L}{d_h} \right)} \frac{d_h}{l_m}.$$
(5.75)

If the pressure loss of the mixing device can be described with convective effects only, which means $C_f = 0$, Eq. 5.75 is simplified to

$$ME_I = \frac{2}{\zeta} \frac{d_h}{l_m}.$$
(5.76)

With Eq. 5.76, the pressure loss coefficient, the hydraulic diameter and the mixing length describe the mixing effectiveness. This presents an opportunity to compare different mixing processes and find optimum arrangements for the specific application.

An important parameter for mixers in general by the dissipated energy ε during the mixing process. In stirred vessels, the dissipated energy is the stirring power divided by the stirred mass in the vessel. For static mixers, the dissipated energy is calculated from the pressure loss, the volume flow and the mass in the static mixer, see Section 5.3.2 and [383] Chapter 8.

$$\frac{P}{\rho V} \overset{\text{stirred vessel}}{=} \varepsilon \overset{\text{static mixer}}{=} \frac{\Delta p_v \cdot \dot{V}}{m} = \frac{\Delta p_v \cdot \bar{w}}{\rho l_m} \tag{5.77}$$

With $l_m = \bar{w} t_m$ and substituting the pressure loss from Eq. 5.77 into Eq. 5.74 results in

$$ME_I = Re \frac{v}{\varepsilon t_m^2} = \frac{\bar{w} d_h}{\varepsilon t_m^2}. \tag{5.78}$$

According to Zlokarnik [383], Chapter 8, there are three different types of micro mixing within liquids: molecular diffusion, laminar deformation below the Kolmogorov length scale, and mutual enclosing of fluid lamellae with different composition, which leads to the growth of micro mixed volume, see Section 5.3.2. The third process, the enclosing of relatively large volumes ("engulfment" acc. to [124]) is the limiting process with the characteristic time t_E, see also Eq. 5.49

$$t_E = 17.3 \left(\frac{v}{\varepsilon}\right)^{1/2}, \tag{5.79}$$

which is in the range of the characteristic time of turbulent flow structures, the Kolmogorov time scale $t_K = (v/\varepsilon)^{1/2}$. Introducing Eq. 5.79 into Eq. 5.78 and rearranging leads to the comparison of two characteristic times.

$$\frac{t_E}{t_m} = 17.3 \sqrt{\frac{ME_I}{Re}} \tag{5.80}$$

The "engulfment" time t_E serves as an independent time scale to compare different mixing times of various mixers. A high ratio represents a short mixing time and good mixing compared to static, turbulent mixers [128].

The characterization of mixing devices with the mixing effectiveness was recently proposed and is still under discussion. Open points to discuss are certainly the definition and measurement of the mixing length combined with a proper definition of mixing quality or segregation intensity. Additionally, the definition and measurement of the pressure loss length must be addressed in future experimental and numerical investigations.

5.4.7 Summary of Convective Micromixers

The investigation of the mixing behavior of convective micromixers has shown several important results. Firstly, it has been shown by comparison of centrifugal to viscous forces in simple T-shaped micromixers that higher Re numbers lead to an enhanced mixing process. The minimum Re numbers for convective micromixers to work lie above 200 (approximately). Using a pH-based color reaction and the iodide-iodate reaction, it has been shown that the characteristic mixing time of convective micromixers is inversely proportional to the Re number. The mixing performance of convective micromixers $MP = d_h/l_m$, the dimensionless mixing length, is constant for a given geometry with values between 0.15 and 0.8. For Re numbers up to 1 000, very short mixing times in the range of 100-400 µs for aqueous solutions can be achieved.

Mixing in static mixers is caused by pressure driven flow. Introducing the dimensionless pressure loss, the mixing effectiveness is formed to $ME_I = MP/Eu$. A low pressure loss with short mixing length leads to high mixing effectiveness, a very effective homogenization of a concentration or temperature field. Mixing devices treat process streams with given flow rate. Introducing the dimensionless flow rate, which is equivalent to the Reynolds number $Re = \bar{w}\, d_h/\nu$, the mixer effectiveness is formed to $ME_I = Re\, MP/Eu$. With higher Re number, the mixer effectiveness increases, and a higher flow rate is homogenized consuming low pressure loss within a short channel. This set of dimensionless numbers characterizes the mixing process as well as the mixing device.

In engulfment flow, initial fluid compartments formed in the inlet and junction engulf each other and generate a multitude of thin fluid lamellae by stretching the swirling flow. Thinning ratios for lamellae of over 500 are achieved, leading to lamella width below 1 µm. This occurs in time frames of several 100 µs. In these thin lamellae, diffusive mass transport is extremely fast. Very short diffusion times are achieved, superimposing the convectional mass transport.

Considering the mean specific energy dissipation $\bar{\varepsilon}$ in the convective micromixers, values of up to several 100 000 W/kg were achieved in the investigated devices. These are extraordinarily high energy dissipations, which explains how the thin fluid lamellae are generated in such very short times. For comparison, the mean specific energy dissipation in conventional batch reactors lies in the range of 10 W/kg.

5.5 Mixing and Chaotic Advection

5.5.1 Concept of chaotic advection

Chaotic advection is a large field in mixing theory, embracing all kinds of mixing processes, as displayed in Figure 5.1. Advection is often synonymously used for convection and originates from meteorology, where it indicates main lateral fluid movements. One of the first textbooks about chaotic advection was written by Ottino [125], who also introduced the term into chemical engineering [384]. A review article on the fluid dynamics of chaotic advection, mixing, and turbulence from Ottino [353] gives many examples of typical flow structures with chaotic behavior. For conventional mixing devices and processes, a good overview was edited by Aref and El Naschie in [385]. A thorough introduction to the mathematical background and formulation can be found in Sturman et al. [354] with the concept of linked twist maps (LTM), see Figure 5.23.

The concept of chaotic advection for microfluidic devices was introduced for complex channel structures with mixing via secondary flow effects or the split-and-recombine processes. The former is used by the prominent herringbone mixer

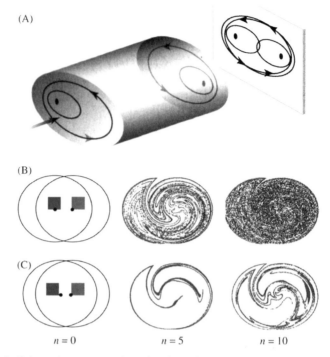

Fig. 5.23. A: Schematic representation of a channel type micromixer constructed from the concatenation of basic mixing elements; B: The linked twist map (LTM) mechanism causes the flow to mix completely after passing through five periodic elements of the mixer; C: The LTM conditions are not satisfied and the flow exhibits islands, which result in poor and incomplete mixing, acc. to Sturman et al. [354], see also Figure A.21.

from Stroock et al. [320] or by meandering channels, while the latter is employed by the caterpillar mixer [357, 386]. For low and intermediate Re numbers, twisted microchannels and 3-dimensional L-shaped microchannels in various forms cause chaotic advection. The modification of the channel wall with ribs and grooves brings chaotic advection to low Re number flow. These mixers are often used for lab-on-chip applications, where good mixing at low Re numbers needs to be achieved and pressure loss is not the main issue.

The functions in Figure 5.24 measure the separation of the stable and unstable manifolds normal to the unperturbed separatrix, the separation surface or interface between the components. The separation is thus expressed as functions of the parameter specific location on the separatrix. The upper separatrix of the open flow is portrayed here, more details are given in Aref et al. [385].

A short summary of designing optimal micromixers with the help of chaotic advection and related theories is given by Ottino and Wiggins [387], where symmetrical and asymmetrical motion in channels play a major role. The role of an active mixer and a meandering serpentine channel is described by Stremler et al. [388] for

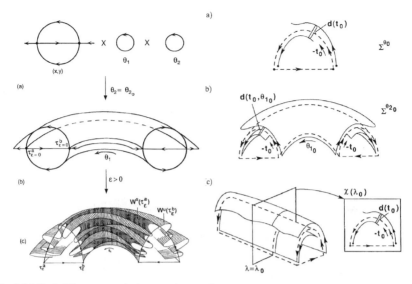

Fig. 5.24. Left: The unperturbed separation surface (separatrices) of two-frequency open flow in (a) the 4D autonomous representation, and (b) the 3D Poincaré map, whose boundaries are formed by global stable and unstable manifolds, W^s and W^u, of persistent normally hyperbolic invariant manifolds (here only the upper chaotic tangle in c)).
Right: The distance function d for (a) near-integrable 2D maps derived from 2D single-frequency velocity fields, (b) near-integrable 3D maps derived from 2D two-frequency velocity fields, and (c) near-integrable sequences of 2D autonomous maps derived from 2D aperiodic velocity fields, acc. to Aref et al. [385].

applications of chaotic advection at the micro-scale. Within the convective laminar mixers, two fluid streams come in contact, and their interface splits up into many lamellae or filaments by vortex flow, which can also be described by the idea of striation dynamics and fluid stretching. Here, the concept is described briefly to treat convective micromixers with a striation model and stochastic description of the striation thickness distribution.

5.5.2 Geometry and flow regimes

The local concentration field is analyzed and typical characteristic point of fluid motion are determined. Two typical points are of special interest, the elliptic and the hyperbolic point, see also Section 5.4.2. The elliptic point is located in the middle of a vortex and promotes the stability of the concentration field and slow mixing, see Figure 5.25. The flow and concentration field is stretched in parallel circular lines around the elliptic point. In engulfment flow in the T-mixer, this point is located directly on the component interface and facilitates the enrolling of the two components. The hyperbolic point is a saddle point of the flow field with in- and outflowing components. It promotes for diverse stretching of the compartments and fast mixing. In the T-mixer, the hyperbolic point is also located at the interface of the components and does not dramatically enhance the mixing process. The periodic flow regime in the T-mixer leads to temporal migration of the characteristic points. The Dean flow in the L-mixer, see Section 5.4.1, has two elliptic points in the middle of the channel, and two hyperbolic points at the inner and outer wall of the mixing channel.

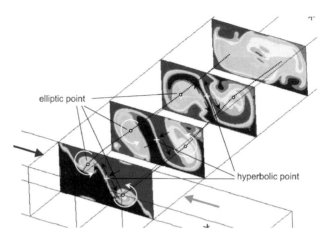

Fig. 5.25. Typical points of the concentration field in a T-mixer (T600×300×300), Re = 300, see also Figure A.20.

The design of mixing devices is assisted by analysis of these characteristic points. As demonstrated for the T-mixer with engulfment flow, an elliptic point should be located on the interface to promote stretching and rolling up of the compartments. A hyperbolic point should be located offside of the component interface, preferably in undisturbed regions to promote diverse stretching. A future task would be to design channel structures, which serve for typical characteristic point at defined locations, see also Figure 5.17.

This design concept also depends on the Re number in the mixing channel, which determines the flow field. With high Re numbers, flow pulsations develop and lead to stochastic treatment of flow and concentration field. A stochastic approach with diffusion on deterministic fluid dynamics is presented by Gombert et al. [381]. The Wiener process is employed to represent diffusive motion of a tracer and is used to determine the concentration fine structure in a micromixer, see Figure 5.14.

Based on the above method of striation design, a short outline is given for a more detailed treatment of the mixing process in T-mixers and other convective micromixers. In Figure 5.26, the mathematical transition is displayed from a complex three-dimensional concentration profile to a one-dimensional mean concentration on a cross section. This helps to indicate key areas of enhanced mixing and reduces analytical effort.

Chaotic advection does not only play a role in single-phase flow mixers. Some research groups are working on multiphase flow mixing, either the mass transfer between two almost immiscible phases (extraction) or the mixing within a bubble or

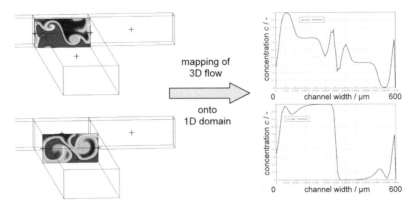

Fig. 5.26. Mapping of the flow, three-dimensional concentration profile onto a one-dimensional concentration function on a cross section.

a meniscus in a two-phase flow. The mass transfer between two immiscible phases can be described by the two-film theory, see also Figure 2.14. Other models are the penetration and surface renewal model [89], Chapter 1.5. The mass transfer in bubble flow in a capillary was experimentally described by Tice et al. [389] for the mixture of aqueous solutions in oil. The slug flow in microchannels enhances the mixing process in one phase due to the Taylor dispersion. Gavriliidis et al. [390] proposed a CFD model for Taylor flow characteristics and axial mixing. The slug flow causes a repeated mixing of the entire contents of a bubble, which is also handled by chaotic advection theory. Most application of this mixing process can be found in analysis systems, however, the fluid extraction may be based on this mechanism. The Jensen research group at MIT have described the production of monodisperse silica nanoparticles in a multiphase flow system [391].

5.5.3 Fluid lamellae and chemical reactions

Another geometric reduction approach starts from a two-lamellae distribution in the mixing channel, which is rotationally stretched and folded into a spiral. The map function gives a lamellae distribution, depending on the geometry and the energy dissipation into the mixing channel. The energy dissipation into the fluid flow creates vortices, which are used for interface stretching and enlargement, see Figure 5.27. This lamellar structure is the basis for the reaction-diffusion calculation, see Eq. 5.3, and further selectivity evaluation. The local energy dissipation rate ε_l is correlated to the striation thickness distribution, i.e. the concentration field with lamellar, distributed structure. The striation thickness distribution is numerically simulated and can be indirectly measured by selectivity of parallel reactions and particle size distribution of precipitation, see Section 6.5.

   The temporal development of a chemical reaction can be simulated in defined lamellae distributions, such as the iodide-iodate reaction, described in Section 6.3.6. Since the basic relationship between 1D diffusive mixing and iodine formation is

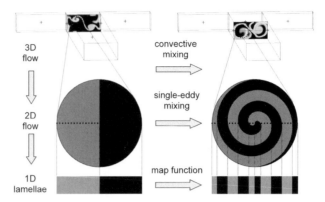

Fig. 5.27. Geometric reduction approach starting from two-lamellae filling the entire mixing channel through single eddy mixing to a one-dimensional lamellae distribution.

interesting, it is sufficient to directly use the iodine concentration as a measure for mixing quality, where a lower iodine concentration means faster mixing. For the simulations, the typical concentrations and reaction rates from Section 6.3.6 are used; the necessary reaction-diffusion equations are similar to the system described in Section 6.3.7, with two additional equations for the reactant species.

In a first step, the concentration development is simulated for a two-lamellae microchannel ($2w = 10$ μm), see Figure 5.28, left side. Iodine I_2 is only formed in the right part of the channel, where iodide/iodate diffuses into the acid lamella. In the left lamella, iodine formation is actually zero, since all penetrating H^+ ions are almost instantaneously neutralized. Figure 5.28

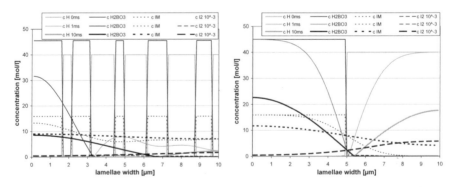

Fig. 5.28. Evolution of species concentrations for the iodide-iodate test reaction in a 10 μm wide microchannel with two symmetric lamellae (left diagram) and smaller, but asymmetric lamellae (right diagram). The iodine concentration c(I2) is scaled with 10^3; the iodate concentration, $c(IO_3^-)$ is not plotted, since it is similar to the iodide concentration.

The right diagram in Figure 5.28 shows a simulation for more than two, asymmetric lamellae (i.e. with varying width) in a 10 μm channel. Since the mean lamella width is approximately one fifth of the width in the two-lamellae case, the final iodine concentration is lower than expected. However, asymmetric conditions have a special side-effect on parallel-competitive reactions, which shows up if the iodine formation is compared for the symmetric and the asymmetric case in terms of varying channel width b, see Table 5.4. Although the characteristic diffusion time grows with the square of the characteristic diffusion length, the ratio between iodine concentrations for symmetric and asymmetric distribution decreases with rising channel width. One reason for this behavior is the imbalance in the concentration distribution: some lamellae with H^+ have too few $H_2BO_3^-$ in their vicinity, can not be neutralized, but instead, are consumed in the Dushman reaction. Therefore, the iodide-iodate test system is not only a measure for the characteristic diffusion length, but also for asymmetries in the lamella width, at least to some extent.

Table 5.4. Final iodine concentration for the iodide-iodate reaction system carried out in a 1D diffusive mixer under symmetric and asymmetric lamellae distribution

Channel width	Final iodine concentration		Ratio sym./asym.
	symmetric	asymmetric	
w (μm)	$c(I_2)$ (mol/l)	$c(I_2)$ (mol/l)	(–)
1	5.30×10^{-8}	1.04×10^{-8}	5.10
10	5.30×10^{-6}	1.03×10^{-6}	5.14
100	0.33×10^{-3}	0.09×10^{-3}	3.67
1 000	1.59×10^{-3}	1.08×10^{-3}	1.47

Hence, it is not advisable to characterize the mixing process with a single test reaction, since every reaction has its own peculiarities. Instead, a set of reactions based on different principles should be used, such as the diazo-coupling (Section 6.3.7), the iodide-iodate reaction (Section 6.3.6), and barium sulfate precipitation (Section 6.5).

5.6 Design and Fabrication of Silicon Micromixers

Microstructured mixing devices for test purposes have to fulfill many requirements concerning the test fluids, optical observation, reproducible experimental data, and affordable fabrication techniques. IMTEK has its own clean room for educational and research purposes with the main focus on silicon technology. Hence, most devices in this work are fabricated with silicon bulk micromachining techniques, which deliver reproducible, exact silicon microstructures with a glass lid for optical observations. These devices have also been tested by research and cooperation partners in Karlsruhe, Jena, and Bremen, to mention just a few.

The harshest environment in this work for the silicon micromixers and microreactors is given by particle precipitation and competitive coupling reaction. With the analysis of precipitation and the influence of mixing, general device requirements for particle precipitation can be derived. Also, the behavior of particle-laden flow, fabrication technology, and chemistry result in the following design rules for the mixing device and peripheral equipment.

Fast generation of interfacial area: Mixing considerations and the simulations for barium sulfate precipitation (Sec. 6.5.1) show that fast generation, along with extensive thinning of fluid lamellae, is the key to yielding small particles. The nuclei are formed in the interfacial region, hence, it is the primary design goal to increase the interfacial area between the reactant fluids as quickly as possible, following the notion that "good mixing" means "rapid stretching and folding of the interfacial area".

Proper location of characteristic points: Each channel element creates elliptic and hyperbolic points at certain locations for different flow regimes describes by chaotic advection. An elliptic point must be located on an interface between two components and roll up the interface. A hyperbolic point must be located in an unmixed region to attract remote material.

Binary mixing: The investigated reactions only require the mixing of two distinct reactant flows. Additionally, cascaded mixing potentially enables control of the particle size distribution or reaction yield and selectivity. They also reduce the risk of fouling or blocking via dilution of the reaction flow with additional water.

Suitability for particle-laden flows: The greatest obstacle to a larger application of micromixers for industrial application is their susceptibility to clogging and fouling. Since the mixing devices deal not only with barium sulfate or pigment synthesis, but also for other reagents, special attention must be paid to material compatibility and particle deposition.

Robustness against aggressive media: Mixing devices for chemical processes must be able to cope with aggressive media. Particularly for experiments with barium sulfate and pigment synthesis, the devices and peripheral equipment must be resistant against corrosive mineral acids (sulfuric acid, hydrochloric acid) and organic solvents (acetic acid). Silicon generates a natural oxide layer on its surface, which also can be passivated by other means, such as silicon nitride or nickel sputtering.

Additionally, temperature control and optical observation of the mixing process are important for reaction and device control. Finally, the mixing structures should be suitable for silicon microfabrication, a well-established, fast and accurate fabrication technology. The implemented anisotropic dry etching technique requires that all 3D structures are generated by extrusion of 2D outlines (round pipes, for instance, are not possible) or stacking of such structures (silicon wafer bonding). The following section describes the fabrication of in-plane channel structures for particle precipitation and aerosol generation, and parallel micromixers with two-wafer design for high flow rates.

5.6.1 Microstructured mixers for liquid phase precipitation

The fabrication of the designed micromixers is performed by silicon bulk micromachining. The fundamentals of this standard technology for microstructured devices are explained in detail by Madou [392] or Menz et al. [393]. The cost situation is crucial for the successful application of microstructured devices. Many of the microfabrication techniques are suitable for mass production, hence, high costs are typical for prototypes and demonstrators. Lawes [394] gives an overview of the cost situation for high aspect ratio, microstructured devices, especially in relation to cost reduction possibilities.

The 2D layout is transferred into 3D structures via photolithography and subsequent anisotropic dry etching of silicon wafers (here: *Advanced Silicon Etching (ASE)*) or deep reactive ion etching (DRIE), resulting in almost vertical side walls. The common etch depth for most structures is set to 300 µm in a 525 µm thick mono-crystalline silicon wafer. The etched wafers are fused with Pyrex glass wafers via anodic bonding, and diced into chips with single mixing devices. A schematic overview of the fabrication steps is displayed in Figure 5.29, left side. A typical photolithography mask for a complete wafer is shown in Figure 5.29, right side, with the mask nomenclature in the corners. The wafer silhouette contains alignment structures for the wafer flat, which is rotated for 3° to avoid long channel structures in a crystal plan of the wafer. The crosses and small lines between the channel structures are saw markings to assist chip dicing.

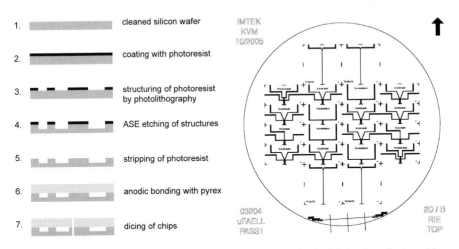

Fig. 5.29. Fabrication process: Left: Scheme of process steps for the fabrication of mixer chips on one wafer; Right: Mask layout for photolithography for 16 mixer chips 20×10 mm^2 and four mixer chips 20×20 mm^2.

Two examples of chip mask layouts and fabricated silicon chips are given in Fig. 5.30. The nomenclature of the chip identification (T-3-6x20) consists of the mixer type (T-mixer), the inlet channel width in 100 μm (300 μm), the mixing channel width in 100 μm (600 μm) and the mixing channel length in 100 μm (2 mm).

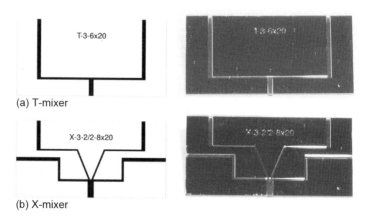

(a) T-mixer

(b) X-mixer

Fig. 5.30. Mask layouts and final chips for particle precipitation.

To keep the inlet and outlet channels short, the mixer chips should be as small as reasonable with respect to handling. Furthermore, to allow connection of all mixer types with the peripheral equipment using only one chip mount, the positions of up to four inlets and the outlet are the same for each mixer type. For this reason, the lateral dimensions of all convective mixer chips are set to 20×10 mm^2 with a thickness of 1.05 mm due to the silicon-Pyrex stack, see Fig. 5.30. In addition, the geometric parameters of all opposite inlets are designed to be equal to ensure symmetric flow conditions.

A fluidic mount is designed to connect the inlet and outlet channels of the mixing devices with the peripheral system. This chip mount is made of PEEK, which has excellent resistance against inorganic media like hydrochloric acid. The chip mount incorporates a cavity with fluidic connections to control the chip temperature by heating the back side with a liquid. The assembly of the chip and chip mount is completed by a silicone sealing with U-shaped cross section, Fig. 5.31 (a), in which the chip is inserted, put into the mount cavity, and fixed with a lid, see Fig. 5.31 (b). Due to the contact pressure from the lid, the silicone is pressed against both the walls and the chip and provides a tight seal. The sealing is fabricated by casting of silicone (Elastosil M from Wacker, Germany) into molds with a dummy block as chip replacement.

5.6.2 Microstructured mixers for aerosol generation

The microreactors described in Section 6.4 are fabricated in silicon with dry etched channels (DRIE process) and covered with glass lids, see Figure 5.32 right side, and [133].

(a) chip inserted in silicone sealing   (b) assembled chip mount

Fig. 5.31. Chip with silicone sealing and fully assembled chip mount.

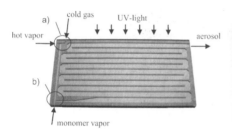

Fig. 5.32. Mixing device for aerosol generation; Left: silicon chip (20×40 mm^2 area) with two mixers (a) and (b) for homogeneous and heterogeneous condensation, and a meander as photo reactor for later use in photo-polymerization (not used in this study); Right: Chip mount from PEEK with assembled chip and fluidic connections.

The inlet channels cause an unwanted heat loss and are prone to particle deposition, therefore, short inlet channels with few bends are preferable. This requires fluidic connections in the plane of the chip with effective sealing, which is achieved with a silicone gasket placed around the chip edges. The fluidic chip mount is fabricated from PEEK due to its low heat conductivity ($\lambda = 0.25$ W/m^{2}K) and good chemical resistivity, see Figure 5.32 right side. The fluidic connections are located in-plane via access holes in the gasket, which also holds the clamped silicon chip.

In a second fabrication lot, the silicon structures serve as a master mold for epoxy-casted polymer devices. To prepare the silicon chips, their structured surface is covered with a thin gold layer sputtered with a thickness of approx. 100 nm. This thin layer prevents the attachment of silicone (Elastosil from Wacker, Germany), which is used to form the negative mold from the silicon chip. The silicone is degassed before filling onto the silicon chip and is fixed in a metal frame. To yield polymer chips with low heat conductivity, epoxy resin is degassed and filled into the mold. After hardening within a day, the epoxy chip can be easily removed from the silicone mold and has a well-defined surface. The polymer chip can also be covered by a glass lid and allows optical observation. Engler et al. [395] describe the entire mold fabrication process together with economic calculations.

The fabrication process and experimental setup for high-throughput mixing devices is demonstrated in Section 5.6.3. Due to the parallelization of the mixing elements, both inlet channels are located on one layer, where the components come in contact. The mixing channel is positioned on another layer to avoid channel crossings. The fabrication steps are similar to those mentioned above, except that two wafers must be structured and bonded together to form the entire mixing device.

5.6.3 Microstructured mixers with two wafers

The cross section of the two-wafer layout for the mixer chips is schematically shown in Figure 5.33.

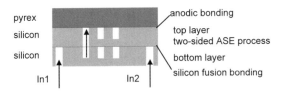

Fig. 5.33. Chip architecture consisting of two microstructured silicon wafers (525 μm thick), processed from both sides, anodic bonded, covered with a Pyrex glass lid.

The silicon wafers are structured using the DRIE process (ASE process), bonded, and diced to the mixer chips at IMTEK clean room facilities. Details from the photolithographic masks of the mixing elements are shown in the Figure 5.34. The dark gray channels are located on the bottom wafer and mainly form the inlet manifolds. The transition hole is fabricated from both sides to obtain a fluidic connection between the wafers. The light gray channels are located on the top wafer, consisting mainly of the mixing channel and the outlet manifold. The flow regimes and mixing process can be observed through the transparent Pyrex lid with an optical microscope.

Four different mixing element combinations were designed, fabricated, and tested. The geometrical details from the mask layout can be seen in Figure 5.34. The four mixing configurations have been fabricated on 20×20 mm^2 mixer chips for experimental investigations. Additionally, the mixing configurations are designed for a parallel arrangement on a single mixer chip. Figure 5.35 shows the mask layout of 16 parallel Tangential mixers on one mixer chip.

The inlet manifold is very important for an equal distribution of the liquid to each mixing element. The two fluid streams enter the chip through three inlet holes and are distributed in the manifold. The mixing elements are arranged in four groups with four mixing elements each, giving 16 mixing elements in total. Their location allows a symmetrical design of the inlet manifold to supply the four groups. The width and depth of the inlet channels to the mixing elements is held constant at 300 μm each. With an analytical calculation the pressure loss of each inlet channel is determined for a certain channel length. This length is adjusted for an equal pressure loss in each

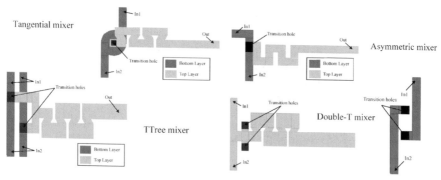

Fig. 5.34. Mask layout of four single mixing elements; Top left: Tangential mixer consisting of the circular mixing chamber with the transition hole, a 90° bend into the top layer, and three successive 120° U-mixers, width 300 µm; Bottom left: TTree mixer with two parallel T-shaped mixers, a second successive T-shaped mixer, and three 120° U-mixers with a width of 600 µm. Top right: Asymmetric mixer consisting of a 90° bend (In1), a T-mixer through the transition hole, a 90° bend into top layer, and three successive 90° U-mixers forming the meandering mixing channel, width 300 µm; Bottom right: Double-T mixer with two successive T-shaped mixing sections and three 120° U-mixers with a width of 600 µm.

inlet channel. The design process of the channel manifold is described in more detail by Kockmann et al. [43].

The mixing channel with the meandering U-mixers is identical in all mixing elements, however, the length of the straight outlet channel was varied for the different positions. A short length for the outer mixing elements and a longer channel for the inner elements ensures an equal outlet pressure loss. Optical investigation of the color distribution and similar mixing quality indicated an almost uniform fl uid distribution in the inlet and outlet manifolds, see Section 4.3.2.

The designed structures have been fabricated using the DRIE process (ASE). Some SEM images are shown in Figure 5.36. A completely assembled mixer chip, $20 \times 20 \times 1.6$ mm^3, is shown in Figure 5.37 together with a coin for scaling purpose. The outlet hole is clearly visible in the middle of the chip.

The mixer chip is mounted on a fl uidic connector block, sealed with a silicone membrane, and can be inspected with an optical microscope, see Figure 5.38. The two liquids are supplied from pressure vessels, pass a fl ow meter, and enter the fl uidic chip. For the measurement of the mixing quality with the help of the iodide-iodate reaction, see Section 6.3.6, the mixed liquid is collected in a cuvette, which allows the UV-spectroscopic measurement of the iodine concentration.

5.7 High Throughput Mixing Devices with Microchannels

Fast mixing with a short mixing time is necessary for chemical processes with enhanced heat transfer and the high selectivities of fast parallel chemical reactions.

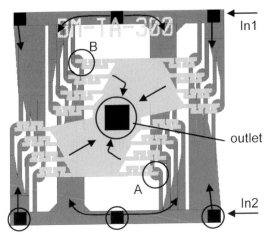

Fig. 5.35. Chip layout of the Tangential mixer chip with 16 mixing units. Dark gray: inlet distribution manifold, Light gray: mixing channels and outlet manifold. The arrows indicate the fl ow directions. The outlet channel is located in the middle of the chip. Details A and B, see Figure 5.36.

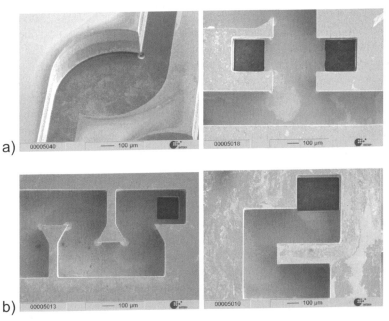

Fig. 5.36. SEM images of DRIE fabricated silicon structures (ASE) before bonding, channel depth 300 μm; a) left: Tangential mixer with inlet channels and mixing chamber; right: fi rst section of the mixing channel of the Double T mixer; b) transition hole with meandering channels, left: bend angle of 120°; right: bend angle of 90°.

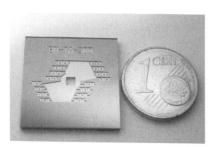

Fig. 5.37. Assembled micromixer chips for high mass flow rates; Left: Image of a complete mixer chip with 16 Tangential mixer units, consisting of two microstructured silicon wafers covered with a Pyrex lid. The coin serves for scaling. Right: Mixer chip integrated within an aluminum fluidic mount with fluidic connections on the back side.

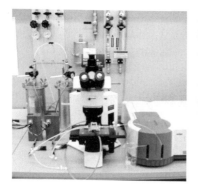

Fig. 5.38. Experimental set-up, Left: pressure vessels, flow meters, microscope with fluidic measurement cell, and UV spectrometer; Right: silicon chip in fluidic mount with fluidic and electrical connections (for pressure measurement in the fluidic mount).

Microstructures offer a short diffusion length between the components and, therefore, a fast mixing time. This has been shown for various applications in analytical equipment by Nguyen and Wu [357], and for laboratory application by Hessel et al. [13]. Recently, the application of microstructured equipment has been used in the production of chemicals, see Markowz et al. [322], and the corresponding process development described by Bayer et al. [396]. For this step, not only are the mixing characteristics important, a high throughput and mass flow rate at tolerable pressure losses are also critical points. To achieve high mass flow rates either the number of channels can be increased, the channel cross section can be enlarged, or the flow velocity in the channels can be accelerated. The first method is used very often, for example, in the concept of numbering-up described by Hessel et al. [13]. The main problem with numbering-up is achieving equal fluid distribution in the inlet and outlet manifolds, see Matlosz [113]. Larger channel cross sections are contradictory to miniaturization benefits and limited by the employed fabrication process, such as the thickness of the metal foil or the silicon wafer. Together with the increased fluid velocity, larger channels lead to higher Re numbers and the occurrence of vortices. Laminar diffusive mixing is assisted by vortices and convective effects, see Chapter 3.

Based on the results of convective mixing in T-shaped micromixers, this section presents the design and characterization of mixing elements and devices suitable for high mass flow rates. The flow of a liquid (water) is numerically simulated with the software package CFD-ACE+ of ESI Group and optimized specifically for the mixing quality, mixing length, and actual pressure loss. Optimized convective mixing elements are systematically combined in series to a complete mixing channel and arranged parallel (16 mixing elements) on a single mixer chip. The designed structures are fabricated from two silicon wafers by deep reactive ion etching (DRIE) with rectangular channels (300 to 600 μm characteristic hydraulic diameter) and covered with Pyrex glass, see Section 5.6.3. The mixing of two media with different pH value, combined with the pH-indicator Bromothymol Blue, complements the simulation results. The pressure loss in the experiments was below 1 bar and, therefore, relatively low for the high mass flow rates achieved. To quantitatively describe the mixing process, the parallel-competitive iodide-iodate reaction according to Fournier et al. [397], Guichardon and Falk [398], and Guichardon et al. [399] gives an indication of the mixing time in the investigated structures. Many research groups use this parallel reaction for characterizing the mixing performance of conventional equipment including Bourne [124], Loebbecke et al. [400], and of microstructures for analytical purposes by Trippa and Jachuk [401].

5.7.1 Numerical simulation of mixing elements

Beginning with the numerical and experimental results for basic mixing elements, such as a T-shaped or 90° bend mixer, see Kockmann et al. [402] and Section 5.4, the results of a numerical design study (CFD-ACE+ of ESI Group) are presented with the goal of efficient micromixers. The design criteria are the mixing performance, i.e. the mixing quality α_m after a certain length in the mixing channel behind the mixing element itself, the mass flow rate $\dot{m}$, and the corresponding pressure loss Δp. Another aim is the parallelization of the elements on a single mixer chip that requires a compact architecture on two wafers with at least two different channel layers and connecting transition holes. Due to the silicon wafer thickness of 525 μm, the channel depth is limited to 300 μm. The channel width is limited to aspect ratios around 1, due to the effectiveness of the centrifugal forces for the mixing process, see also Section 3.2. The chip dimensions are set to 20×20 mm^2 to make use of a standardized fluidic chip mounting. For the analysis, the Re number is determined inside the inlet channel:

$$\text{Re} = \frac{\dot{m}\, d_{h,\text{in}}}{A_{\text{in}}\, \eta} \tag{5.81}$$

Besides the T- and U-shaped elements, a tangential mixing element is investigated with a rotating and stretching effect, see Figure 5.39, left side. The two fluids enter the mixing chamber tangentially and leave it perpendicularly to a different channel level. A 90° bend guides the fluid into the mixing channel, where viscous damping leads to the resulting laminar flow. Here, the necessity of the two-wafer design is visible for the proper arrangement of the two inlet flows and the resultant mixing flow. Other mixing elements are described in the following section. In Figure 5.39, right side, the mixing quality α_m in the Tangential mixer is compared with the T-shaped micromixer for similar mass flow rates due to different channel cross sections. The rotating flow leads to a large interface and, consequently, a high mixing quality even for low mass flow rates. The engulfment flow in the T-shaped micromixer increases the interface and the mixing quality. This result is also achieved from experiments with chemical reactions, see next section.

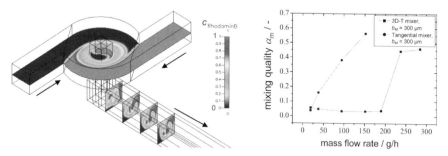

Fig. 5.39. Left: Concentration profile in a Tangential mixer with outlet mixing channel, $300 \times 300 \ \mu m^2$, 100 g/h total mass flow rate, representing a Re number of 108.3 in the inlet channel; Right: Comparison of the mixing quality in the T-shaped mixer and the Tangential mixer dependent on the mass flow rate.

The large sizes of the complex mixing structures have limited the numerical simulation to the number of finite volumes and the consumed calculation time. The numerical model of the Tangential mixer involves approx. 758 000 grid points and requires almost approx. 89 hours on a PC (Intel Pentium P4 with 1.6 MHz) to calculate, hence, the simulation of a combination of various mixing elements is very expensive, if at all possible. The numerical simulation of a complete mixer with 16 parallel elements, plus inlet and outlet manifolds, is nearly impossible with current capabilities.

The numerical diffusion of the mixing simulations allows only a qualitative comparison of the concentration fields and the mixing quality. The actual grid size is sufficient for the calculation of the flow field and stream lines, but does not allow the correct determination and display of the concentration gradients. A grid refinement study was only possible for simple mixing devices and gave no additional insight, see Engler [127] or Kockmann et al. [128]. More information on numerical diffusion in CFD simulation of microchannel flow can be found in Hardt and Schönfeld [403] with an analytical model of the concentration profile in multi-laminated flow.

5.7.2 Experimental results and discussion

The mixer chips are designed for liquid flow with a high throughput, hence, the achieved mass flow rate is measured for a given pressure drop over the chip and the entire test apparatus. The results are displayed in Figure 5.40 for the four mixer configurations. The pressure loss over the mixer increases quadratically with the total mass flow rate in all mixer configurations. The regression curve to the power of 2 correlates very well with the measured values and indicates the convection inside the mixing device. A more detailed description is given by Engler [127]. Due to the complex geometry of the mixing elements, it is not reasonable to determine a single friction factor or pressure loss coefficient for the entire micromixer. The pressure is limited to approx. 1 bar by the tightness of the fluidic connections, the chip sealing, and the pressure supply from the laboratory network.

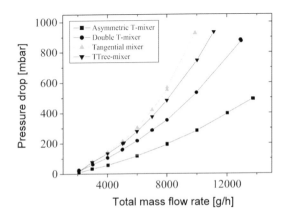

Fig. 5.40. Pressure drop versus total mass flow rate in the investigated mixing configurations with 16 parallel mixing elements.

The asymmetric mixer shows the least pressure loss with the highest throughput of approx. 14 kg/h due to the simple arrangement. The Tangential mixer produces the highest pressure loss and obtains a mass flow rate of only 10 kg/h in the investigated pressure range.

To complement the mixing results from the numerical simulations, the mixing of two fluids was visualized with the pH-indicator Bromothymol Blue for good optical contrast. The neutralization reaction of an alkaline solution of di-sodium hydrogen phosphate (pH8) with deionized water (In1) and Bromothymol Blue (pH = 7, light gray, In2) results in a mixture with a pH value of 7.5, which is indicated by a blue color (dark gray). The concentration of the reactants is very low and has no significant influence on the fluid properties. The mixing process for two aqueous solutions with a 1:1 ratio is clearly visible for various mass flow rates in a TTree mixer, see Figure 5.41. Other mixer configurations show similar behavior.

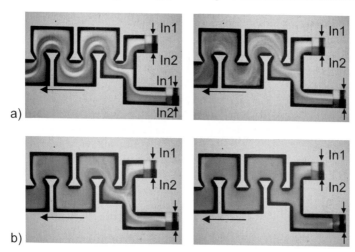

Fig. 5.41. Flow and mixing visualization in a single TTree mixer with four different mass flow rates: a) In = 11 g/h, Re_{in} = 11.9; In = 86 g/h, Re_{in} = 93.1; b) In = 250 g/h, Re_{in} = 270.7; In = 755 g/h, Re_{in} = 817.6, see also Fig. A.8.

The Re numbers are determined for the inlet channel, because the hydraulic diameter and average velocity are not constant in the various cross sections in the U-shaped mixing channel. For low Re numbers, the two streams are clearly separated near the entrance and the transition hole. In the following course of the mixing channel, the dark gray areas are growing, indicating slight vortex mixing coupled with diffusion. For higher Re numbers, the liquid is still separated in the contact element, the T-mixer, see Figure 5.41. The curved channels cause a swirling flow and increase the mixing quality, indicated by the homogeneous gray shading. For high Re numbers of approx. 800, mixing is almost complete before the liquid enters the meandering channel.

The mixing situation in a Tangential mixer is displayed for an aqueous solution in Figure 5.42. The inlet streams are already well mixed in the tangential mixing chamber at relatively low Re numbers. The meandering channel increases the mixing quality, however, shading variations and striations are still visible in the outlet manifold, which indicates incomplete mixing in that region.

To characterize the mixing, two specific parameters have been determined: the mixing time and the segregation index. The mixing time gives an indication of when the mixing is completed. The color change of the Bromothymol Blue solution is observed optically to determine the mixing length, where no more color change is visible. This indicates complete mixing at a certain length in the mixing channel. The mixing length is divided by the average velocity in the mixing channel to calculate the mixing time, see Figure 5.43.

The mixing time clearly decreases below 1 ms for all investigated mixer configurations. The curves follow almost the same trend, indicating the same mixing characteristics as convective mixing. The Tangential mixer achieves short mixing

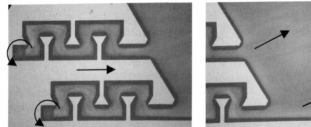

Fig. 5.42. Flow and mixing visualization in a Tangential mixer with a mass flow rate of In = 1040 g/h, $Re_{in} = 70.4$.

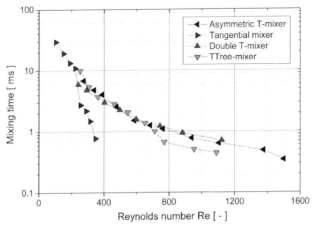

Fig. 5.43. Logarithmic mixing time for various mixer configurations over the Re number in the inlet channel Re_{in}. Due to the optical determination of the mixing length, the mixing time has an accuracy of approx. ±10 %.

times at relatively low Re numbers due to good premixing in the tangential mixing chamber.

The quantitative description of the mixing quality is a complex process and has been previously documented, see for example, Engler et al. [382]. Based on this prior work, the competitive-parallel iodide-iodate reaction (Villermaux/Dushman reaction acc. to Fournier et al. [404]) is used to determine the mixing quality, see Section 6.3.6. The iodine concentration from the mixing process is measured by UV-spectroscopy and transformed to a segregation index, which is inversely proportional to the mixing quality given in Figure 5.11. The experimental results are displayed in Figure 5.44 for the four investigated mixing configurations. A direct comparison between the inverse of the mixing quality and the segregation index is not possible due to the different nature of the concentration variation and the competitive-parallel chemical reaction, however, the curves follow the same trend for various Re numbers and different geometries. The concentration variation is the first indicator for mixing characterization.

For Re numbers lower than 250, the Double T-mixer shows the best mixing quality, due to the fine splitting of the liquids. For Re numbers from 250 to 600, the Tangential mixer shows good mixing performance, which is then surpassed by the asymmetric T-mixer. Nevertheless, the differences between the various configurations are small; their mixing characteristics follow the same trend. Other conventional mixers, such as heavily stirred vessels, achieve a segregation index from approx. 0.1 to 0.003, depending on the equipment setup (see Fournier et al. [404] or Guichardon and Falk [398]. Mixing processes in microchannels produce a segregation index of 0.01 and higher, see Trippa and Jachuk [401].

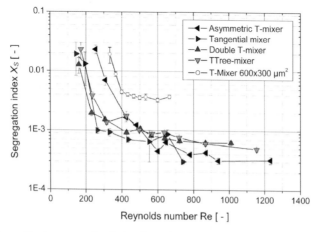

Fig. 5.44. Logarithmic segregation index from the iodide-iodate reactions in various mixing configurations over the Re number Re_{in}. The mixing quality is inversely proportional to the segregation index X_S.

5.7.3 Performance investigation

To provide a more detailed description for the potential of convective micromixers, they were compared to existing micromixers, so, suitable comparable parameters had to be chosen. In this work, three characteristic performance issues were utilized, see also Section 5.4.6:

1. the mixing time t_m characterizing the mixing performance,
2. the mass flow rate $\dot{m}$ characterizing the throughput,
3. the pressure loss Δp characterizing the energy loss of the device.

The mixing time gives an idea of how "effective" the mixing process is, in comparison to parallel processes, for example, chemical reactions. Usually, a short mixing time is desired being faster than other processes. The throughput is an important issue for mixing processes, as only devices with a considerable throughput can be economically applied, which is especially the case for chemical production purposes. Finally, the pressure loss is a measure of the energy necessary for the mixing processes in flow mixers.

In summary, the asymmetrical T-mixer is the most effective micromixer of the investigated configurations. With a relatively low pressure loss of less than 500 mbar, a mass flow rate of more than 15 kg/h is homogenized with a mixing time shorter than 0.5 ms and a segregation index X_s below 0.0005. Successful internal numbering-up has been highlighted with optimized convective micromixers.

5.7.4 Injection micromixers

Further development for simple parallelization of microstructured mixing elements leads to jet-in-cross flow mixing elements, which are well-known from combustion engineering. In microfluidic devices, injection micromixers are used to produce emulsions [28] or to mix gaseous streams [85]. Here, the jet-in-cross flow principle is combined with convective mixing in curved channels, see Figure 5.45 right side.

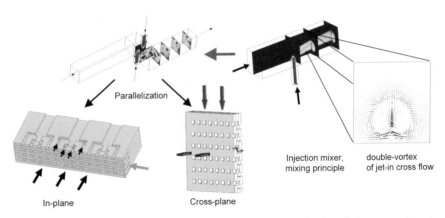

Fig. 5.45. Left: implementation of injection micromixers, parallelization of elements and stack setup; Right: mixing principle of one channel injection mixer. See also Figure A.12.

Two basic mixer geometries are designed producing an in-plane and cross-plane arrangement, adding a second component to the leading component flowing through a regular-porous, microstructured device, see Figure 5.45. First simulation results indicate fast mixing with low pressure loss, resulting in a high mixing performance for various mixing ratios, gas/liquid mixing, or the different fluid properties of the components to be mixed. The main applications planned are for fuel gas injection into combustion chambers and for high throughput chemical reactors.

6

Chemical Reactions and Reactive Precipitation

This chapter gives an introduction to chemical engineering methods for continuous flow reactor design. Special attention is payed to the wall mass transfer and analytical solutions. Some examples are given for reactions of interests, such as parallel competitive-concurrent reactions. Finally, a focus sheds light on particle technology in microchannels.

6.1 Chemical Reactor Engineering

The design of chemical reactors is based on material and energy balances as well as on chemical reaction kinetics. The material balance must be set up for all species participating in the reactions. The material balance for a component A_i can be formulated in the following manner, see also Eq. 2.1:

$$
\begin{bmatrix} \text{accumulation} \\ \text{of } A_i \text{ in} \\ \text{the system} \end{bmatrix} = \begin{bmatrix} \text{rate of flow} \\ \text{of } A_i \text{ into} \\ \text{the system} \end{bmatrix} - \begin{bmatrix} \text{rate of flow} \\ \text{of } A_i \text{ out of} \\ \text{the system} \end{bmatrix} + \begin{bmatrix} \text{rate of production} \\ \text{of } A_i \text{ within} \\ \text{the system} \end{bmatrix}
\tag{6.1}
$$

$$
\frac{\partial n_i}{\partial t} = \dot{n}_{i0} - \dot{n}_i + L_{p,i}
$$

with n_i as the number of moles of species A_i in the system at time t and $L_{p,i}$ as the rate of A_i production. For constant temperature and concentration of the chemical species, the production rate of A_i corresponds to the product of the reaction volume V and the A_i transformation rate R_i.

$$
L_{p,i} = R_i \cdot V = \dot{n}_i - \dot{n}_{i,0}
\tag{6.2}
$$

The gross reaction correlation for simple reactions is the sum of the product of the species A_i with their stoichiometric coefficients v_{ij} of reaction j.

$$
0 = \sum_{ij} v_{ij} \cdot A_i
\tag{6.3}
$$

The transformation rate R_i of A_i is the sum of the rates r_j of the reactions j with participating A_i and molar density ρ_m.

$$R_i = \rho_m \sum_j \nu_{ij} r_j \tag{6.4}$$

Parallel reactions are often studied in microreactors to characterize their performance. The parallel reaction with two competitive products, the desired product P and the side product P_S, is termed the parallel-competitive reaction, resulting from the educts A, B, and C. The educt B reacts with one of the other educts, but with different velocity or reaction rate k_2.

$$\nu_A A + \nu_{C1} C \xrightarrow{k_1} \nu_P P$$

$$\nu_B B + \nu_{C2} C \xrightarrow{k_2} \nu_{PS} P_S \tag{6.5}$$

A prominent example of this reaction class is the iodide-iodate reaction, also known as the Villermaux-Dushman reaction, see Section 6.3.6. Another often-studied system is the competitive-consecutive reaction, where the product P of the first reaction reacts further with a second reactant B to an undesired side product P_S.

$$\nu_A A + \nu_{C1} C \xrightarrow{k_1} \nu_{P1} P$$

$$\nu_B B + \nu_{P2} P \xrightarrow{k_2} \nu_{PS} P_S \tag{6.6}$$

Both reactions are important for many industrial systems in fine chemistry or pharmaceuticals, where several reaction steps lead to the desired product. To compare different reaction conditions, the ratio of the product to the key educt $\nu_P P / \nu_A A$ is defined as the selectivity S of the reaction

$$S = \frac{\text{moles of the desired product}}{\text{transformed moles of the key educt}} = \frac{\nu_P [P]}{\nu_A [A]} = \frac{\nu_P c_P}{\nu_A c_A}, \tag{6.7}$$

and the ratio of the desired product P to the maximum possible output $P/P_{\max}$ is defined as the yield Y of the reaction

$$Y = \frac{\text{moles of the desired product}}{\text{moles of the maximum possible product}} = \frac{\nu_P [P]}{\nu_P [P,\max]} = \frac{\nu_P c_P}{\nu_P c_{P,\max}}. \tag{6.8}$$

In a continuous flow system, such as a microreactor with negligible species accumulation in the system ($dn_i/dt = 0$), the reaction rate r_j for a key species i is expressed as a combination of Eqs. 6.1 to 6.4

$$r_j = \frac{L_i}{\nu_i \cdot \rho_m} \frac{1}{V} = \frac{\dot{n}_i - \dot{n}_{i,0}}{\nu_i \cdot \rho_m} \frac{1}{V}, \tag{6.9}$$

with $n_{i,0}$ as the inlet species flow rate. On the other side, the reaction rate r_j depends on the concentration of the participating reacting species. This correlation can be described by a power law [135]

$$r_j = \pm k_j \cdot \prod_i c_i^{m_i}, \tag{6.10}$$

with the reaction order m_i and a negative sign for educts. The notation for species concentration of component A is given in brackets $[A]$, where the reaction rate can be written as

$$r_j = \frac{1}{\rho_m \cdot V} \frac{dn_A}{dt} = \pm k_j \frac{1}{V} \cdot ([A]^{m_A} \cdot [B]^{m_B} \cdot \dots) \tag{6.11}$$

If the stoichiometric coefficients $|v_i| = 1$, and if for each reactant a first order reaction can be assumed, the reaction rate of the parallel-competitive reaction for Eq. 6.5 can be rewritten as

$$R_1 = \rho_m \left(-1 k_1 c_A c_C - 1 k_2 c_B c_C \right). \tag{6.12}$$

The Gibbs free energy G determines the direction of the reaction. When the reaction reaches its equilibrium ($R_i = 0$), the free energy of the reactant mixture has a minimum value. Eq. 6.12 then simplifies to

$$k_1 c_A c_C = -k_2 c_B c_C, \quad \text{and} \quad K_C = \frac{k_1}{k_2} = -\frac{c_B}{c_A} \tag{6.13}$$

In many cases, the equilibrium constant K_C can be determined with the second law of thermodynamics.

$$K_C(T) \cong K(T) = \exp\left(\frac{-\Delta G^0}{RT} \right) \tag{6.14}$$

The temperature derivative of Eq. 6.14 with $\Delta G^0 = \Delta H^0 - T \Delta S^0$ leads to the van't Hoff equation.

$$\frac{d \ln K}{dT} = \frac{d}{dT} \left(\frac{-\Delta G^0}{RT} \right) = \frac{-\Delta H^0}{RT^2} \tag{6.15}$$

The integration starting from the standard temperature ($T_0 = 298\ K$) to the actual temperature T with constant reaction enthalpy leads to the expression

$$\ln \frac{K}{K(298)} = \int_{298}^{T} \frac{-\Delta H^0}{RT^2} dT. \tag{6.16}$$

The reaction rate constant k_j of the reaction j is independent of the reactant concentrations and composition of the mixture, but is dependent on the temperature, described by Arrhenius law:

$$k_j = k_{0,j} \cdot \exp\left(\frac{-E_{a,j}}{RT} \right) \tag{6.17}$$

with the reaction rate coefficient at standard conditions or frequency factor $k_{0,j}$ and the activation energy $E_{a,j}$. For most reactions, the activation energy lies between 40 and 300 kJ/mol, which suggests an increase of k by a factor of 2 to 50 with a temperature rise of 10 K [30].

6.1.1 Heat and mass transfer with chemical reactions

The mass balance for chemically reacting flow is determined by three factors: diffusive and convective mass transfer, see Eq. 5.3, as well as a species source or sink by chemical reaction. With the reaction kinetics, defined in Eq. 6.10, the reaction-diffusion-convection equation is given by

$$\frac{\partial c}{\partial t} + (\vec{w} \cdot \text{div})\, c = \text{div}\,(D\,\text{grad}\,c) \pm k\,c^m. \tag{6.18}$$

Often, this equation can be reduced to one dimension z in channel direction or two dimensions with z and x in radial direction. Due to the reaction kinetics, a third time scale has to be considered together with the convection and diffusion time scale.

The energy balance with chemical reactions can be derived from the first law of thermodynamics, Eqs. 2.33 and 2.38, including the reaction enthalpy.

$$\frac{\mathrm{d}\,E_{sys}}{\mathrm{d}t} = \dot{Q} - \dot{W} + \dot{n}_{in}E_{in} - \dot{n}_{out}E_{out} \tag{6.19}$$

In words: the energy change within the system equals transferred heat, technical work, and the energy flow in and out of the system. The technical work $\dot{W}$ consists of mechanical work $\dot{W}_S$ by machines, such as stirrers or pumps, and displacement work $\dot{n}_i p \widehat{V}_i$ with the molar volume $\widehat{V}_i$ of the reactant A_i. Introduced into the energy equations, this gives

$$\frac{\mathrm{d}\,E_{sys}}{\mathrm{d}t} = \dot{Q} - \dot{W}_S + \sum_i \dot{n}_i \left(E_i + p\widehat{V} \right)_i \bigg|_{in} - \sum_i \dot{n}_i \left(E_i + p\widehat{V} \right)_i \bigg|_{out}. \tag{6.20}$$

The energy of a system encompasses internal, kinetic, and potential energy as well as all other forms of energy, including electric, magnetic, sound, or light. In most chemical reactors, the internal energy $\widehat{U}_i$ dominates the other energy forms and can be combined with the flow energy to the molar enthalpy $\widehat{H}_i = \widehat{U}_i + p\widehat{V}_i$. This simplifies the energy equation to

$$\frac{\mathrm{d}\,E_{sys}}{\mathrm{d}t} = \dot{Q} - \dot{W}_S + \sum_i \dot{n}_{i,0} \cdot \widehat{H}_{i,0} - \sum_i \dot{n}_i \widehat{H}_i \tag{6.21}$$

with the subscript "0" for the inlet conditions. The enthalpy difference between inflowing and product components is often described with the reaction enthalpy ΔH_r, see also Eq. 4.10. Dealing with chemical reactions, molar values for heat and concentrations are applied to be compatible with the stoichiometry. In mechanical engineering and heat transfer, the specific parameters are related to the mass or the volume of a component and commonly used in design calculations. Here, the concentration c is defined as the mass ratio of a component to the total mass, hence, the molar density ρ_m is included in some equations.

6.1.2 Characteristics of continuous flow reactors

In continuous flow reactors, the mixing situation and residence time in the microchannels belong to the most important characteristics. In an ideal mixed reactor, the temperature and concentrations are uniformly distributed within the reactor volume. In the steady state, the material balance Eq. 6.1 simplifies to

$$0 = \dot{n}_{i,0} - \dot{n}_i + L_{P,i} = \dot{n}_{i,0} - \dot{n}_i + R_i \cdot V. \tag{6.22}$$

In an ideal mixed reactor, the transformation rate R_i is constant and gives the necessary reactor volume

$$V = \frac{\dot{n}_{i,0} - \dot{n}_i}{-R_i}. \tag{6.23}$$

The ideal tubular reactor has no backmixing due to the plug flow of the fluid and uniform radial composition and temperature. This is the case in turbulent flow, which occurs often in conventional equipment. The plug-flow reactor PFR is treated here for comparison. The reactants are continuously consumed when they flow through the reactor. The material balance can be arranged for a differential element with $\Delta V \to 0$,

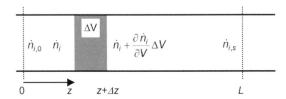

Fig. 6.1. Ideal tubular plug-flow reactor (PFR), see also Figure 3.1 and 6.3.

see Figure 6.1. The reaction rate equals the species consumption in a volume element.

$$-\frac{d \dot{n}_i}{dV} + R_i = 0 \tag{6.24}$$

Integration yields for the reactor volume

$$V = \int_{\dot{n}_{i,0}}^{\dot{n}_{i,s}} \frac{d \dot{n}_i}{R_i}. \tag{6.25}$$

With the conversion X for the key reactant A_i

$$X = \frac{\dot{n}_{i,0} - \dot{n}_{i,s}}{\dot{n}_{i,0}}, \tag{6.26}$$

the volume of the reactor can be written as

$$V = \dot{n}_{i,0} \int_0^{X_s} \frac{d X}{-R_i} = \dot{V}_0 c_{i,0} \int_0^{X_s} \frac{d X}{-R_i}. \tag{6.27}$$

In a plug-flow reactor, all fluid elements leaving the reactor have the same residence time and the same history. In reality, the velocity distribution leads to a residence time distribution (RTD) of all entering fluid elements, which affects the reactor performance, the product selectivity, and yield. The RTD depends on the axial mixing, recirculation, backmixing, and on other fluid dynamic effects. Axial mixing in a tubular reactor can be described by the dispersion model, which considers a plug flow with superimposed lateral dispersion. The latter is described by means of a constant effective axial dispersion coefficient D_{ax}, which has the same unit as the molecular diffusion coefficient D_m, but is much larger due to radial velocity distribution, vortices, and eddies. The mass flow by dispersion is described similar to diffusion by Fick's law

$$J = -D_{ax} \cdot A \cdot \frac{\partial c}{\partial z}. \tag{6.28}$$

The mass balance of a non-reacting tracer with the concentration c gives the RTD from the dispersion model. For constant density and mean flow velocity w, the mass balance can be written as Fick's second law

$$\frac{\partial c}{\partial t} = -\bar{w}\frac{\partial c}{\partial z} + D_{ax}\frac{\partial^2 c}{\partial z^2}. \tag{6.29}$$

For steady reacting flow conditions, the above equation can be written as

$$0 = -\bar{w}\frac{\partial c_i}{\partial z} + D_{ax}\frac{\partial^2 c_i}{\partial z^2} + \sum_j v_{ij}r_j. \tag{6.30}$$

Introducing the dimensionless variables $t^* = t/t_P$, $t_P = L/\bar{w}$, $z^* = z/L$, $c^* = c/\bar{c}_0$, $\bar{c}_0 = n_{0,i}/V$, and the dimensionless Bodenstein number $\text{Bo} = w \cdot L/D_{ax}$ gives the non-dimensional differential equation of Fick's second law from Eq. 6.29

$$\frac{\partial c^*}{\partial t^*} = -\frac{\partial c^*}{\partial z^*} + \text{Bo}\frac{\partial^2 c^*}{\partial z^{*2}}. \tag{6.31}$$

The Bodenstein number Bo is the ratio of the axial dispersion time $t_{ax} = L^2/D_{ax}$ and the mean residence time t_P, which is identical to the space time for reaction mixtures within the reactor. For small Bo numbers, the axial dispersion time is short compared to the mean residence time, indicating high backmixing within the reactor. For larger Bo numbers, no dispersion occurs and plug flow is assumed. Axial dispersion can be neglected for Bo > 100. Clicq et al. [405] investigated the axial dispersion in nanoscale channels with Bo numbers higher than 10^5 in analytical and high-throughput applications. A second dimensionless number is important for dispersion in channel flow, the Fourier number Fo' in mass transfer, similar to the Fourier number in heat transfer, see Eq. 4.21.

$$\text{Fo'} = \frac{D_m t}{x^2} \tag{6.32}$$

The characteristic length x points in radial direction and can be replaced by the "radius" of the channel, here $d_h/2$.

$$\text{Fo'} = \frac{4\,D_m\,t}{d_h^2} \tag{6.33}$$

With their high surface-to-volume ratio, microreactors provide enhanced heat transfer and also increased mass transfer due to the small dimensions . Similarly, the reduced size decreases the necessary residence time compared to large scale reactors. The residence time is often a key element in designing chemical processes, as different kinetic processes must be adjusted accordingly to achieve the desired product qualities.

To determine the residence time distribution, a tracer is injected as a Dirac pulse with an amount n_{inj} at the reactor entrance. For a continuous flow system, the outlet concentration distribution to an inlet pulse concentration can be described with

$$C(t^*) = -\frac{1}{2}\sqrt{\frac{\text{Bo}}{\pi t^*}}\,\exp\left(\frac{(1-t^*)^2\,\text{Bo}}{4t^*}\right), \tag{6.34}$$

which is displayed for various Bo numbers in Figure 6.2.

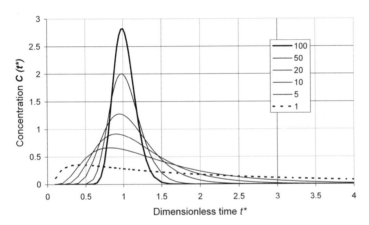

Fig. 6.2. Residence time distribution predicted by the dispersion model with the Bo number $\text{Bo} = w \cdot L/D_{ax}$.

A broad RTD from small Bo numbers reduces the reactor performance with positive reaction order, particularly for high reaction order m. The reaction selectivity S and yield Y of the target product are even more sensitive in parallel-competitive, competitive-consecutive, or more complex reactions with intermediates. The axial dispersion D_{ax} in tubular reactors depends on the flow situation described by the Reynolds number $\text{Re} = w \cdot d_h/v$ and the fluid properties described by the Schmidt number $\text{Sc} = v/D_m$ with the molecular diffusion coefficient D_m. The combination of both, the Péclet number $\text{Pe} = \text{Re} \cdot \text{Sc} = w \cdot d_h/D_m$, is used to describe the molecular diffusion in a fluid element. The axial Péclet number $\text{Pe}_{ax} = w \cdot d_h/D_{ax}$ with the axial dispersion is similar to the molecular Péclet number Pe_m, but should not be confused. The Bo number correlates with the axial Pe number, see also Eq. 5.30,

$$\text{Bo} = \text{Pe}_{ax} \frac{L}{d_h}. \tag{6.35}$$

The flow in straight microchannels with typical hydraulic diameters between 10 and 500 μm is often laminar and has a parabolic velocity profile. Hence, the molecular diffusion in axial and radial direction determines the RTD, see Figure 6.3.

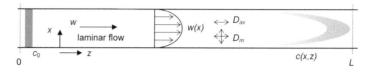

Fig. 6.3. Flow situation in a laminar flow microchannel, axial spreading of an initial concentration pulse by the laminar flow profile.

Radial diffusion diminishes the spreading effect of the parabolic velocity profile, while the axial diffusion increases the dispersion [406]. The axial dispersion of laminar flow in an empty circular tube is called Aris-Taylor dispersion and can be estimated with the help of the dimensionless numbers,

$$D_{ax} = D_m + \frac{w^2 d_h^2}{192\, D_m} \quad \text{for} \quad \frac{L}{d_h} \geq 0.004 \frac{w\, d_h}{D_m} \tag{6.36}$$

leading to

$$\frac{1}{\text{Pe}_{ax}} = \frac{1}{\text{Re} \cdot \text{Sc}} + \frac{\text{Re} \cdot \text{Sc}}{192}. \tag{6.37}$$

Laminar flow with velocity and concentration profile must be developed for correct application of these correlations in fluid dynamics and mass transfer. The axial Pe_{ax} number for turbulent flow in circular pipes is described with the empirical correlation

$$\frac{1}{\text{Pe}_{ax}} = \frac{3 \cdot 10^7}{\text{Re}^{2.1}} + \frac{1.35}{\text{Re}^{1/8}}. \tag{6.38}$$

Returning to laminar flow, Eq. 6.37 can express the ratio of the axial diffusivity D_{ax} to the molecular diffusion coefficient D_m

$$\frac{D_{ax}}{D_m} = 1 + \frac{\text{Pe}}{192}. \tag{6.39}$$

Introducing the axial diffusivity D_{ax} of laminar flow into the Bodenstein number Bo gives the following correlation

$$\text{Bo} = \frac{\bar{w} L}{D_{ax}} = \frac{\bar{w} L}{D_m + w^2 d_h^2 / (192\, D_m)}. \tag{6.40}$$

For low molecular diffusion coefficients $D_m \to 0$ of liquids, the denominator of the above equation can be simplified and leads to

$$\text{Bo} \cong \frac{192 \, D_m \cdot L}{\bar{w} d_h^2}. \tag{6.41}$$

The channel length L divided by the mean velocity $\bar{w}$ gives the mean residence time t_P. Introducing the residence time into the above equation gives an expression with the Fourier number Fo' in mass transfer, see Eq. 6.33.

$$\text{Bo} \cong \frac{192 \, D_m \cdot t_P}{d_h^2} = \frac{192 \, \text{Fo'}}{4}$$

$$\cong 48 \, \text{Fo'}. \tag{6.42}$$

This correlation shows the relationship between axial and radial mass transfer in laminar flow and is very important for the design of microchannel reactors. The Fo' number in mass transfer is similar to the Fourier number Fo in heat transfer, see Section 4.1.2.

For potential applications of microreactors, a comparison with fixed-bed reactors assist in locating optimal parameters and application limits. Renken and Kiwi-Minsker [30] compare the dispersion in a fixed-bed reactor with that in a microchannel reactor with the same reactor cross section porosity $\varepsilon = A_{channel}/A_{total}$. The particle diameter in the bed is correlated to the hydraulic diameter of the microchannel with $d_P = 1.5(1 - \varepsilon)d_h/\varepsilon$. In straight laminar flow, the axial Pe_{ax} number is higher in the microreactor than in the bed over the range of $3 < \text{Re} \cdot \text{Sc} < 100$. This represents a low axial dispersion and a narrow RTD in this range, where the microreactor exhibits better performance than a fixed-bed reactor with the same porosity ε. The Pe_{ax} number of the microreactor has a maximum due to the relatively short radial diffusion time $t_D = d_h^2/D_m << L/w = t_P$ compared to the space time, the mean residence time. Then, axial dispersion is determined by the second term in Eq. 6.37 to $\text{Pe}_{ax} \cong 192/\text{Re} \cdot \text{Sc}$. The minimal dispersion (maximum Pe_{ax}) is obtained by derivation of Eq. 6.37 at $\text{Re} \cdot \text{Sc} = \sqrt{192} = 13.86$. Together with reaction kinetics and residence time the axial dispersion determines the characteristics of continuous flow reactors for design tasks. A more detailed description follows in Section 6.3.

6.1.3 Temperature control in microchannel reactors

Heat transfer plays an important role in reaction engineering since the reaction transformation rate is influenced by the temperature and many reactions are accompanied by heat release (exothermic) or consumption (endothermic). The activation energy E_a of the reaction is delivered from the temperature increase of the reactants. An insufficient heat transfer may lead to a temperature increase, which raises the transformation rate of the reaction and allows the reaction run away. Insufficient temperature control may result in hot or cold spots, where reaction conditions change, and diminishes the reaction performance. Also, reactor construction material or catalysts may be damaged or destroyed by hot spots resulting in irreversible reactor destruction. Aside from heat transfer, operational parameters are important for the temperature management of a reactor, which is known as "parametric sensitivity". Small channel diameters enhance the heat transfer and may overcome some heat transfer limitations. To determine the necessary channel diameter, the reaction kinetics, enthalpy, and the heat transfer performance need to be estimated to avoid "runaway" and "parameter sensitivity". In small channels, the temperature and composition are assumed to be constant in a cross section. The wall temperature is constant due to the high heat conductivity in the reactor walls and a high surface-to-volume ratio $a_V = \Delta A/\Delta V \approx 4/d_h$. A laminar and fully developed velocity and temperature profile can be assumed as well as almost constant fluid properties. Neglecting energy contributions from flow friction and technical work in the channel, the energy change in a channel element ΔV is determined by the reaction heat $R_1 \Delta H_r$ of the leading component, the heat transfer coefficient α at the channel wall, the cross section A, the surface-to-volume ratio a_V, and the heat capacity flux $\dot{m}\, c_p$

$$\frac{d\,T}{dV} = \frac{d\,T}{A \cdot dz} = \frac{R_1 \cdot \Delta H_r}{\dot{m} \cdot c_p} - \frac{\alpha\, a_V\, (T - T_W)}{\dot{m} \cdot c_p}. \tag{6.43}$$

The reaction enthalpy ΔH_r is determined according to Eq. 4.10, while the reaction rate R_1 is determined from the material balance of the leading component and its conversion along the channel axis.

$$\frac{d\,X}{A \cdot dz} = \frac{-R_1}{\dot{n}_{1,0}} \tag{6.44}$$

The temperature profile along the channel axis can be derived from the heat balance along the straight channel with constant cross section A

$$\frac{d\,T}{dz} = \frac{R_1 \cdot \Delta H_r}{\bar{w} \cdot \rho \cdot c_p} - \frac{4\,\alpha\,(T - T_W)}{d_h \cdot \bar{w} \cdot \rho \cdot c_p}, \tag{6.45}$$

which gives, with the space-time correlation $dt_P = dz/\bar{w}$ and the adiabatic temperature rise

$$\Delta T_{ad} = -c_{1,0} \cdot \rho_m \cdot \Delta H_r/(\rho c_p), \tag{6.46}$$

the following correlation:

$$\frac{d\,T}{dt_P} = \frac{-R_1}{c_{1,0}} \Delta T_{ad} - \frac{4\alpha\,(T - T_W)}{d_h \cdot \bar{w} \cdot \rho \cdot c_p}. \tag{6.47}$$

Microreactors often operate under isothermal conditions due to the high heat transfer where temperature changes dT/dt_P can be neglected. The mass balance can be simplified with the above assumptions to

$$\frac{d\,X}{dt_P} = \frac{-R_1}{c_{1,0}}. \tag{6.48}$$

The energy balance can also be formulated to match Eq. 6.48 for isothermal conditions

$$\frac{4\,\alpha\,(T - T_W)}{d_h \cdot \rho \cdot c_p\, \Delta T_{ad}} = -\frac{R_1}{c_{1,0}}. \tag{6.49}$$

With the exception of a zero-order reaction $m = 0$, the reaction rate R_1 varies along the channel length z. Eq. 6.49 can only be satisfied with varying $(T - T_W)/d_h$ along z, which is not feasible. However, the reactor can be designed with an acceptably small temperature difference $(T - T_W)$, which is demonstrated here for a first-order reaction $-R_i = kc_1$. The rate constant k is related to the rate constant k_W at wall temperature T_W:

$$k = k_W \left(\frac{k}{k_W} \right) = k_W \cdot \exp \left(\frac{E_a (T - T_W)}{R \cdot T \cdot T_W} \right) \tag{6.50}$$

Rearranging Eq. 6.49 with Eq. 6.50 for the hydraulic diameter gives an estimation of the maximum diameter for sufficient small temperature difference:

$$\frac{1}{d_h} = \frac{\rho \cdot c_p}{4 \cdot \alpha} \cdot \exp \left(\frac{E_a (T - T_W)}{R \cdot T \cdot T_W} \right) \cdot \frac{\Delta T_{ad}}{(T - T_W)} \cdot k_0 \exp \left(\frac{E_a}{R \cdot T_W} \right) \tag{6.51}$$

The heat transfer coefficient α and the hydraulic diameter can be combined to the Nußelt number $Nu = \alpha \cdot d_h / \lambda$, see Section 4.1.3 with Eq. 4.30. Now, the minimum hydraulic diameter can be estimated with

$$\frac{1}{d_h^2} = \frac{\rho \cdot c_p}{4 \cdot Nu \cdot \lambda} \cdot \exp \left(\frac{E_a (T - T_W)}{R \cdot T \cdot T_W} \right) \cdot \frac{\Delta T_{ad}}{(T - T_W)} \cdot k_0 \exp \left(\frac{E_a}{R \cdot T_W} \right) \tag{6.52}$$

An estimation of Renken and Kiwi-Minsker [30] yields a channel diameter of approx. 100 μm allowing a maximum temperature increase of 3 K for a strong exothermic gas-phase reaction.

For the thermal behavior of a channel reactor, three different parameters or numbers are important:

$$\text{Arrhenius number: } \gamma = \frac{E_a}{R \cdot T_W} \tag{6.53}$$

$$\text{heat production potential: } S' = \frac{-\Delta T_{ad}}{T_W} \cdot \frac{E_a}{R \cdot T_W} \tag{6.54}$$

$$\text{time ratio of reaction and cooling: } N = \frac{t_R}{t_C} = \frac{1}{k(T_W) \, c_{1,0}^{m-1}} \cdot \frac{4 \cdot \alpha}{\rho \cdot c_p \cdot d_h} \tag{6.55}$$

The characteristic reaction time for a first-order reaction is based on the inlet conditions. It is assumed that the inlet temperature T_0 is equal to the wall temperature T_W. With these boundary conditions, the axial temperature profile is strongly influenced by the time ratio $N = t_R/t_C$ of reaction and cooling. The temperature is very sensitive at a minimum ratio N_{min}, where the reactor moves into an unstable regime. For safer reactor operation, a minimum ratio N_{min} can be estimated by an empirical function with the heat production potential S':

$$N_{min} = \left(\frac{t_R}{t_C} \right)_{min} = 2.72 \cdot S' - B \sqrt{S'}, \tag{6.56}$$

where the parameter B is dependent on the reaction order m:

- $B(m = 0) = 0$,
- $B(m = 0.5) = 2.6$,
- $B(m = 1) = 3.37$, and
- $B(m = 2) = 4.57$.

A parameter study by Renken and Kiwi-Minsker [30] reveals that a channel diameter of approx. 200 µm is necessary for an effective and safe reactor operation under highly exothermic conditions (gas phase reaction with partial oxidation of o-xylene) with ΔT_{ad} up to 400 K and wall temperatures up to approx. 620 K. With the application of microreactors, low and homogeneous temperatures can often be achieved to optimize the selectivity of the desired product from a complex reaction under harsh conditions. Particularly, the concentrations of the educts can be increased to enhance the volume productivity or space-time yield.

6.2 Wall Mass Transfer and Surface Reactions in Microfluidic Systems

Mass transfer to and from reactive surfaces in microchannel devices is essential to evaluating and quantifying heterogeneous catalytic reactions or analytical detection of biological molecules in surface plasmon resonance (SPR) devices. As an introduction, Homola et al. [407] show in their review characteristics and applications of microstructured SPR sensors. The major application fields for SPR sensors are biochemical analysis [408] of proteins, DNAs, and other larger molecules as well as surface characterization in the nanometer range [409]. Catalytic reaction rates are controlled by the mass transfer to and from the catalytic wall as well as the transport within the porous catalytic material. The small length scales lead to laminar flow, and the correspondingly high surface-to-volume ratio suggests that an understanding of transport processes and reactions on surfaces is necessary.

Many applications of microreactors are dependent on the accurate prediction of wall-fluid interaction at interfaces. The steady-state, diffusion-limited transport to a surface in a pressure-driven flow is known as the Graetz problem. Yoon et al. [410] describe analytical and numerical solutions of mass transfer in electrochemical microreactors. They suggest possibilities for renewing the mass transfer boundary layer by removal of the wall layer via multiple outlets, adding fresh material through multiple inlets, or create a spiraling, transverse flow with the integration of herringbone ridges along the channel walls. The situation in a channel with symmetrical wall is displayed in Figure 6.4 with entrance and fully-developed region for mass transfer. The mass transfer along a channel is usually divided into two main operating regimes: the "entrance region", implying the presence of a mass transfer boundary layer, and the "fully developed region", with bulk depletion of the sample [180].

In this section, the term boundary layer strictly applies to mass transfer boundary layers and not to momentum boundary layers, which are not often present since the flows are fully developed laminar flows as a consequence of the small dimensions in straight microchannels. In the context of microchannel wall mass transfer, the

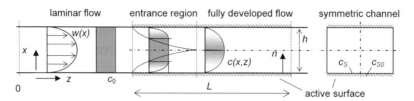

Fig. 6.4. Channel flow with mass transfer in a certain region.

entrance region has been studied in detail since it is relevant to most analytical SPR systems. Two-compartment models, approximating the mass transfer in the entrance region, have provided additional insight into transport processes in SPR systems by taking into account surface saturation over time and reaction-limited conditions.

From a practical point-of-view, operating a device in the entrance regime, as is often the case in SPR, leads to low sample capture fractions. For example, in a typical SPR experiment operating with a flow rate of 5 µl/min through a channel of cross-section 50×500 µm^2 over a 1 mm long detection zone, the capture fraction f, being the total mass fraction of a specific analyte captured at the surface, is estimated to be 7 % less than diffusion limiting conditions. The remaining 93 % of the analyte flows in the bulk region outside the capture region and is lost for detection purposes.

In general, microfluidic devices aim to minimize sample consumption, yet operating in the entrance region goes against this goal. However, when channels reach the micro-scale and the regime becomes "fully developed", the capture fraction increases greatly, minimizing the detection losses. In catalytic microreactors, a smaller hydraulic diameter d_h increases the conversion rate.

Therefore, the main goal of this section is to summarize and discuss expressions for relevant operation regimes of flow-through microfluidic devices involving surface reaction. Particular emphasis is placed on understanding transport processes in the fully developed region of the Graetz problem. Illustrative experimental examples and kinetic data from microfluidics literature are given by Gervais and Jensen [411]. The overall analytical procedure provides criteria for designing and operating continuous microfluidic sensors and microreactors with catalytic walls based on convection, diffusion, and reaction of multiple species.

6.2.1 Continuum transport model

The general conservation equations (Eq. 6.18), including a bulk reaction term, take the form

$$\rho_m \cdot \frac{\partial c_i}{\partial t} = -\frac{1}{A} \, \mathrm{div} \cdot \vec{N}_i + R_i \tag{6.57}$$

where the molar flux $\vec{N}_i$ of the component i consists of the diffusion and convection term

$$\vec{N}_i = \rho_m \cdot A \left(-D_{m,i} \cdot \mathrm{grad} \, c_i + \vec{w} \right). \tag{6.58}$$

Here, c_i is the volume concentration of the ith species, $D_{m,i}$ is the molecular diffusivity of the solute, $\vec{w}$ is the fully developed velocity profile, and R_i is the volumetric rate of species generation in the bulk. Introducing the coordinate in radial direction x and axial direction z, the convective and diffusive bulk transport (Eq. 6.57) can be written as

$$\frac{\partial c(x,z,t)}{\partial t} = D_m \frac{\partial^2 c(x,z,t)}{\partial x^2} + D_m \frac{\partial^2 c(x,z,t)}{\partial z^2} - w(x) D_m \frac{\partial c(x,z,t)}{\partial z}. \tag{6.59}$$

For all surfaces, the boundary conditions on the molar flux $\vec{N}_i$ represent the balance between flux to the surface and surface reaction.

$$\vec{n} \cdot \vec{N}_i = R_{s,i} \cdot V_{s,i} \tag{6.60}$$

The surface unit vector $\vec{n}$ is perpendicular to the surface and points channel inward. R_{si} is the inward flux of species i into the bulk, such that surface adsorption indicates a negative R_{si} term.

Solutions to these equations have been developed for various geometries [180], however, only flat plates will be considered here. The lithographic, embossing, and ablation techniques used to create most microfluidic systems produce broad and shallow channels, which are desirable in most analytical applications requiring measurement of the surface-bound species. The fully developed three-dimensional laminar axial velocity profile w(height h, width b) can be represented as a simple function w(height h) except close to the channel side walls. For studies of mass transfer in channels with cross sections of different geometries, such as triangular channels produced from anisotropic etching of silicon or semi-elliptical channels obtained through isotropic etching in glass, solutions for simple cases have been provided in literature for the fully-developed region [212].

To further simplify the analysis, constant diffusivity D_m is assumed for all species. For larger molecules from biological applications, the diffusivity can be non-linear and highly dependent on molecule size, shape, and charge. Bimolecular surface reactions involving reversible binding of species are considered as being sufficient to describe the majority of surface association/dissociation reactions. The rate equation of the surface concentration c_s takes the form

$$\frac{\partial c_s}{\partial t} = k_{on} c_w (c_{s0} - c_s) - k_{off} c_s \tag{6.61}$$

where c_w is the bulk concentration in the channel near the reactive wall. The concentration of surface bound species per area c_s is a function of surface position. The concentration c_{s0} is the total number of binding sites per area. The association and dissociation rate constants of the bimolecular reaction are k_{on} and k_{off}, respectively (with units $M^{-1}s^{-1}$ and s^{-1}).

Assuming that the channel is broad and shallow ($w \gg h$), and that bulk reaction is negligible compared to surface reactions ($R_i = 0$), the transport equations and boundary conditions simplify to a two-dimensional problem. It is useful to scale the equations in order to reveal the dimensionless parameters governing the system. The dimensionless parameters are the axial position $z^* = z/(h\,\text{Pe})$, the channel height $h^* = x/h$, the normalized transversal diffusion time $t^* = D_m t/h^2$, relative adsorption capacity $\varepsilon_A = c_0 h \rho_m/c_{s0}$, and bulk concentration $c^* = c_i/c_{i0}$. Additionally, the flow velocity and the reaction rate are represented in their dimensionless form $Pe = \bar{w} h/D_m$ and $k_{on} c_{s0} h/(\rho_m D_m) = DaII_s$, the Péclet number and Damköhler number of second kind for surface reactions, respectively. The resulting dimensionless transport equation (Eq. 6.59) takes the general form:

$$\frac{\partial c^* (h^*, z^*, t^*)}{\partial t^*} = \left(\frac{\partial^2 c^* (h^*, z^*, t^*)}{\partial h^{*2}} + \frac{1}{Pe^2} \frac{\partial^2 c^* (h^*, z^*, t^*)}{\partial z^{*2}} \right)$$
$$-w(h^*) \frac{\partial c^* (h^*, z^*, t^*)}{\partial z^*}, \tag{6.62}$$

The bulk transport in a channel is distinguished between mass transfer to a single wall (asymmetrical case) or to both walls (symmetrical case) as described by the following equations, respectively.

$$\frac{\partial c^*}{\partial t^*} = \left(\frac{\partial^2 c^*}{\partial h^{*2}} + \frac{1}{Pe^2} \frac{\partial^2 c^*}{\partial z^{*2}} \right) - 6 h^* (1 - h^*) \frac{\partial c^*}{\partial z^*}, \tag{6.63}$$

$$\frac{\partial c^*}{\partial t^*} = \left(\frac{\partial^2 c^*}{\partial h^{*2}} + \frac{1}{Pe^2} \frac{\partial^2 c^*}{\partial z^{*2}} \right) - \frac{3}{2} (1 - 4 h^{*2}) \frac{\partial c^*}{\partial z^*}. \tag{6.64}$$

The dimensionless rate equation for surface reactions is given by

$$\frac{\partial c_s^* (z^*, t^*)}{\partial t^*} = \varepsilon_A \, DaII_s \left[c^* (h^* = 0, z^*, t^*) \, (1 - c_s^* (z^*, t^*)) - \bar{K}_D \, c_s^* (z^*, t^*) \right]. \tag{6.65}$$

$\bar{K}_D = k_{off}/(k_{on} c_0)$ is the dimensionless equilibrium dissociation constant and $w(h^*)$ is the normalized fully developed laminar flow velocity profile depending solely, in two-dimensional flow, on the normalized height parameter h^*. The boundary conditions of the dimensionless equations for both walls (symmetrical case) are

$$\left. \frac{\partial c^*}{\partial h^*} \right|_{h^* = 0} = 0 \text{ (symmetry),} \tag{6.66}$$

$$\left. \frac{\partial c^*}{\partial h^*} \right|_{h^* = 1/2} = -\frac{1}{\varepsilon_A} \frac{\partial c_s^*}{\partial t^*} \text{ (reaction at walls),} \tag{6.67}$$

for one wall with reactive surface (asymmetrical case)

$$\left. \frac{\partial c^*}{\partial h^*} \right|_{h^* = 0} = -\frac{1}{\varepsilon_A} \frac{\partial c_s^*}{\partial t^*} \text{ (reaction at wall 1),} \tag{6.68}$$

$$\left. \frac{\partial c^*}{\partial h^*} \right|_{h^* = 1} = 0 \text{ (insulation at wall 2),} \tag{6.69}$$

and the initial and geometrical boundary conditions

$$c^*(h^*, z^*, t^* = 0) = 0, \quad c^*(z^*, t^* = 0) = 0, \tag{6.70}$$

$$c^*(h^*, z^* = 0, t^*) = 1, \quad \left.\frac{\partial c^*}{\partial z^*}\right|_{z^* = \infty} = 0. \tag{6.71}$$

Eqs. 6.62 and 6.65 are coupled by the species flux balance at the wall

$$\left.\frac{\partial c^*(h^*, z^*, t^*)}{\partial h^*}\right|_{h^* = [0,1]} = -\frac{1}{\varepsilon_A}\frac{\partial c_s^*(z^*, t^*)}{\partial t^*}. \tag{6.72}$$

The above set of equations can be analytically solved for special cases, numerical solutions are given by Gervais and Jensen [411]. To interpret these results, it is important to know what the dimensionless parameters stand for.

6.2.2 Physical meaning of dimensionless parameters

The dimensionless parameters in Eqs. 6.62 to 6.72 simplify the correlations and show important characteristics of the physical system. The relative absorption capacity $\varepsilon_A = c_0 h \rho_m / c_{s0}$ corresponds to the relative density of species between bulk and fully saturated surfaces, It is a measure of surface adsorption capacity relative to the bulk concentration and matches the units of surface concentration and bulk concentration. A small ε_A indicates a high relative surface capacity leading to a longer saturation time. In cases of large Damköhler numbers, $DaII_s \gg 1$, the transport to the surface is strongly diffusion-limited. While for $DaII_s \ll 1$, the transport becomes limited by the reaction at the surface. Therefore, a flat concentration profile can be expected across the channel.

The diffusion/convection length scale z^* is also known as the inverse of the Graetz number $Gz = Peh/z$. It is an important parameter in this section, as it controls the mass transfer transition from the entrance region to the fully developed region. At $z^* = 1$, the time scale required for diffusion across the channel height is in the same range as the time scale required for crossing a distance z at velocity $\bar{w}$. For $z^* < 1$, the convection time scale is shorter than the diffusion time scale and a portion of the species in the channel is not able to "reach" the reactive surface and leaves the active region resulting in a mass transfer boundary layer. For $z^* > 1$, bulk species have sufficient time to diffuse to the reactive surfaces and no mass transfer boundary layer develops. In the case of a reaction on both the top and bottom walls, the diffusion time scale becomes $\bar{t}_d = h^2/4D$, as a molecule is, at most, half a channel height away from a reactive wall.

6.2.3 Graetz problem and wall mass transfer

The non-linear convection/diffusion/reaction problem summarized in Eqs. 6.62 to 6.72 can be solved numerically (e.g., by using finite element methods) to provide an accurate solution to a specified set of parameters. Analytical solutions, when feasible, provide insight into the relationship between parameters (space, time, rate constants, etc.) and dependent variables (bulk and surface concentration, capture fraction, etc.).

No complete analytical solutions exist for the whole problem due to the non-linear surface reaction term. Chattopadhyay and Veser [412] propose a 2D boundary-layer model coupled with reaction kinetics with surface and gas-phase reactions to investigate the intrinsic safety of microreactors in hydrogen/air combustion. Gervais and Jensen [411] segmented the microchannel and its related solution space into physically relevant parameter regimes, for which analytical solutions exist. In particular, the transport equations can be decomposed in two accurate analytical solutions for $z^* \gg z^*_{crit}$ and $z^* \ll z^*_{crit}$, where z^*_{crit} represents the critical value, from which the two analytical models have the same relative error compared to the numerical solution. These solutions can be subdivided once more depending on whether the surface Damköhler number is finite (partially reaction-limited) or infinite (fully diffusion-limited).

In most applications involving high relative adsorption capacity (low ε_A), the bulk concentration reaches a steady state value long before the surface has been significantly saturated. The transport in the bulk can then be assumed to be in pseudo-steady state in relation to the surface at all times. Moreover, in most microfluidic applications, the axial convection is much faster than the axial diffusion, i.e. Pe $\gg 1$. Pe is usually on the order of $10^2 - 10^4$ in lab-on-a-chip applications. As a result axial diffusion can be neglected and the simplified, scaled transport problem, known as the Graetz problem in heat transfer, then takes the form:

$$\frac{\partial^2 c^* (h^*, z^*)}{\partial h^{*2}} = \text{Pe}(h^*) \frac{\partial c^* (h^*, z^*,)}{\partial z^*}. \tag{6.73}$$

The coupled, nonlinear boundary condition from the kinetic rate equation, Eq. 6.65, implies that the system, described with Eqs. 6.72 and 6.73, must generally be solved numerically. However, in cases where $c_s \ll c_{s0}$ (at sufficiently small times), the bimolecular surface reaction, Eq. 6.61, reduces to a pseudo first order reaction and the boundary condition becomes linear:

$$\frac{\partial c^* (h^*, z^*, t^*)}{\partial h^*}\bigg|_{h^*=0} = \text{DaII}_s \, c^* \, (h^* = 0, z^*, t^*). \tag{6.74}$$

The linear system, which is also relevant to practical cases, can be solved analytically using the separation of variables method to obtain an eigenfunction expansion of the solution. The exact form of the eigenfunction expansion has been formulated for circular pipe and parallel plate geometry with rapid reaction at both walls. In the following, the exact form of the eigenfunction expansion is introduced for the case of parallel plates with reaction at only one wall, including the first five expansion coefficients and eigenvalues for the fully diffusion-limited case. The scaled Graetz problem for a first order reaction is described by

$$\frac{\partial^2 c^*}{\partial h^{*2}} = 6h^* (1 - h^*) \frac{\partial c^*}{\partial z^*},$$

$$\text{or} \tag{6.75}$$

$$\frac{\partial^2 c^*}{\partial h^{*2}} = \frac{3}{2} (1 - 4h^{*2}) \frac{\partial c^*}{\partial z^*},$$

for the asymmetrical and symmetrical case with both reacting walls or a single wall, respectively. The boundary conditions for both cases are given by the Eqs. 6.66 and 6.68. The surface reaction of first order is described by temporal derivative of the dimensionless concentration c_s^* and the DaII$_s$ number

$$\frac{\partial c_s^*}{\partial t^*} \approx \varepsilon_A \, \text{DaII}_s \, c_w^* \tag{6.76}$$

for an unsaturated surface $c_s^*(z^*,t^*) \ll 1$ and a pseudo first order reaction. The general form of the solution is a convergent series of exponential functions

$$c^*(h^*,z^*) = \sum_{i=1}^{\infty} a_i G_i(\lambda_i, h^*) \, \exp\left(-\frac{\lambda_i^2 z^*}{6}\right) \tag{6.77}$$

with the basis function $G_i(\lambda_i, h^*)$. For the symmetrical case with two active walls, the basis function is given by

$$G_i(\lambda_i, h^*) = \exp\left(-\lambda_i h^{*2}\right) M\left(\frac{1}{4} - \frac{\lambda_i}{16}, \frac{1}{2}, \lambda_i h^{*2}\right), \tag{6.78}$$

the so-called Kummer M function, a confluent hypergeometric function. More details can be found in Abramowitz and Stegun [413] and Deen [180] Chapter 9.

These analytical solutions converge slowly: after a five term expansion, the maximum relative error on the concentration remains ~ 5 %. However, the complete set of functions is only necessary for the complete two-dimensional concentration profile must be mapped. For most practical applications, the saturation time scale of the substrate or the fraction of mass bound to the surface is of primary interest, rather than the detailed concentration profile in the fluid channel. In these cases, the general solution can be separated according to different parameter regimes and analytical solutions calculated with respect to the velocity-weighted, or perfectly mixed vessel ("mixed-cup"), bulk concentration c_b^*:

$$c_b^*(z^*,t^*) = \int_A c_b^*(h^*,z^*,t^*) w(h^*) dh^*$$

$$= \sum_{i=1}^{\infty} A_i \exp -\frac{\lambda_i^2 z^*}{6} \tag{6.79}$$

where $w(h^*)$ is the normalized velocity profile. The coefficients A_i are calculated from the integral

$$A_i = \int_0^1 a_i G_i(\lambda_i, h_*) w(h^*) dh^* \tag{6.80}$$

with eigenvalues and solution function from mathematical reference books [414].

6.2.4 Wall mass transfer in microchannel reactors

For heterogeneous catalytic reactions, the walls of microstructured reactors are coated with a porous catalyst material. For fast and strongly exo- or endothermic

reactions, the mass transfer to the wall has to be considered for the reactor performance. The velocity and concentration profile at the entrance develops along the
channel axis to a stagnant profile in fully developed flow. For laminar flow conditions, the length of the fluid dynamic entrance is given by

$$L_e \leq 0.06 \cdot \text{Re} \cdot d_h, \tag{6.81}$$

see also Section 3.1.6 and 4.2.2. The mass transfer can be described by the same
basic correlations, but must be scaled by the fluid properties. These are expressed
as the ratio of the momentum transfer to the mass transfer defined as the Schmidt
number $\text{Sc} = v/D_m$. The scaled entrance length for the concentration profile is given
by

$$L_e \leq 0.05 \cdot \text{Re} \cdot \text{Sc} \cdot d_h. \tag{6.82}$$

For gases, the Sc number is approximately unity, hence, the fluid dynamic entrance
length is in the order of the concentration entrance length. In liquids, the Sc number
is in the order of $O(10^3)$ and higher, where the entrance length differs significantly
for fluid dynamics and mass transfer.

In the entrance zone, the concentration gradients are very high, as is the corresponding transfer coefficient $k_D = \dot{n}/(\rho_m A \Delta c)$, which decreases with the developing
flow. In the fully developed zone, the mass transfer has a constant value comparable
with the laminar heat transfer and Nu number, see Section 4.1.3. As a dimensionless
number, the Sherwood number $\text{Sh} = k_D d_h/D_m$, the constant value can be expressed
as follows:

$$\text{Sh} = B \left(1 + 0.095 \frac{d_h}{L} \, \text{Re} \cdot \text{Sc} \right)^{0.45} \tag{6.83}$$

The constant B is dependent on the geometry and is given for the following geo-
metries [30]: $B(\text{circle}) = 3.66$, $B(\text{ellipse with length/width} = 2) = 3.74$, $B(\text{parallel}$
plates$) = 7.54$,

- $B(\text{rectangular with length/width} = 4) = 4.44$,
- $B(\text{rectangular with length/width} = 2) = 3.39$,
- $B(\text{square}) = 2.98$,
- $B(\text{equilateral triangle}) = 2.47$,
- $B(\text{sinusoidal channel}) = 2.47$, and
- $B(\text{hexagonal channel}) = 3.66$.

The constant B corresponds to the asymptotic Sherwood number for constant wall
concentrations $\text{Sh}_\infty = 3.66$ in a circular channel, which is valid for fast reactions,
where the components are rapidly consumed at the wall. In the case of slow reactions, the wall is saturated with the components, and the concentration only varies
along the channel length. This situation is similar to the case of constant wall heat
flux in convective heat transfer. The Sherwood number for circular tubes becomes
$\text{Sh}'_\infty = 4.36$ for slow reactions.

The ratio of the reaction rate to the rate of wall mass transfer is expressed with
the second Damköhler number DaII,

$$\text{DaII} = \frac{k_S}{k_D} = \frac{k_S \cdot d_h^2}{2 \cdot D_m}, \tag{6.84}$$

for a first order reaction and laminar flow. An empirical correlation of the Sh number as a function of the DaII number is given by

$$\frac{1}{\text{Sh}} = \frac{1}{\text{Sh'}_\infty} + \frac{\text{DaII}}{\text{DaII} + 1.979} \left(\frac{1}{\text{Sh}_\infty} - \frac{1}{\text{Sh'}_\infty} \right). \tag{6.85}$$

Renken and Kiwi-Minsker [30] compare the mass transfer characteristics of a microreactor with a packed bed consisting of particles with a diameter of 1 mm, see Section 6.1.2. For gas flow (Sc = 1) and fast reactions (DaII $\geq$ 100), the mass transfer coefficient k_D in the microreactor with different channel diameters from 20 to 1 000 μm is always higher than in the packed bed with Re numbers between 1 and 10.

6.2.5 Wall adsorption and saturation time scales

The total capture fraction f of a bed or a surface with length L represents the fraction of species adsorbed on the surface after one passage

$$f(L) = \frac{c_0 - c(L)}{c_0} = 1 - c_b^*(L). \tag{6.86}$$

When the reaction is fully diffusion-limited, solutions of Eq. 6.86 are dependent only on the diffusivity; for reaction-limited situations, only the reaction rate constant is involved. For intermediate cases, the saturation time will depend on both diffusion and reaction rates, making it possible to express it as a function of DaII_s. This parameter and the diffusion/convection length scale z^* are the two main parameters to characterize surface capture and transport phenomena in microfluidic devices. The possible regimes of entrance and fully-developed flow are summarized next and will be discussed in detail in the following sections.

In the *entrance region* with $z^* < z^*_{crit}$ and for the infinite DaII_s numbers of very fast reactions with slow diffusion, the capture fraction according to Eq. 6.86 can be given as a fully analytical solution. The surface capture fraction is a function of the dimensionless length.

$$f(z^*) = 1 - \exp\left(-\frac{3n_w}{\Gamma(1/3)} \left(\frac{3}{2} z^* \right)^{2/3} \right)$$

$$= 1 - \exp\left(-1.467 \cdot n_w z^{*2/3} \right) \tag{6.87}$$

Here, n_w denotes the number of active walls, $n_w = 1$ for asymmetrical adsorption at one wall and $n_w = 2$ for symmetrical adsorption at two walls. The surface concentration is a function of the time and the channel length

$$c_s^*(z^*,t) \approx 0.98 \, \frac{\varepsilon_A \, t}{t_D \, z^{*1/3}} \, c_b^*(z^*) \qquad (6.88)$$

with the diffusion time scale $t_D = h^2/D_m$. The saturation time scale is defined herein as the time required to saturate 95 % of the available binding sites for the species.

$$t_{0.95} \approx 0.97 \, \frac{h^2}{\varepsilon_A \, D_m(1+\tilde{K}_D)} \, z^{*1/3} \qquad (6.89)$$

with the dimensionless equilibrium dissociation constant $\tilde{K}_D = k_{off}/(k_{on}c_0)$, see Eq. 6.65. This equation is valid for $\text{DaII}_s(1 - c_s^*)z^{*1/3} \gg 1$ for high flow rates in long channels, see Eq. 6.61. The rate constants k_{on} and k_{off} are replaced by effective rate constants

$$k_{on}^{eff} = \frac{k_{on}}{1 + 0.95\text{DaII}\left(1 - \langle c_s^* \rangle\right) z_L^{*1/3}},$$

$$\qquad (6.90)$$

$$k_{off}^{eff} = \frac{k_{off}}{1 + 0.95\text{DaII}\left(1 - \langle c_s^* \rangle\right) z_L^{*1/3}},$$

which includes the almost saturated surface. Applications for this case are the surface plasmon resonance sensor (SPR) or catalytic wall reactions in large channels with high flow rates. The concentration profiles for the symmetrical and asymmetrical case are given in Figure 6.5 from numerical and analytical solutions for the entrance region.

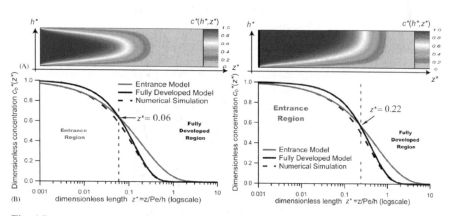

Fig. 6.5. Numerical simulation of the Graetz problem for parallel plates and diffusion limited reaction at two walls (symmetrical reaction, left side) and at only one wall (asymmetrical reaction, right side), adopted from Gervais and Jensen [411]; (A) Numerical simulation of the concentration profile; (B) Plot of the normalized bulk concentration.

In the *fully-developed region* with $z^* > z_{crit}^*$, diffusive transport is fast enough compared to convective transport, hence, the mass transfer boundary layer disappears. The critical length z_{crit}^* is numerically estimated by Gervais and Jensen [411] with $z^*_{crit} = 0.06$ for the symmetrical ($n_w = 2$) and $z^*_{crit} = 0.22$ for the asymmetrical case ($n_w = 1$).

For arbitrary $DaII_s$ numbers and an unsaturated surface $c^*_s \ll 1$, the capture fraction according to Eq. 6.86 can be given by a fully analytical solution for pseudo first order reactions.

$$f(t) = 1 - \exp\left(-n_w \bar{k}_{d/r}(DaII_s)z^*\right) \tag{6.91}$$

The effective diffusion/reaction transport coefficient $\bar{k}_{d/r}$ of the Graetz problem is a function of the DaII number. Numerical solutions for the symmetrical and asymmetrical case are given by Gervais and Jensen [411].

$$\bar{k}_{d/r,\text{sym}}(DaII_s) = 1.6304 \ (1 - \exp(-0.68 \cdot DaII_s)) + \exp(-6.1/DaII_s) \tag{6.92}$$

$$\bar{k}_{d/r,\text{asym}}(DaII_s) = 1.4304 \ (1 - \exp(-0.68 \cdot DaII_s)) + \exp(-5.77/DaII_s)$$

The surface concentration c^*_s is a function of the time and the channel length

$$c^*_s(z^*,t) \approx \frac{\bar{k}_{d/r}\varepsilon_A t}{t_D} \ \exp(-n_w \bar{k}_{d/r}z^*) \tag{6.93}$$

This correlation is applicable for heat and mass transfer in long, narrow channels. Interested readers are referred to Shah and London [212].

If the $DaII_s$ number is greater than one for *rapid reactions*, $DaII_s \gg 1$, the surface capture fraction is given by

$$f(t) = 1 - \exp\left(-n_w \bar{k}_{d/r}\left(\frac{L}{Pe\,h} - \frac{w_{eff}\,t}{\bar{w}\,\bar{t}_d} + \frac{1}{\varepsilon_A(1+\bar{K}_D)}\right)\right). \tag{6.94}$$

with the number of active walls n_w. The effective velocity of the propagating saturation front w_{eff} is a function of the mean flow velocity

$$w_{eff} = \frac{\varepsilon_A \bar{w}}{\varepsilon_A + n_w(1+\bar{K}_D)^{-1}} \tag{6.95}$$

The time for almost complete surface adsorption is a function of the effective mean bulk velocity

$$t_{1.00} = \frac{L}{w_{eff}} = \frac{L}{\bar{w}} + \frac{L\,n_w}{\varepsilon_A\,\bar{w}\,(1+\bar{K}_D)}. \tag{6.96}$$

For *slow reactions* or for improved wall mass transfer, $DaII_s \ll 1$, the surface capture fraction is expressed by the error function erf(x), the integral over the Gaussian error function, see also Section 4.1.2.

$$f(t) = 1 - \frac{1}{2}\,\text{erfc}\left(\sqrt{\frac{\pi}{4}} \cdot z^*_{eff}\,\sigma\right) \quad \text{with} \quad \sigma = n_w \bar{k}_{d/r}(DaII_{eff}) \tag{6.97}$$

Here, the moving concentration front coordinate z^*_{eff}

$$z_{eff}^* = z^* - \frac{w_{eff}}{\bar{w}} t, \qquad (6.98)$$

is defined with the moving front velocity w_{eff}, see Eq. 6.95.

The effective DaII number $DaII_{eff}$ is the $DaII_s$ number at the point of half saturation $c_{eq}/2$ of the surface

$$DaII_{eff} = \frac{k_{on}(c_{s0} - c_{eq}/2)h}{D} = DaII_s \frac{0.5 + \bar{K}_D}{1 + \bar{K}_D}. \qquad (6.99)$$

Numerical simulations of propagating fronts are given by Gervais and Jensen [411] and indicate a good correlation between analytical solutions and numerical results, see Figure 6.6.

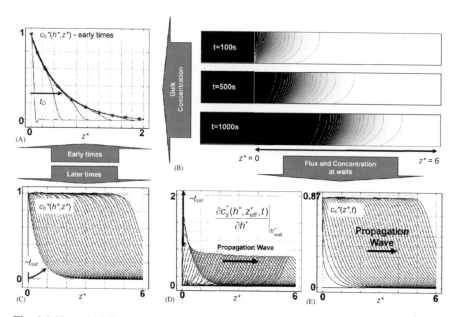

Fig. 6.6. Numerical flow simulation of a molecule reacting with the wall of a sensor, adopted from Gervais and Jensen [411]. (A) Bulk concentration profile for short times ranging from 0 to 0.15 s; the star-dotted line is a fit of the fully developed model for $DaII = 0.14$. (B) FEM flow simulation for three different time points: approximately the saturation time scale at 100 s and above the saturation time scale with 500 and 1 000 s in the wave propagation regime. (C) Bulk concentration $c_b^*(h^*,t)$ as a function of h^* for 50 to 2 000 s. Each step of the curves has a distance of 50 s. (D) Normalized mass flux at the wall with time steps acc. to C. (E) Normalized surface concentration $c_s^*(h^*,t)$ as a function of h^* for 50 to 2 000 s.

During propagation of the concentration front, five different time scales for the transport process have to be considered: diffusion t_D, reaction t_R, convection or residence time t_P, saturation t_{sat}, and wave propagation t_{wp}. Gervais and Jensen discuss the influence of the time scales for different microfluidic applications, where typical time scales must be considered.

Transport problems involving flow through devices, surface reactions, and adsorption are wide spread in micro process technology. It may be difficult to select the appropriate model to characterize device performance as devices and processes vary greatly in geometry and operation range. The analysis and modeling in this section focus on fully characterized transport to surfaces in the absence of a boundary layer (laminar flow) and proposes a simple approach to obtain the diffusion/reaction mass transfer coefficient $\bar{k}_{d/r}(\text{DaII}_s)$. Application of these coefficients allows accurate mapping of the bulk and surface concentrations along the channel and helps to determine the expected time for a surface to saturate.

As the field of micro process engineering evolves, fluidic modules and elements have to be integrated on a very large scale, both in series and in parallel, and process times should be controlled carefully. Saturation time scales and capture fractions in flow through reactions will be important parameters to optimize.

6.3 Design Criteria for Microchannel Reactors

Chemical reactor design is governed by different constraints, such as material properties, available geometries, and fabrication technologies, as well as economical constraints. At the same time, typical time and length scales have to be considered for fluid dynamics, mixing, mass and heat transfer, as well as for the relevant reaction kinetics. An overview of scale-up issues concerning reactor design for pharmaceuticals and fine chemicals production is given by Caygill et al. [415] with emphasis on single-phase and multi-phase systems. Major issues resulting from an industrial inquiry point on foam creation, mixing, heat transfer time, and solids suspension. In order to provide satisfactory design, the authors recommend to gain detailed knowledge of the difficulties, either from literature or extensive experimentation.

Dimensionless numbers and their order-of-magnitude assist in estimating the reactor performance, although detailed calculation of the convection-reaction-diffusion system (Eq. 6.18) is often very laborious, if possible at all. A good introduction to microreactor design is given by Emig and Klemm [33] Chapter 16, based on their experience of various reactor concepts. The textbooks of Levenspiel [416], Smith [417], and Hartmann and Kaplick [418] are some introductory literature on the design and integration of chemical process design, which is not treated in detail here. The role of transport phenomena in chemically reacting flow systems is described by Felder and Rousseau [419] Belfiore [420], as well as Rosner [421], specifically addressing design aspects for reacting flow processes.

6.3.1 Equipment design process

The detailed design of chemical process devices comprises the dimensioning of physical or chemical processes, the transfer processes within the device, the dimensioning and shaping of active areas, and the involved functional and structural elements. Together with material choice and dimensioning according to mechanical loads and stresses, and fabrication method selection, the interfaces, connections, and the housing are designed, resulting in the shape and structure of the process device. An introduction to process and equipment design is given in Section 2.2.2.

To reduce the number of variations from the problem-solving process, some criteria can be analyzed first:

- Is the function performed and fulfilled with a simple structure?
- Do intrinsic effects achieve the desired physical process?
- Should the active area be shaped by simpler fabrication process or would a different fabrication technique form a suitable active area?
- Can the device be operated in a safe manner?

Whether the solution obtained is appropriate to the specified task will be evaluated according to the selection criteria of quantity, quality, and cost of the product. Here, the specification with quantifiable and measurable targets is very helpful. The equipment design in the detail study is illustrated in Figure 6.7. The design of microstructured devices highlights some characteristics that will be described in the following.

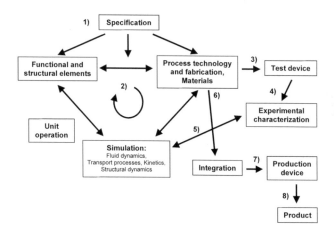

Fig. 6.7. Iterative design process starting with the specification, initially resulting in a test device. With improved knowledge and refined simulation methods, the production equipment is designed, fabricated, and integrated into the entire periphery to produce the desired product.

Beginning with the specified targets, the actual design process moves between the necessary prerequisites according to the functionality of the system, the available fabrication possibilities, the associated materials, and the simulation tools with different modeling focus. The entire process is highly interactive and interdisciplinary. The design process starts with a well-defined specification (1), which leads to the design iteration (2) for microsystems between the functional and structural elements, the fabrication technology with related materials, and the simulation. The test device is fabricated (3) and tested experimentally (4). The close relationship between experiment and simulation enhances the design quality of successive devices (5). To find new pathways for structuring, fabrication, or applicable physical effects, creativity techniques are very helpful, described in Section 2.5. In Section 1.5, an overview of scaling opportunities can be found to assist the design process.

With improved structures and fabrication technologies (6), the final production device is fabricated with proper integration of various functions and elements (7). The entire chemical process within the microstructured device leads to the desired product (8) fulfilling the given specification, which is the final indication of the device's suitability. This method may slightly vary for different applications, but follows the same basic principle.

Gerlach and Dötzel [201] list many functional elements according to their mechanical, thermal, fluidic, optical, electrical, magnetic, electronic, chemical, and/or biological nature. The fabrication techniques can be roughly divided into surface and bulk methods and are dependent also on the appropriate materials determined by the mechanical, thermal, and chemical requirements. The simulation begins with the calculation of mass and energy balances, followed by fluid dynamics and employed transport processes. Eventually, calculation of reaction kinetics or the structural behavior are necessary.

The pressure loss in the chemical reactor is one of the key factors of the reactor performance. With equations from Section 3.1.3, the pressure loss over a long straight channel with circular cross section is expressed as

$$\frac{\Delta p}{L_C} = C_f \frac{\eta \cdot w}{2 \, d_h^2}. \tag{6.100}$$

The friction factor $C_f = 64$ of a circular tube determines the friction coefficient in this equation, other geometries are given in Table 3.1. The pressure loss determines the viscous dissipation within the channel, expressed by the Eckert number Ec, see Section 3.1.3. Additionally, the pressure loss determines the specific energy dissipation ε within the channel as the driving force for convective mixing, see Section 5.4.5. The specific energy dissipation ε correlates with the pressure loss Δp and volume flow rate $\dot{V}$ divided by the fluid mass m_R in the reactor and can be expressed for straight channels with constant cross section A and length L_C by

$$\varepsilon = \frac{\Delta p \dot{V}}{m_R} = \frac{\Delta p \, \bar{w}}{L_C \, \rho}. \tag{6.101}$$

The mean residence time t_P within a reactor with long straight channels is given by

$$t_P = \frac{L_C}{\bar{w}}. \tag{6.102}$$

The space time t_{ST} of a reactor is defined as the ratio of the fluid volume V_R in the reactor to the volume flow rate $\dot{V}_0$ at the reactor entrance and can be simplified for reactions with constant volume to

$$t_{ST} = \frac{V_R}{\dot{V}_0} = \frac{L_C}{\bar{w}}. \tag{6.103}$$

Aside from the mean values, the residence time distribution RTD is important for a complete and effective reaction within the microstructured device, see Section 6.1.2, hence, the reaction kinetics must be known with rate constant k and reaction order m. The appropriate combination and arrangement of reaction kinetics with the transport processes of species, heat, and mass is essential for successful application and operation of reactor devices.

6.3.2 Reaction kinetics and transport processes

The reactor performance is expressed with the production of the product $\dot{n}_i$ as the conversion of the key reactant

$$X = \frac{\dot{n}_{i,0} - \dot{n}_{i,s}}{\dot{n}_{i,0}} \tag{6.104}$$

or related to the volume flow rate $\dot{V}_0$ through the reactor as the space-time yield

$$Y_{ST} = \frac{\dot{n}_{i,0} - \dot{n}_{i,s}}{\dot{V}_0}, \tag{6.105}$$

see also Eqs. 6.24 to 6.27. Including the reaction kinetics with the rate correlation, see Section 6.1,

$$r = k \cdot c^m, \tag{6.106}$$

dimensionless numbers are formed to determine the characteristic time and length scales of chemical reactors.

The typical time and length scales in chemical reactors are compared to determine operating ranges and limits. Besides chemical kinetics, the residence time and its distribution in a device, as well as the heat and mass transfer characteristics have to be treated and adjusted for successful operation and efficient device performance. The flow situation plays a major role due to the different transport phenomena in single phase and multiphase flow reactors. Catalytic reaction are treated in the next section, including effective transport coefficients, which allows the lumped treatment of complex coupled transport phenomena with reasonable effort. Often, too many parameters are not exactly known and an order-of-magnitude estimation is the appropriate tool for designing the equipment. Therefore, experimental evaluation and verification of design assumptions are essential for the development of microstructured devices.

The single-phase flow fluid dynamics in a microchannel can be described with the Aris-Taylor dispersion, see Section 6.1.2, expressed with Bodenstein number Bo for axial mass transfer. The heat transfer of laminar flow is given by a constant Nußelt number Nu, see Section 4.1.3. The reaction kinetics are described by Eqs. 6.10 to 6.17 and must be related to the heat and mass transfer characteristics. The reactor design is described following Emig and Klemm [33] Chapter 16 and accomplished by further details. The following steps are necessary for the successful design of microstructured chemical reactors:

1. The channel diameter d_h or characteristic dimension h or b are determined with the aim to avoid radial heat and mass transfer limitations. High transfer coefficients can be achieved either within small channels, with curved channels, or with internals, such as obstacles, which induce convective vortices and enhanced dissipation. Fabrication restrictions, pressure loss, and the desired flow rate must be considered when choosing the channel diameter. The second Damköhler number has to be lower than 1, i.e. DaII < 1, see Section 6.2.1, allowing fast and sufficient mixing in respect to the chemical reaction.

2. For heat transfer considerations, the temperature change from the reaction heat ΔH_r has to be controlled appropriately. The characteristic time of heat conduction $t_H = d_h^2/a$ with the temperature diffusivity $a = \lambda(\rho c_p)$ must be shorter than the characteristic reaction time $t_R = c^{1-m}/k$. For strong exothermic reactions, the adiabatic temperature rise ΔT_{ad}, see Eq. 6.47, from the reaction must be in the range of the temperature difference between reactants and cooling medium in the reactor. The ratio of reaction heat to the adiabatic temperature rise ΔT_{ad} via the heat capacity of the reactants is expressed by the third Damköhler number

$$\text{DaIII} = \frac{r \cdot \Delta H_r}{\dot{m} \, c_p \cdot \Delta T_{ad}}. \tag{6.107}$$

see Brötz [422]. This ratio must be in the range of one to provide isothermal conditions for the reaction.

3. In exothermic reactions, the reaction heat must be removed quickly from the reactants by convective cooling. The ratio of the reaction heat to the convective heat transfer is expressed by the fourth Damköhler number

$$\text{DaIV} = \frac{r \cdot \Delta H_r}{k \cdot a_V \cdot \Delta T_{log}} \tag{6.108}$$

where k and a_V denote the overall heat transfer coefficient and specific surface, respectively, see Section 6.1.2. A ratio less than one indicates that the reaction heat can be transported away from the reaction channel by sufficient heat transfer. A higher temperature difference ΔT_{ad} and larger specific surface a_V lead to enhanced heat transfer and lower DaIV number.

4. The necessary reaction volume V_R is determined with the reaction loading inside the reactor. If no heat and mass transfer limitations are present, the necessary reaction volume V_R can be determined by simple integration over the channel length or the produced species, see Eq. 6.25.

$$V_R = \int_{\dot{n}_{i,0}}^{\dot{n}_{i,s}} \frac{\mathrm{d}\,\dot{n}_i}{R_i}. \tag{6.109}$$

For an isothermal reaction with constant volume, the necessary reaction volume V_R divided by the volume flow rate $\dot{V}_R$ gives the mean residence time t_P, the time required to perform the reaction within the reactor. The Fo' number is determined for the system. For Fo' > 1, the case of an ideal flow reactor is valid. If Fo' < 1, the channel diameter must be decreased or the influence of the molecular diffusion must be accurately determined, see Eq. 6.30.

5. The channel length L_C and the channel number N_C are determined from the necessary reaction volume V_R for the given reaction loading. The channel length depends not only on the reactor volume, but also on the necessary residence time for the reacting flow. The ratio of the residence time to the characteristic reaction time is expressed by the first Damköhler number DaI, which must be greater than one for a complete reaction.

$$\text{DaI} = \frac{t_P}{t_R} = \frac{v_i k \cdot c_i^{m-1} \cdot L_C}{\bar{w}} > 1 \tag{6.110}$$

The fluid must have sufficient time to react within the reactor channels, although, this is not a necessary condition, because in some cases the mixed fluid can also react outside of the reactor. The flow velocities within the reactor channels are typically in the range of $0.01 \leq \bar{w} \leq 5$ m/s for liquids and $0.1 \leq \bar{w} \leq 50$ m/s for gases resulting in a reasonable pressure loss. The number of channels N_C is combined with the channel length over the necessary space loading

$$t_P \cdot \text{DaI} = \frac{V_{C,\text{tot}}}{\dot{V}_{in}} = \frac{V_C \cdot N_C}{\dot{V}_{in}} = \frac{b \cdot h}{\dot{V}_{in}} \cdot (N_C \cdot L_C). \tag{6.111}$$

The channel diameter is represented by the height h and width b of the channel. Often, the channel height is determined by the fabrication technology, but also controls the typical transfer characteristics, such as heat transfer for cooling or wall mass transfer for catalytic reactions. Hence, the channel width b can be varied without too many constraints to achieve high flow rates.

6. In general, the channels can be arranged in a single device with many channels and appropriate flow distribution system (internal numbering-up), or microstructured devices with a certain channel number can be set up in parallel (external numbering up), see Section 6.3.5. The economical optimization is controlled by fabrication restrictions for microstructured devices and must balance out between fabrication cost (investment cost) and operation cost coming from pressure loss, installed pumping power, the risk of blocking and fouling, and related cleaning and maintenance costs, see Section 6.3.4. The pressure loss over the entire device is the main iteration parameter for the channel diameter. If the pressure loss is too high, the diameter is enlarged without losing the necessary transport characteristics for heat and mass transfer.

This design method is also valid for fluid/fluid reactions in emulsion or for gas-liquid reactions. Eventually, the mass transfer between the phases has to be considered, see Section 3.3. A good estimation is also the transport length within the fluid segments, bubbles or droplets, and the comparison of the characteristic times. With the above described steps, main geometry and process parameters are determined for successful application of microstructured equipment.

6.3.3 Heterogeneous catalytic reactions in microchannel reactors

Heterogeneous catalysis is based on the activity of a solid catalyst material, often distributed in a nanoporous carrier material. There are two major possibilities to fasten the carrier within the microchannel, either at the channel wall or as immobilized pellets in the channel. The immobilized pellets form a fixed bed and can be treated with related methods from conventional equipment, see Renken and Kiwi-Minsker [30]. The fabrication and pellet filling of a microreactor for adsorption is described by Henning et al. [423] and references therein.

The discussion here concentrates on a catalyst within a porous layer attached at the wall. Two geometrical parameters have to be considered, the catalyst layer thickness δ_{cat} and the channel diameter d_h. The thickness determines the internal mass transfer within the porous structure, and the diameter governs the external mass transfer from the bulk flow to the wall, see Section 6.2.3 on the Graetz problem. The reactive heat is produced close to the wall and can be transferred directly to the cooling medium flowing in a parallel channel. The main heat transfer limitation originates from the porous layer of the carrier substrate, often consisting of oxides with low thermal conductivity λ_{cat}. Sufficient heat transfer is important for stable operation of the reactor, see Emig and Klemm [33] Chapter 7.

The equipment design procedure can be performed in analogy to the scheme mentioned in the last section concerning the channel diameter d_h and the catalyst layer thickness δ_{cat}. The Mears criterion [33], with limitation from the second Damköhler number DaII, determines the upper limit of the channel diameter

$$\frac{d_h \cdot r_{eff}|_S}{D_m\, c_{in}} < \frac{0.05}{m} \tag{6.112}$$

where the effective reaction rate at the catalytic surface $r_{eff}|_S$ is determined from the effective rate coefficient

$$k_{eff} = k_s \cdot \beta\, a_V / (k_s + \beta\, a_V) \tag{6.113}$$

including the mass transfer within the catalytic layer. The Mears criterion limits the effective reaction rate $r_{eff}|_S$ of heterogeneous reactions to 5 % of the chemical reaction rate without mass transfer limitations. In flat wide channels with low aspect ratio, the channel height h determines the transfer characteristics in the channel flow. The following correlation determines the upper limit of the catalyst layer thickness δ_{cat} by internal mass transfer limitation,

$$\frac{\delta_{cat}^2 \cdot r_{eff}|_V}{D_{cat}\, c_{in}} < \begin{cases} 0.7 & \text{for } m = 0 \\ 0.07 & \text{for } m = 1 \\ 0.03 & \text{for } m = 3 \end{cases}, \tag{6.114}$$

according to the Weisz-Prater criterion [33]. The reduced heat transfer due to the insulating effect of the catalyst layer is determined with the Anderson criterion [33]

$$\frac{\Delta H_r \cdot r_{eff}|_V\, \delta_{cat}^2}{\lambda_{cat}\, \Delta T_W} < 0.15\, \frac{R \Delta T_W}{E} \tag{6.115}$$

giving a limit of the layer thickness to the allowable driving temperature gradient ΔT_W. This equation is similar to the definition of the fourth Damköhler number DaIV in Eq. 6.108. The correlations illustrated above for reactor design are valid for cases, where

- no mass transfer limitations occur, DaII $\ll 1$,
- ideal flow reactor can be assumed with Fo' > 1,
- isothermal conditions are achieved from adjusted DaIII and DaIV numbers, and
- constant volume of the reagents can be assumed.

With the effective rate coefficients k_{eff}, the necessary residence time t_P or reactor volume V_R can be determined in analogy to Eqs. 6.109 to 6.111. With the above conditions, the channel length L_C and the channel number N_C can be determined from the necessary reactor volume. The fabrication method as well as the flow distribution are very important for the equipment design.

6.3.4 Scale-up and economic situation

The above section on the reaction kinetics and related transport phenomena indicates the prominent role of dimensionless numbers, especially the ratio of characteristic time scales, which are integrated in the various Damköhler numbers for residence time, mixing, fluid heating, and heat transfer. Here, the Damköhler numbers are summarized for equipment and process design as discussed by Damköhler [424] for scale-up purposes.

$$\text{DaI} = \frac{\text{chem. component increase}}{\text{convective comp. increase}} = \frac{v_i\, r}{-\text{div}\,(c_i\,\vec{w})} \propto \frac{v_i\, r\, L_C}{c_i \bar{w}} = \frac{t_P}{t_R} \qquad (6.116)$$

$$\text{DaII} = \frac{\text{chem. component increase}}{\text{diffusive comp. increase}} = \frac{v_i\, r}{\text{div}\,(D_i \text{grad}\, c_i)} \propto \frac{v_i\, r\, d_h^2}{c_i D_m} = \frac{t_m}{t_R} \qquad (6.117)$$

$$\text{DaIII} = \frac{\text{chem. heat generation}}{\text{heat capacity flow}} = \frac{\Delta H_r\, r}{\text{div}\,(\rho\, c_p\, \Delta T_{ax}\,\vec{w})} \propto \frac{\rho_m\, \Delta H_r\, r\, L_C}{\rho\, c_p\, \Delta T_z\, \bar{w}} \qquad (6.118)$$

$$\text{DaIV} = \frac{\text{chem. heat generation}}{\text{heat transfer}} = \frac{\Delta H_r\, r}{\text{div}\,(\lambda\, \text{grad}\, T_{rad})} \propto \frac{\rho_m\, \Delta H_r\, r\, d_h^2}{\lambda_W\, \Delta T_W} \qquad (6.119)$$

with the temperature difference along the channel ΔT_z and the temperature difference between the bulk flow and the wall ΔT_W. For a successful and complete reaction within a long straight channel and feasible temperature control, the following conditions have to be fulfilled:

DaI > 1 for complete reaction within the channel

DaII < 1 for sufficient fast mixing and radial mass transfer

DaIII ≈ 1 for sufficient low temperature change along the channel

DaIV < 1 for sufficient heat transfer in the channel

These conditions must be adjusted to the actual problem with relevant boundary conditions, in particular, the heat transfer and the combined mass transfer situations must be managed in catalytic processing.

The main goal of engineering activities is to achieve high performance in various technical systems. In communication and information technology, the miniaturization of electronic equipment has led to an incredible performance increase during the last decades. Moore's law of doubling the number of circuits in electronic devices in 18 months still holds after decades. In process technology, the scale-up of chemical production or power plants has lead to high energy efficiencies and affordable consumer products, including compact discs or food products.

For example, around 1920 cryogenic air separation units could produce approximately 1.3 t/h oxygen with 98 - 99 % purity. In the middle of last century, the largest air separation units delivered approx. 5.2 t/h oxygen with 99 % purity, 50 years later the largest air separation units are supplying customers with approx. 65 t/h oxygen with 99.5 % purity and higher. As the throughput has increased, the specific energy consumption has decreased from approx. 1.5 kW/kg oxygen to approx. 0.42 kW/kg oxygen. At the same time, the consumer specific supply was also addressed with small and adjusted plants for flexible production satisfying costumer demand.

The economy of scale in the bulk chemical and petrochemical industries can be expressed with the help of the plant capacity C, which is given by the volume or mass flow rate $\dot{m}$ of the desired product. Often, the cost situation of a plant or a certain product is given on an annual basis. The annual processing cost A_{P1} of a product from plant 1 with capacity C_1 is related to a reference plant 0 with the following correlation

$$A_{P1} = A_{P0} \cdot \left(\frac{C_1}{C_0} \right)^{\kappa} \tag{6.120}$$

with the scale-up exponent κ. Experience from scale-up projects with bulk chemicals, such as ethanol, ethylene oxide, or sulfuric acid shows a κ value in the range of 0.6 to 0.7. Often, a value of 2/3 is assumed, see Baerns et al. [425]. For example, the increase of the mass flow rate with the factor of 2 will lead to a cost increase of $2^{0.67} = 1.59$, which provides enormous potential for cost savings. In the case of batch equipment, Roberge et al. [36] specified the scale-up exponent with 0.3 in a reactor size range between 0.1 to 10 m^3. The typical decline of production cost with increasing capacity is displayed in Figure 6.8. Besides capacity increase, the introduction of new technology can lead to production cost savings, for example, turbo compression in ammonia production [33] or microreactors in specialty chemistry [36].

On the other hand, some branches of the chemical industry are not subject to the economy of size, such as the pharmaceutical industry or fine chemicals, where flexibility, regulations, as well as the product price and quality are the important factors. However, all areas share the common need for high throughput in a desired process to lower the operational cost. The work load of the installed capacity determines process efficiency and product profitability to a great extent [33]. The product asset increases with increasing work load or capacity utilization.

A closer look at the cost structure reveals the different roles of various cost elements. The annual processing cost A_{P1} with annual production rate R_1 can roughly be estimated for a plant with comparable technology and production rate R_0 with the correlation

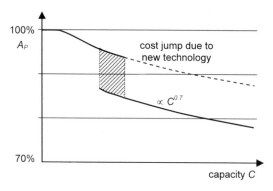

Fig. 6.8. Correlation of production cost with plant capacity, acc. to Emig and Klemm [33]. The production cost include resources, energy, catalyst, chemicals, depreciation, wages, maintenance, and repairs.

$$A_{PI} = \eta_1 \cdot C_{FC0} \left(\frac{R_1}{R_0}\right)^{0.7} + \eta_2 \cdot c_L N_0 \left(\frac{R_1}{R_0}\right)^{0.25} + A_{U0} \cdot \left(\frac{R_1}{R_0}\right) \qquad (6.121)$$

where the factors η_1 and η_2 can be obtained from the data available in a particular company [41] Chapter 9. The fixed-capital cost C_{FC} consists of the plant investment minus the annual depreciation, depending on the local tax law. The labor-related cost is dependent on the number of workers N_0 multiplied by their earnings c_L plus an overhead factor η_2. The utility cost A_U includes raw materials, energy, or chemicals. The above correlation does not include every influence factor, which might influence the cost situation, but provides a guideline for preliminary estimations.

Introduction of a new technology, refurbishment (revamp) of an existing plant, or the design and construction of an entirely new plant begins with a financial investment leading to first product sales after start-up and production onset. The financial situation of a company at this point is displayed by the cash-flow analysis, the temporal examination of income and expenses of the company related to the project. The typical cash flow analysis is given in Figure 6.9 for an investment project. Roberge et al. [36] performed an economical study investigating the capital expenditures CAPEX and the operational expenditures OPEX for continuously operating plants with microstructured equipment in comparison with batch plants. For large scale production, a gain in yield and safety are the main drivers for microstructured devices, however, the gain must be significant in order to cope with the capital expenditure associated with a new technology.

At the beginning of the project, the expenditure during the research and development stage is normally relatively small. The cash flow curve is on the negative side (expense) with low decline. The expenses will usually include some preliminary process design and a market survey. Project change will not effect the cost situation very much, but have a major impact on the projects success. Once the decision has been made to go ahead with the project, detailed process-engineering design will commence, and the rate of expenditure starts to increase. The rate is increased further when equipment is purchased and construction begins. There is no return on

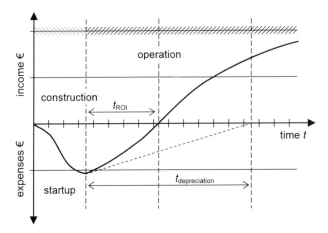

Fig. 6.9. Cash flow analysis of an investment project, see Baerns et al. [425] Chapter 13 and Perry's Handbook [41] Chapter 9.

this investment until the plant begins production. Even during startup, there is some additional expenditure, but most of the project costs are collected. Once the plant is operating smoothly, an inflow of cash is established, however, depreciation starts lowering the book value of the plant. Up to the point where the cash flow curve touches the baseline, the invested capital is at risk. This is called the breakeven point at t_{ROI}, where the manufacturing process returns its capital. The acronym ROI means Return On Invest. Beyond the breakeven point after t_{ROI}, any cash flow above the horizontal baseline is in excess of the return on the invested capital with interest rate. The situation including taxes and interests is much more complicated and described in more detail in [41] Chapter 9.

The cash flow curve is influenced by many factors and each time period has its own role. A shorter time startup period gives a short time-to-market for the desired product and an earlier breakeven point. Lower project cost will also lead to an earlier breakeven point. Higher product sales and assets due to better product quality or faster time-to-market will also lead to an early breakeven point. Here, the introduction of microstructured equipment can help to increase the profitability of existing plants or to introduce new plant concepts. On the other hand, it is economically unreasonable to substitute a moderately successful, but depreciated plant with new and expensive technology.

6.3.5 Scale-up method with equal-up principle

The process and plant design procedure begins with a product idea or product formula, which is tested in the laboratory with stirred beakers or standard calorimeters. Here, the chemical receipts and protocols are developed, which must be transferred into a technical process. In the conventional way, a miniplant in the laboratory is set up to gain more information of the process, transport and chemical kinetics, material properties, and initial feedback about potentially successful reactor concepts and unit operations for separation and product purification, see Section 2.2.2. The most promising route is tested in a pilot plant, the last step before a complete production plant with target capacity. The length scales and capacity are increased from millimeter to meters and from ml/min to tons/hour or day, respectively. This process is called scale-up and a well-established procedure in chemical engineering, see Zlokarnik [32]. The main steps of this procedure are displayed on the left side of Figure 6.10, see also Figures 1.3 and 2.4.

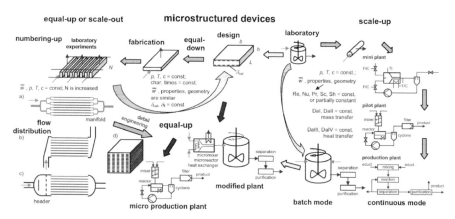

Fig. 6.10. Equal-up, scale-out, and scale-up procedures for chemical engineering purposes with microstructured devices, see also Figures 1.3, 2.3, and A.4. Development of microstructured devices and partly integration of microstructured devices.

With miniaturization of chemical equipment, laboratory development of chemical receipts becomes feasible and leads to continuous processing on laboratory scale. The variety of microfabrication technologies and related material choice enables the design of microchannel devices tailor-made for single chemical processes. The existing data from laboratory developments or existing production plants form the basis of microstructured equipment design as well as new process routes to be tested in such devices, indicated by the arrows from left to right in Figure 6.10. Channels are the basic element of microstructured equipment, often with rectangular cross section. Depending on the process conditions, such as temperature, pressure, and fluid properties, the channel cross section, catalyst layer (if necessary), and minimum number are determined on the basis of available fabrication technology and device material. This process is called equal-down, because the small channel dimensions are determined according to the necessary heat and mass transfer and relevant chemical kinetics. The channel dimensions should be as small as necessary or as large as reasonable for safe and efficient operation. Often, this data provides a rough estimation, however, only experimental tests will give an accurate insight into the physical and chemical processes within continuously-operated, microstructured devices.

A laboratory device is fabricated and tested with the chemical system to yield experimental data. This data is compared with design assumptions and preliminary simulation results. In case of larger discrepancies or insufficient device performance, it may be necessary to redesign the process or device is necessary on the basis of the comparison between experimental and design data. Critical review of experimental data is the key issue for successful redesign of microstructured devices. In case of successful experimental tests, the next design step is to layout a device or a number of devices handling the desired product capacity.

To increase the mass flow rate or volume throughput of a device consisting of several microstructured elements, internal numbering-up of these elements increases the entire flow rate. This simple approach increases only the number of channels, platelets with microstructures or devices to enlarge the capacity, see Figure 6.10 left side. The flow distribution and correct integration must be considered and is the most critical point for successful implementation. Depending on the size of the entire device, various fluidic manifolds or flow headers can be applied. The flow distribution on 8 parallel channels is shown in Figure 6.10, case a). The fluidic manifold is similar to the channel network for micro heat exchangers, see Section 4.2.4. Wada et al. [426] propose a fluidic manifold for 16 microchannels with a width of 300 μm for a two-phase flow microreactor similar to the manifold in Figure 6.10, lower left side. The equal distribution is obtained via high pressure loss in a narrow channel section at the entrance of the reactor channel. Pressure loss in channel segments is the key to a uniform flow distribution in the manifold.

For more than 20 or 30 channels, the dendritic fluidic manifold, case a), is difficult to design, and high flow velocities in the channels lead to flow maldistribution due to vortex formation in the bends. Flow distribution according to cases b) and c) are able to uniformly feed a multitude of channels. The design of case b) was investigated and optimized by Tonomura [129] and Commenge et al. [427]. The flow distributor in case c) is similar to flow headers in conventional tube bundle devices. The velocity of the inlet flow is directed to the side walls by a central plate. Between the inlet plate and the microchannel entrance, a flow grid or straightener across the entire inlet section causes a small pressure loss and leads to uniform flow velocity over the entire header cross section, see Rebrov et al. [428]. With this arrangement, all channels are facing the same fluid velocity and are supplied with uniform flow rate. This concept is known from wind tunnels to produce a uniform, homogeneous turbulence field in an aerodynamic test section.

The appropriate flow manifold depends on the number of channels as well as the shape of the entire device. Microstructured devices for high flow rates often consist of a stack of microstructured plates, case d). Experimental experience and proper integration of microstructured elements in a conventional apparatus are essential in order to design and fabricate this plate stack. The equal-up process starts from the process or product to be realized, and the main effects and parameters are identified for miniaturization. These key parameters have been tested in the equal-down step and now determine the design of the industrial reactor. The laboratory design is equal to the production design with respect to key geometries, fluid dynamics, mixing, reaction kinetics, and heat management. The experimental results from the laboratory reactor clearly indicate, which dimensions cause the benefits of the small length scale (e.g. the flow boundary layer δ_{fl}). Other design parameters from fabrication have to be transferred to the production device, such as shape and structure of the active surface [12], e.g. the catalyst layer thickness δ_{cat}. The key parameters of the production device, including channel height h or wall thickness s, are equal to the laboratory equipment, however, the geometry of the active surface or the fabrication techniques may differ, for example, the microchannels have the same cross section or only the same height, but differ in length.

To introduce microstructured equipment into industrial production plants, different strategies have been successful. A new plant with microstructured devices allows the design of key components as well as peripheral equipment according to special demands, such as pulsation free pumping, proper filtering, or temperature control. The integration into an existing plant with given process conditions must consider more boundary conditions, but allows a stepwise integration. For example, a plant with a stirred vessel batch reactor can be upgraded with a continuous flow microreactor, as shown in Figure 6.10, lower middle, as modified plant. The microreactor with heat exchanger is positioned upstream of the vessel and feeds the former reactor vessel. The vessel is used as a buffer tank, which utilizes downstream steps for separation and purification of the desired product. The microstructured device is more likely to be accepted by operators, when as much of the conventional equipment remains in the process as possible.

Industrial application of microstructured equipment is a long-term target, which requires strategic thinking and appropriate action. Design strategies and creativity techniques have been mentioned in Section 2.5 including new aspect for micro process engineering. With the equal-up procedure, comparable process conditions can be realized in laboratory devices to rapidly optimize many parameters and conditions, as well as in production devices with high flow rates, reproducible process conditions, and controllable safety requirements. The equal-up design strategy and the consequent implementation of microstructures allows enhancement of device throughput from laboratory to production scale without the risk and cost of the conventional scale-up procedure.

6.3.6 Iodide-iodate reaction for selectivity engineering

The quality of a chemical reaction transformed in a technical process is evaluated and measured by the selectivity S, see Eq. 6.7. In fine chemicals and pharmaceuticals production, the selectivity is often the major issue in cost-effective plant operation, see Cybulski et al. [370] Chapter 4. The authors coined the issue of "selectivity engineering", where proper control and manipulation of the "micro-environment", like lamellae engineering, and of the "macro-environment", like using other additives, catalysts, or solvents, is essential to increase the output of a chemical reaction. Here, engineering of the micro-environment is aimed at and evaluated by the selectivity S of a parallel-competitive reaction, which is used to characterize the mixing process, see Eq.6.5.

For this purpose, the iodide-iodate-reaction by Villermaux/Dushman has been used. Although this reaction has some drawbacks, see [374], it can be applied to extract the mixing time of any arbitrary convective micromixer in a convenient way.

The reaction is based on competition for H^+ between an acid-base neutralization and the Dushman reaction:

$$H_2BO_3^- + H^+ \rightleftharpoons H_3BO_3 \tag{6.122}$$

$$5I^- + IO_3^- + 6H^+ \rightleftharpoons 3I_2 + 3H_2O. \tag{6.123}$$

While the neutralization (6.122) can be considered instantaneous, the Dushman reaction (6.123) is fast, but has a reaction rate some magnitudes below that of the neutralization. The produced iodine I_2 is in equilibrium with the iodide ion according to the correlation

$$I_2 + I^- \rightleftharpoons I_3^-. \tag{6.124}$$

The amount of created I_3^- is spectroscopically measured with two characteristic absorption peaks in the UV range at 352 and 286 nm. The implementation of the iodide-iodate reaction is well documented by Guichardon et al. [398, 399]. Its implementation in micromixers is described by Panić et al. [400]. The experimental procedure is straightforward: an acid solution is mixed with a solution containing H_2BO_3, I^- and IO_3^- in appropriate concentrations, see Table 6.1. In areas with sufficient amount of H_2BO_3, the acid is instantaneously neutralized, and the Dushman reaction does not occur. If a local excess of H^+ is present due to inadequate mixing, the reaction (6.123) forms iodine, I_2. Hence, the final amount of iodine is a measure for mixing quality: the less I_2 has been formed, the more homogeneous and fast the mixing process was.

Table 6.1. Applied concentrations for the iodide-iodate reaction [127]

Substance	1 l of Solution 1	1 l of Solution 2	Resulting Concentration [mol/l]
NaOH	45.5 ml 1.0 M	-	≈ 0
H_3BO_3	2.81 g	-	$[H_2BO_3^-]_0 = 0.045$
KI	2.64 g	-	$[I^-]_0 = 0.0159$
KIO_3	0.71 g	-	$[IO_3^-]_0 = 0.0033$
H_2SO_4	-	40 ml 0.5 M	$[H^+]_0 = 0.04$

The absolute concentration of iodine depends on the initial reactant concentrations. To yield a concentration-independent scale for mixing quality, Fournier et al. [404] define the so-called segregation index X_S, which relates the actual present iodine to the potential of iodine formation.

$$X_S = \frac{Y}{Y_{ST}} \tag{6.125}$$

The yield Y is the ratio of the amount of acid consumed by reaction (6.123) and the total acid amount

$$Y = \frac{2(n_{I_2} + n_{I_3^-})}{n_{H_0^+}} = \frac{4([I_2] + [I_3^-])}{[H^+]_0}, \tag{6.126}$$

where $[H^+]_0$ is given in Table 6.1. The reference value Y_{ST} is the yield of iodate I_3^- generated in a totally segregated state with infinitively slow mixing. Acid is consumed in proportion to the local concentration of borate and iodide/iodate ions, hence, $Y_S T$ is related to the concentration of iodate.

$$Y_{ST} = \frac{6\,[IO_3^-]_0}{6\,[IO_3^-]_0 + [H_2BO_3^-]_0} \qquad (6.127)$$

The segregation index X_S is the main outcome of the reaction, which is calculated from absorption measurements of the mixed solutions. The larger X_S, the more undesirable iodine is created and the longer is the mixing time. This is a simple illustration and may not be true for all mixers. In reality, the underlying mixing mechanism with non-stoichiometric mixed areas must be considered, as well as the interfacial length and lamella thickness, see Section 5.5.3. As a consequence, the segregation indices measured for completely different mixers need not necessarily be comparable, even if the same solutions are used. However, for convective mixers, the segregation index is generally proportional to the mixing time, as will be shown in the following.

The iodide-iodate reaction was implemented in three different T-mixers with rectangular cross section, and the resulting segregation index is displayed in Figure 5.13. Further results are shown in Figure 6.11, where the segregation index X_S is drawn over the second Damköhler number $DaII = t_m/t_R$. $DaII$ compares the characteristic timescales of mixing to reaction, see also the definition of a characteristic mixing time from Figure 5.20 and Eq. 5.58. The characteristic reaction time of the iodide-iodate reaction for the applied concentrations is determined to $t_R \approx 0.3$ s by solving the underlying ordinary differential equations, see [127]. As expected for small Damköhler numbers, see [398], the segregation index is proportional to the mixing time, confirming the accuracy of Eq. 5.58.

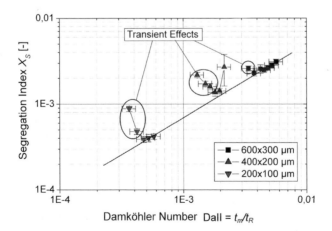

Fig. 6.11. Segregation index X_S of the iodide-iodate reaction for three different T-shaped micromixers over the Damköhler number $DaII$, see also Figure 5.13.

The results of Figure 6.11 can be used as a calibration curve for the correlation of mixing time t_m in seconds to segregation index X_S:

$$t_m = 0.38 \cdot X_S \tag{6.128}$$

This equation holds for results of the iodide-iodate reaction with concentrations as shown in Table 6.1, used in convective micromixers.

The deviations of the measured values from the proportional correlation in Figure 6.11 originate from arising transient effects inside the T-mixers, see Section 5.4.2. These transient effects create local unmixed regions, which increase the segregation index, which is one of the drawbacks of the iodide-iodate reaction mentioned above. As shown in Section 5.7.2 and with Figure 5.13, convective micromixers with meandering channels show no increase of the segregation index. As a result, simple T-mixers need to be improved to counter-act the transient effects, see Section 5.4.3.

6.3.7 Competitive-consecutive reactions

The basic principle of competitive-consecutive test reactions is the combination of a fast primary reaction with a slow secondary reaction,

$$\nu_A A + \nu_{C1} C \xrightarrow{k_1} \nu_{P1} P$$
$$\nu_B B + \nu_{P2} P \xrightarrow{k_2} \nu_{PS} P_S$$

see Eq. 6.6, which only occurs if both reactant A and primary product P are available. If mixing proceeds too slowly, the primary reaction is retarded and more secondary product P_S is formed than with faster mixing, hence, the selectivity S can be used as a measure of mixing time.

A widely used test reaction of this type is the coupling of 1-naphthol and diazotized sulfanilic acid [429], which belongs to a class of reactions with industrial importance. In the primary reaction, the sulfanilic acid couples with 1-naphthol; the rate constant is given in Bourne et al. [429] as $k_1 = 11\,500$ m^3/mol s, a very fast reaction. The primary product can then couple again with the remaining sulfanilic acid, however, this reaction is relatively slow with $k_2 = 1.85$ m^3/mol s.

Similar to Eq. 6.18, the one-dimensional reaction-diffusion equations for the competitive-consecutive system with product P and side product P_S are:

$$\frac{dc_A}{dt} = D_A \frac{\partial^2 c_A}{\partial x^2} - k_1 c_A c_B \tag{6.129}$$

$$\frac{dc_B}{dt} = D_B \frac{\partial^2 c_B}{\partial x^2} - k_1 c_A c_B - k_2 c_B c_P \tag{6.130}$$

$$\frac{dc_P}{dt} = D_P \frac{\partial^2 c_P}{\partial x^2} + k_1 c_A c_B - k_2 c_B c_P \tag{6.131}$$

$$\frac{dc_{PS}}{dt} = D_{PS} \frac{\partial^2 c_{PS}}{\partial x^2} + k_2 c_B c_P \tag{6.132}$$

After this set has been discretized in relation to spatial coordinate x, the reaction system can be simulated for one-dimensional diffusive mixing by integrating the resulting first order ordinary differential equations; details on discretization and implementation are given by Kastner [49]. To evaluate the impact of mixing time on product selectivity

$$Sp = \frac{c_P}{c_{PS}},\tag{6.133}$$

this simulation is carried out for a two-lamellae diffusive mixer with varying channel width $2b_l$ where b_l is the width of one lamella. A similar simulation was presented in Section 5.5.3 for the iodide-iodate reaction.

Table 6.2. Relationship between product selectivity S_P and characteristic diffusion time $t_D = b_l^2/2D$ for competitive-consecutive coupling between 1-naphthol and diazotized sulfanilic acid. The process time t_P is the time span until all sulfanilic acid is consumed

Lamella width b_l [µm]	Char. diffusion time t_D [ms]	Process time t_P [ms]	Selectivity S_P [–]
50	2 500	2 755	0.128
5	25	32	0.776
0.5	0.25	0.79	15.787

From the evolution of concentrations over time in the case of a 10 µm wide channel, it can be seen that the secondary product S is only formed in regions containing reactant B and primary product P. Hence, the faster the reactants mix, the less secondary product is formed since most of reactant B is consumed in the primary reaction. The relationship between mixing time, in this case characteristic diffusion time t_D, and selectivity S_P is shown in Table 6.2. Selectivity and channel width are obviously indirectly proportional, with the rough approximation:

$$\frac{S_P'}{S_P} \approx \frac{b_l}{b_l'}.\tag{6.134}$$

Using this competitive-consecutive test reaction, different mixing devices can be compared in relation to their mean lamella width b_l. The reaction gives an experimentally accessible measure of mixing quality.

Dyes and dye pigments are among the oldest products of the chemical industry, in particular the aromatics, which are known since 1860 [430]. The basic molecular formula for this class of compounds is $R - N = N - R'$, where the group $N = N$ is called *azo* and R and R' denote aromatic groups. The synthesis of azo dyes is generally a two step process [430]:

1. **Diazotization:** Starting from an aromatic amine, $Ar - NH_2$, the reactive diazo salt $Ar - N_2X$ (X is an inorganic acid anion) is prepared by nitration:

$$Ar - NH_2 + 2HX + NaNO_2 \longrightarrow Ar - N_2X + NaX + 2H_2O\tag{6.135}$$

2. **Azo-coupling:** the electrophilic -N=N$^+$ group in the diazonium ion couples with a nucleophilic carbon atom of another organic compound, typically by substitution of a hydrogen atom:

$$Ar - N_2X + H - R \longrightarrow Ar - N = N - R + HX \qquad (6.136)$$

The coupling component R is usually also an aromatic compound, resulting in an extended de-localized π-electron system within the azo compound Ar-N=N-R. This enables azo compounds to absorb light in the visible part of the electromagnetic spectrum (the color of azo compounds range from yellow to red), and makes them an important class of industrial interest.

Diazotization is a highly exothermic reaction and, in most cases, also subject to a consecutive-competitive side reaction, which reduces the yield of the desired diazonium ion $Ar - N = N^+$. Hence, diazotization can benefit from microreactor technology due to the excellent heat transfer and good mixing performance of microchannel devices, however, diazotization in micromixers is not subject of this section. Instead, the coupling step of azo compound pigments (i.e. dye particles) in convective micromixers is investigated with the pigment CLA1433 from Ciba, see Section 6.5.

The azo-coupling is performed in aqueous media, resulting in competitive reaction and precipitation. When the solved reactants (i.e. diazonium ion and coupling compound) are mixed into the aqueous liquid, competitive reaction precipitation occurs: the coupling compound can precipitate in water before it has reacted with the diazonium ion to give the desired dye molecule, which precipitates anyway. The occurring processes are:

$$\text{diazonium ion}\,(l) + \text{coupling compound}\,(l) \longrightarrow \text{azo dye comp.}\,(l)$$
$$\text{coupling compound}\,(l) \xrightarrow{\text{H}_2\text{O}} \text{coupling comp.}\,(s) \quad (6.137)$$
$$\text{azo dye compound}\,(l) \xrightarrow{\text{H}_2\text{O}} \text{pigment}\,(s)$$

As precipitation occurs almost immediately, it is not possible to carry out the coupling from the solved state by semi-batch operation in a large stirred vessel; the coupling compound would precipitate mostly unreacted. The generally applied procedure is to first precipitate the coupling compound in such a way that the formed particles are as small as possible. Then the diazonium solution is added to this suspension. Due to their charge and polarization, diazonium ions are soluble in aqueous media. Since even particles of sparingly soluble substances are not static, meaning there is always a dynamic equilibrium between deposition and solution at the particle surface, some of the coupling compound molecules migrate into the liquid phase and react there with a diazonium ion. The formed dye molecule precipitates again due to its low solubility. This procedure has two major drawbacks:

- It takes a long time until most of the molecules have been in a solved state at least once, resulting in process times of up to one or more days.
- A noticeable portion of unreacted coupling compound remains in the particle core, even after long processing times.

Thus, the basic concept is to enhance conversion and reduce process time with the aid of convective micromixers. Due to their rapid mixing, the fast coupling reaction can possibly be accomplished before a larger amount of coupling compound is precipitated. This can be assisted by increased process temperature, accelerating the reaction and reducing the precipitation rate. While long-term heating can be problematic in conjunction with diazo compounds, the short heating time within microreactors avoids significant decomposition. The experimental results of fast mixing within T-shaped micromixers are described in Section 6.5.6.

6.4 Microreactors for Aerosol Generation

This section discusses the application of T-shaped micromixers for the generation of aerosols with nanoscale droplets by mixing a hot vapor-gas mixture with a cold gas. Vitamin E – nitrogen aerosols with droplet diameters from 15 to 50 nm have been produced in T-shaped microreactors with typical channel dimensions from 200 to 400 μm. With numerical simulation of the mixing geometries, three different flow regimes have been observed in these devices: symmetrical vortex flow, S-shaped engulfment flow and mushroom-shaped engulfment flow.

Optimized microreactor geometries were designed and fabricated in silicon with Pyrex glass lids. Special attention was paid to thermal insulation and particle deposition at the channel walls. This concerns not only the mixer chip, but also the design of the fluidic mount, using few bends and corners. The simulated flow regimes are validated and visualized with a liquid phase reaction. The particle deposition of a monodisperse NaCl aerosol shows that particles smaller than approx. 200 nm are probable to pass the microchannels. The Stokes number St has been identified as a useful parameter for describing this behavior.

The droplet size distribution of a vitamin E aerosol generated in the micromixers indicates a passive particle generation by cooling of the vapor-gas mixture in the inlet channel and active particle generation by mixing of the two gas streams. The very high temperature gradients of up to 60 K/mm and approx. $1.3 \cdot 10^6$ K/s are responsible for the narrow distribution of fine particles. The study shows the applicability of microreactors to produce nanosized particles or droplets in a continuous process.

6.4.1 Nanoparticle generation

The generation of nanoparticles, either solid or liquid, leads to new product formulations with special properties in biological, pharmaceutical, or medical applications [431]. Often, nano particles are produced by combustion of the educts [432] or electrical discharge processing [433]. Fabrication in microreactors is a promising route for reproducible and reliable particle properties, such as defined mean particle size with narrow particle size distribution. The production of nanoparticles in microreactors has entered the industrial scale in specialty chemistry [69]. Although particle handling is difficult in microchannels due to the high surface-to-volume ratio, which increases the risk of fouling and blocking the channels [27], the generation and handling of nanoparticles is certainly feasible in microchannels, as has been shown by various applications in micro reaction engineering [13] and Lab-on-Chip applications [391]. The particles are generated in a liquid phase with high shear gradients [434] or in bubbly flow of two immiscible liquids or gas-liquid flow [84]. A three-phase microreactor system produces nanoparticles by fast turbulent mixing in a nozzle [435].

The synthesis of solid nanoparticles from the gaseous phase in microreactors up to now has only been achieved by flame synthesis [432]. The reactor consists of three concentric tubes with a minimum inner diameter of 0.8 mm, which lead into a large combustion chamber. The components are fed through the inner pipe and concentric slits and react within the combustion chamber to produce nanosized powder [436]. This setup supports the concept of using microstructured devices only in positions where benefits can clearly be utilized.

This section describes the generation of aerosols with nanosized liquid particles or droplets. From a vapor-gas mixture, droplets can be generated either by cooling or by mixing. In both cases, the state of a vapor-gas mixture is changed in such a way that the vapor component exceeds the saturation concentration. Droplet generation by mixing is very fast and concerns the entire gas flow, while cooling down a vapor-gas mixture utilizes wall heat transfer, which produces high supersaturation close to the wall and, thus, condensation at the wall. This causes additional wall losses and leads to a broad particle size distribution (PSD). Nanoparticles produced by mixing are very small with a narrow PSD. Here, mixing occurs in well-characterized T-shaped micromixers. Mixing of liquids has already been intensively studied in such devices, with extremely short mixing times below 1 ms [133, 128].

For gaseous mixing in Tmixers, often asymmetrical mixing conditions occur due to different inlet temperature, and hence, viscosity and density. For asymmetrical mixing conditions, three different flow regimes are found by CFD calculation and can be described by typical Re numbers and momentum ratios of the inlet streams. A pH neutralization reaction with Bromothymol Blue (BTB) as a colorimetric pH indicator is performed in the liquid phase, with similar Re numbers and momentum ratio of the inlet flows. The particle loss in the microchannels is determined in deposition experiments with nanoscale sodium chloride particles in a nitrogen carrier stream. For aerosol generation, a gaseous vitamin E - nitrogen mixture close to saturation state, related to 150 °C, is mixed with a nitrogen stream at ambient temperature to produce a vitamin E aerosol with a droplet diameter less than 50 nm.

6.4.2 Asymmetrical mixing in microchannels

Particle formation includes the process steps of nucleation and particle growth from a supersaturated mixture. Nucleation without any starting nuclei is called homogeneous and is the starting point for the investigations. Homogeneous nucleation requires a high supersaturation depending on the properties of the vapor and the liquid, such as the surface tension. Two possible ways to produce a high supersaturation are illustrated in the h, Y-diagram, where the enthalpy h_{1+Y} of a mixture (carrier gas plus vapor $1 + Y$) is given over the concentration Y of the vapor, see Figure 6.12. The thick curved line marks the saturation line with the relative "moisture" $\varphi = 1$, below which the vapor condenses and precipitates as liquid droplets or surface film. The almost horizontal lines indicate isotherms with $T = $ const., which bend downwards at the saturation line.

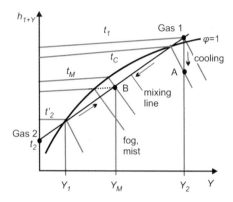

Fig. 6.12. Schematic h, Y-diagram of a vapor-gas mixture with cooling for Gas 1 (vapor content Y_2) and mixing of Gas 1 with Gas 2 (no vapor content).

The cooling of the vapor-gas mixture (point Gas 1 with the vapor content Y_2) down to a temperature below the saturation line results in the condensation of the vapor (point A), either to droplets in the bulk flow or as a film at the cold wall. Alternatively, mixing of a vapor-gas mixture with a cold gas stream (point Gas 2) leads to a supersaturated state (point B) and, consequently, to the formation of droplets in the bulk flow. Rapid mixing of the process streams is fundamental for high supersaturation and homogeneous nucleation in the mixing channel. Both combined will result in almost instantaneous nuclei formation and uniform particle growth, which is reflected in a narrow particle size distribution at the outlet of the mixing device.

As a result of the high spatial resolution of the elements, the fundamental equations of fluid dynamics are resolved correctly, also called direct numerical simulation DNS, see Section 2.4.5. The simultaneously solved species equations must be treated in a different way, because the mass transport is determined by the Schmidt number Sc, the ratio of the kinematic viscosity v to the diffusion coefficient D. In gases, the Sc number has the order of O(1), where the simulated species concentration profiles are on the same time and length scales as the velocity profiles. In liquids where the Sc number has the order of O(1 000) and greater, the diffusion is much slower and occurs on smaller length scales than the fluid dynamics. Therefore, the simulation results of gas-phase mixing are accurate, while the concentration profiles of liquid-phase mixing simulation indicate just a trend or a direction. For better results, a much finer mesh is necessary, which is beyond present computational possibilities. The concentration profiles of liquid mixing are "smeared" in regions with high gradients, which can not be represented by the mesh. This effect is summarized under the issue of numerical diffusion and often treated by sub-grid modeling.

Mixing of two streams with different temperature and concentration leads to mean values of temperature and concentration. In this case, the temperature is more characteristic of gaseous flow due to density and viscosity changes. The vapor concentration has only a small influence on the mixing process, hence, the mixing quality is defined for the temperature field in the mixing channel. To obtain a mixing characterization closer to reality, the velocity-weighted mixing quality α_V in a cross section of the mixing channel is defined as a normalized standard deviation of the temperature field in a certain cross section of the mixing zone, see Eq. 5.9. The temperature T_i at a grid point i is weighted with the corresponding velocity at the grid point i with the corresponding cross section part A, and the mean velocity $\bar{u}$ with the entire cross section A_M, see Section 5.1.2.

To obtain small particles with a narrow particle distribution, the cooling of the vapor by mixing must occur quickly, as slow mixing will allow nucleation to start earlier in some regions than in other. When these parts of the gas are mixed with hot vapor, the nuclei can grow for a longer period than with instantaneous mixing, resulting in larger particles. As growth conditions for different parts of the gas are different with slow mixing, the particle size distribution becomes wider, therefore, the mixers are optimized for a short mixing time in relation to the temperature distribution.

Starting with a simple T-mixer made from straight channels with a $300 \times 300 \ \mu m^2$ inlet cross section and $400 \times 300 \ \mu m^2$ outlet cross section, the mixer geometries are investigated for short mixing times at different inlet conditions. With a fixed channel depth of $300 \ \mu m$, the inlet and outlet channel widths have been varied according to Figure 6.13. Local widening and narrowing of the inlet channels lead to longer mixing times.

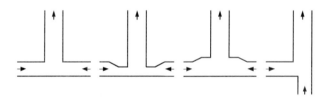

Fig. 6.13. Simulated mixer geometries: simple T-mixer, T-mixer with narrowing of inlets, T-mixer with widening of inlets, 90° mixer (from left to right).

Mixing of cold nitrogen at 20 °C with hot nitrogen at 70 °C and 257 °C causes three different mixing regimes, see Figure 6.14:

1. vortex flow with symmetrical vortices,
2. so-called S-engulfment with S-shaped vortices, identical with engulfment flow described in Section 3.2.3 and 5.4.2, and
3. so-called mushroom-engulfment with mushroom-like vortices.

The point-of-view for the streamlines and temperature-velocity vector plots in Figure 6.14 is in flow direction into the mixing channel. At a fixed channel depth of $300 \ \mu m$, an inlet channel width of $300 \ \mu m$ and a $400 \ \mu m$ wide mixing channel give the shortest mixing times with mushroom-engulfment flow regime. Alternatively, an injection mixer with a 90° junction was studied and produced mushroom-like vortices over a wider range of Re numbers. The larger interface leads to a higher mixing quality α_m.

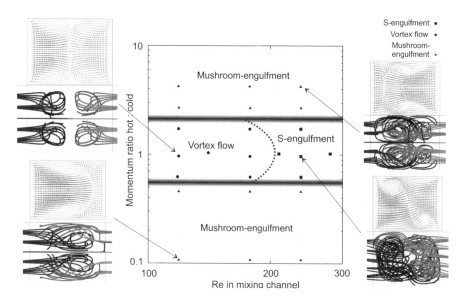

Fig. 6.14. Simulations of flow regimes in T-mixers, see also Figure A.9: flow regimes found in a T-mixer ($300 \times 300 \ \mu m^2$ inlets, $300 \times 400 \ \mu m^2$ mixing channel) showing stream lines and velocities with gray-shaded temperature in a lateral section through the mixing channel; mixing of nitrogen at 20 °C and 70 °C.

As the main result, a combination of straight narrow inlet channels (200 μm width) at the mixing zone with a straight 500 μm wide mixing channel gives the best mixing results for S-engulfment, see Figure 6.14 left. This layout also yields the shortest mixing time with mushroom-engulfment, with an inlet channel width of 300 μm and 400 μm wide mixing channel. For particle processing, the choice of mixer geometries is limited due to the risk of particle attachment to the wall. Therefore, the simple contacting element of a T-shaped mixer with a straight mixing channel is best suited for particle generation in the gas phase.

The development of the various mixing regimes is dependent on the Re number in the mixing channel, as a measure of the normalized velocity, and the momentum ratio between the two inlet flows, see Figure 6.14. With symmetrical inlet flow conditions, vortex flow is established first. At a certain Re number of approx. 180 to 200, a transition to S-engulfment flow is observed. With asymmetrical inlet flow conditions, mushroom-engulfment begin to occur at lower Re numbers. Vortex flow leads to insufficient mixing due to the stable interface between the two components and should, therefore, be avoided. Both engulfment flow regimes lead to close entanglement between the two components and generate fast and effective mixing.

The transition stage between vortex and mushroom-engulfment remains constant at the higher temperature of the hot gas (257 °C). The transition between the vortex and S-engulfment flows shifts to higher Re numbers for the higher hot gas temperature. This effect results from a temperature-dependent density variation and is also predicted by an evaluation of the ratio between centrifugal force and inertial force, see Section 3.2.

6.4.3 Experimental characterization of mixing regimes

An experimental study was performed to verify the simulated mixing regimes in the liquid phase with the help of a pH neutralization reaction. Moreover, the measurement of particle deposition in fabricated silicon microreactors was conducted with a particle-laden gas stream, see Section 4.3.4. A silicon chip for aerosol generation and its fluidic mount are displayed in Figure 5.32.

The mixing regimes are determined in the liquid phase by a pH-induced color reaction, which has been used in previous investigations, see Section 3.2.3. The mixture of a neutral Bromothymol Blue solution (green) and an alkaline buffer solution with pH 8 (clear) changes its color to blue depending on the local mixing quality. More details can be found in [374]. The regions of the different mixing regimes found by simulation, see Figure 6.14, are clearly identified in this control experiment, see Figure 6.15.

In the experimental investigations, the transition to the mushroom-engulfment regime occurs in a smaller band around a momentum ratio of 1, due to the entrance effects of the short inlet channels in the simulation. One inlet channel includes a bend, which presumably disturbs the symmetrical inlet flow, see the top left corner of the mixer chip in Figure 5.32. The difference between engulfment regime (middle of Figure 6.15) and mushroom vortex (lower image of Figure 6.15) clearly can be seen in the symmetry breakup due to the asymmetrical inlet conditions. Additionally, the sharp line between inlet 1 and inlet 2 in the lower image indicates the higher flow rate of inlet 2.

6.4.4 Aerosol generation by homogeneous condensation

The Vitamin E-aerosol was generated by mixing the two starting components of the Vitamin E - nitrogen vapor-gas mixture with pure nitrogen in the first mixing element of the microreactor, see Figure 5.32. The measurement test rig is displayed in Figure 6.16. An inlet temperature for the cold gas $T_{nitrogen} = 25$ °C and the superheated vapor-gas mixture $T_{superheat} = 150$ °C along with a saturation temperature of $T_{vapor} = 130$ °C were set as process conditions. At both inlets, the mass flow ratio and the total mass flow rate are maintained with mass flow controllers (MFC). In a vaporizer, the nitrogen at one inlet is saturated with vitamin E and overheated to reduce condensation within the connection tubes and in the inlet channel. After mixing the hot vapor-gas mixture with cold nitrogen in the microreactor, the generated

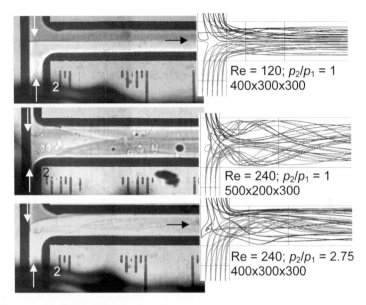

Fig. 6.15. pH neutralization reaction in a T-shaped micromixer (mixing channel width × inlet channel width × channel depth in μm) together with the corresponding simulations: vortex flow (top), S-engulfment (middle) and mushroom-engulfment (bottom), see also Figure A.10.

aerosol is charged in a Kr_{85} charger. The particle number density and particle size distribution are measured in a scanning mobility particle sizer (SMPS) consisting of a differential mobility analyzer DMA and a condensation particle counter CPC.

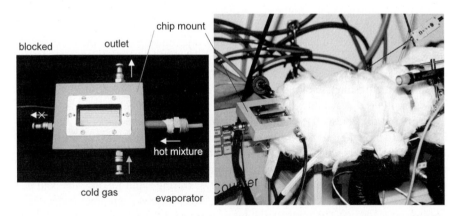

Fig. 6.16. Experimental setup of aerosol generation

The particle size distribution of three different tests is displayed in Figure 6.17. During the experiment the chip temperature rose from 23 to 30 °C. The droplet size distribution clearly exhibits two peaks indicating different droplet generation mechanisms, i. e. a mere cooling of the vapor-gas mixture in the inlet channel and an active supersaturation by mixing of two gas streams. Both mechanisms are sketched in Figure 6.12.

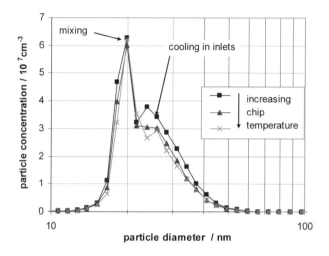

Fig. 6.17. Particle size distribution of Vitamin E-aerosol generated in the microreactor with Re = 440 in the mixing channel ($500 \times 300\ \mu m^2$).

Experiments with a modulated cold gas flow show that droplets with a diameter of approx. 20 nm are generated through rapid mixing of the two fluid streams. Particles with a diameter in the range of 20–50 nm are generated by cooling in the inlet channel. This passive cooling effect diminishes with increased chip temperature, but is still visible at the end of the experiment. The maximum value of the particle size distribution with passive cooling, i. e. no cold gas supplied, is almost identical to the cooling peak in Figure 6.17, which clearly identifies the influence of the inlet channel cooling. From the inlet, the vapor-gas mixture is cooled down from 150 °C to approx. 120 °C at the contact point with the cold gas stream. With the typical inlet flow velocity of approx. 28 m/s (Re = 340) and an inlet length of 1 mm ($\Delta T_l = 30$ K/mm), a temporal temperature gradient of approx. $\Delta T_t = 840000$ K/s is obtained, which leads to high supersaturation. From experimental investigations it is known that the mixing length in the channel is constant, representing three times the hydraulic diameter d_h for Re numbers higher than 350 [127]. During rapid mixing, the cold gas stream is heated from 25 °C to approx. 85 °C, which means a temporal temperature gradient of 1 300 000 K/s with a velocity of more than 22 m/s in the mixing channel (Re = 440). These extremely high temperature gradients are responsible for the small droplet diameter and narrow droplet size distribution, see Figure 6.17.

Further experimental results from the Karlsruhe group, MVM, indicate a clear distinction between two different particle generation mechanisms: convective cooling and mixing. Convective cooling gives a supersaturation profile in the microchannel with different saturation values. These lead to different nucleation and particle growth conditions, resulting in a broad particle size distribution. Mixing is a very fast process with a nearly discrete supersaturation, resulting in a narrow particle size distribution. The analysis of the bimodal particle size distribution is displayed in Figure 6.18. The raw experimental data, given on the right side, weakly correlates the bimodal particle size distribution. The mathematical analysis clearly distinguishes small particles with narrow size distribution from mixing and larger particles with wide size distribution from cooling.

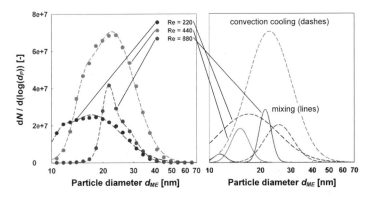

Fig. 6.18. Influence of the Re number on the particle size distribution; left: raw particle size distribution for four different Re numbers; right: bimodal split curve with mixing and cooling influence.

The results from Figure 6.18 also indicate future research and development directions. Temperature control to minimize cooling effects is essential for small aerosol particles with narrow size distribution. High Re numbers lead to small particles with narrow size distribution as well as high volume flow rates. The main issue in mixer design is fast mixing at the first contact of the fluids and, thereafter, undisturbed flow to prevent particles from contacting the walls.

The results of this section are financially supported by the Federal State of Baden-Württemberg under the research project "Microtechnology-Supported Integrated Processes". Thanks go to the group of Prof. Nirschl with Mr. Wengeler and the group of Prof. Kasper with Mr. Heim for their assistance and support in experimental investigations.

6.5 Mixing and Defined Precipitation in Liquid Phase

Continuous precipitation by mixing of two fluids is a simple and important method to produce nanoparticles for many industrial applications in specialty chemistry, pharmaceuticals, or the food industry. Convective mixing and mass transfer determine the nucleation and growth of the nanoparticle, which is displayed in the various particle size distributions. The mixer chips have a straight mixing channel to minimize particle attachment and blocking. Particle nucleation is modeled on classical thermodynamic nucleation theory. The growth of a single nanoparticle is analytically described by the diffusive mass transfer in its vicinity and the Sherwood number Sh. Including the local mixing characteristics of the T-shaped micromixer, the particle size distribution (PSD) is numerically simulated by a finite difference method as a function of the Re number in the mixing channel. The micromixers are also used for an azo-coupling under industrial conditions.

In an excellent analysis of the potential of microreactor technology in the fine chemicals and pharmaceutical industry, Roberge et al. [36] reveal that approx. 50 % of the reactions in this sector could benefit from microreactors. The authors also state that approx. 60 % of these reactions involve a solid phase, making them inappropriate for microreactors due to fouling and blocking issues. Particle-generating processes have been realized in micromixing devices mostly for scientific purposes, such as barium sulfate precipitation, but also for industrial applications like pigment synthesis. Beside the thorough studies of Schwarzer et al. [437, 438] on the precipitation of barium sulfate within T-shaped mixers, there are a few works dealing with particle generation in microchannels [439, 440, 441, 442]. Of particular interest are reports of the production and conditioning of dye pigments in microreactors. Pennemann et al. [443] describe the synthesis of the azo pigment Yellow 12 in a microreactor and the accompanying improvement of its properties. Clariant, a manufacturer of fine chemicals, has patented the production and conditioning (i.e. the improvement of the pigment properties by physical treatment) of diketopyrrolopyrrol pigments in a microreactor [62, 63].

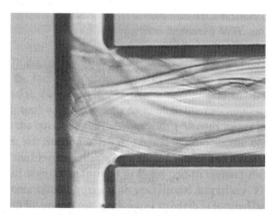

Fig. 6.19. Convective T-mixer with two fluids intertwined by convective engulfment flow (Re = 200) and diffusive mixing between the lamellae. The dark lines indicate generated barium sulfate particles at the interface between the components.

This section investigates the influence of convective mixing in T-shaped micromixers on particle precipitation, see Figure 6.19. For this purpose, a model for coupled mixing-precipitation processes is derived and numerically simulated. The results are compared to experimental data gained from the precipitation of barium sulfate, a widely used model system that has already been applied to convective micromixers by Schwarzer [444]. Furthermore, the suitability of convective micromixing devices for industrial pigment synthesis is evaluated in collaboration with Ciba Specialty Chemicals Inc. (Ciba). Both objectives require micromixers with special properties. Hence, design rules must be established and realized with basic mixing devices, which are subsequently implemented with the aid of silicon bulk micromachining.

6.5.1 Modeling and simulation of particle precipitation

Considering a single particle, precipitation can be divided into three distinct stages: nucleation, growth, and secondary effects. In the beginning, an initial embryonic particle – the nucleus – must be formed, which can then grow further. Already existing particles will grow as long as the surrounding solution is supersaturated. For nucleation, however, there exist two distinct mechanisms. If the embryo is only made up of molecules of the precipitating species, the nucleation is called homogeneous. In heterogeneous nucleation conversely, deliberate particles of arbitrary composition can serve as precipitation nuclei.

If mixing is accomplished only by diffusion at a constant interface area between the two components, the primary particle size distribution PSD, being the PSD without modifications due to secondary processes, is determined by the initial species concentrations. In convective mixing, on the other hand, the fluid lamellae are stretched and folded, consequently leading to an increased interface area and a shorter lamellae width. The first step in understanding the impact of mixing on the PSD is to develop an appropriate model for precipitation at constant interfaces. This is assisted by a short overview of the basic thermodynamic principles for precipitation. Particulate systems are characterized by the particle size distribution PSD, which is subsequently used to derive the population balance equation PBE as the basis for all precipitation simulations.

Precipitation of a solved species only takes place as long as the solubility is exceeded, which is expressed by the supersaturation S. At high supersaturations, homogeneous nucleation is the dominant mechanism [444]. The rate of homogeneous nucleation B_{hom} can be computed according to Söhnel and Garside [445] using the classical nucleation theory. After a critical nucleus has formed, it grows continuously until the activity equilibrium is reached, in other words: $S_a = 1$. This process is composed of two successive steps:

1. transport of single precipitate "molecules" to the particle surface, and
2. incorporation of those "molecules" into the lattice.

At high supersaturations, as usually encountered in technical precipitation processes, the growth rate is determined by the transport process. Incorporation kinetics are only important at small supersaturation values. The transition from integration-controlled to transport-controlled growth takes place at $S \approx 40$ [61].

The transport of the precipitating species to the particle surface is then a rotational symmetrical diffusion problem, and can be described by the diffusion equation in spherical coordinates [446], see also Eq. 5.2:

$$\frac{\partial c}{\partial t} = \frac{1}{r^2} \frac{\partial}{\partial r} \left(D r^2 \frac{\partial c}{\partial r} \right) \tag{6.138}$$

If the bulk concentration of the species changes only slightly with time, the concentration profile around the spherical particle is assumed to be in quasi-steady state, i.e. $\partial c / \partial t \approx 0$. With constant diffusivity D, Eq. 6.138 reduces to an ordinary differential equation

$$\frac{d}{dr} \left(r^2 \frac{dc}{dr} \right) = 2r \frac{dc}{dr} + r^2 \frac{d^2 c}{dr^2} = 0. \tag{6.139}$$

It must be stressed that this approximation is only applicable if

1. the concentration profile around the particle is rotationally symmetric,
2. equilibrium is maintained near the particle surface at all times, and
3. the variation of the bulk concentration with time, dc_∞ / dt, is slow compared to diffusion. Only then can a quasi-steady concentration profile develop and the assumption $\partial c / \partial t \approx 0$ may be used to derive Eq. 6.139.

However, real particles rarely have a spherical shape, they usually have a form of crystal lattice. Thus, the spherical radius R is not appropriate to specify the size of real particles. The linear size L is used instead and is defined as the diameter of a sphere exhibiting the same volume as the arbitrarily shaped crystal. Hence, particle growth is described by the linear growth rate $G = dL/dt$. For mass transfer to the particle surface, the scale-dependent information is included in particle diameter and species diffusivity. Dividing the mass transfer coefficient by D/L gives the corresponding dimensionless Sherwood number [181]. This yields an expression for the linear growth rate G in terms of the Sherwood number Sh and particle size L

$$G = \frac{2 D \text{Sh}}{L \, c_s} \psi, \tag{6.140}$$

with the molar density c_s of the crystal, also called the surface concentration of the particle, and the growth potential $\psi = \text{const.} \cdot (S_a - 1)$ as a linear relationship to the supersaturation. This correlation describes the size-dependent particle growth, which will be applied in simulations of the PSD.

6.5.2 The population balance equation

To derive of the population balance equation PBE, the discrete PSD is depicted in Figure 6.20. The number density concentration n is defined as the number of particles per class range Δl_i and volume of solution V_{sol}. Particles may be directly generated

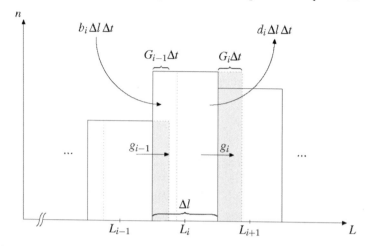

Fig. 6.20. Detail of a histogram particle size distribution for derivation of the population balance equation PBE.

in class i, i.e. they do not grow in from class $i-1$, but are generated by nucleation or agglomeration of much smaller particles. Correspondingly, if particles of class i are involved in cluster formation, they disappear from this class. Both phenomena are accounted for by birth and death rates (number per volume, class range and time), b_i and d_i, respectively. Summation of these four terms and multiplication with the solution volume gives the change in the total particle number in class i. Taking now the limits $\Delta t \to 0$ and $\Delta l \to 0$ yields the well known population balance equation, see Hounslow et al. [447]

$$\frac{\partial n}{\partial t} + \frac{\partial (Gn)}{\partial L} = b - d. \tag{6.141}$$

Since agglomeration and heterogeneous nucleation are not considered here, the death rate d can be discarded, and the birth rate b is identical with the nucleation rate. Expanding the partial derivative $\partial (Gn)/\partial L$ and applying Eq. 6.140 for the growth rate, the left side in the above equation yields

$$\frac{\partial n}{\partial t} + \frac{\partial (Gn)}{\partial L} = \frac{\partial n}{\partial t} + n\frac{\partial G}{\partial L} + G\frac{\partial n}{\partial L} = \frac{\partial n}{\partial t} + \frac{2D\,\mathrm{Sh}}{c_s}\frac{\psi}{L}\left(\frac{\partial n}{\partial L} - \frac{n}{L}\right). \tag{6.142}$$

using Eq. 6.140 for the growth rate G. If SI units and dimensions are used, the involved quantities differ remarkably in their numerical values (typical values: $n = 10^{22}$ m^{-4}, $L = 10^{-9}$ m), making numerical computations difficile and inaccurate. Hence, all variables are normalized with an appropriate reference value:

$$n^* := \frac{n}{n_0} \qquad t^* := \frac{t}{t_0} \qquad L^* := \frac{L}{L_N} \qquad \psi^* := \frac{\psi}{c_0} \qquad b^* := \frac{b}{b_0} \tag{6.143}$$

where L_N is the nuclei diameter and c_0 is the initial bulk concentration. With these definitions, the dimensionless PBE is transformed into

$$\frac{n_0}{t_0}\frac{\partial n^*}{\partial t^*} + \frac{2D\,c_0\,n_0\,\mathrm{Sh}}{c_s\,L_N^2}\frac{\psi^*}{L^*}\left(\frac{\partial n^*}{\partial L^*} - \frac{n^*}{L^*}\right) = b_0 b^*. \tag{6.144}$$

To create reasonable choices for the remaining reference values, two dimensionless numbers, Z_I and Z_{II} are formed by dividing two of the three coefficient factors in Eq. 6.144

$$Z_I = \frac{2D\,c_0\,t_0}{L_N^2\,c_s}\mathrm{Sh}, \tag{6.145}$$

$$Z_{II} = \frac{n_0}{t_0\,b_0}. \tag{6.146}$$

Choosing the reference values properly, the dimensionless numbers will also have magnitudes close to unity. Setting $Z_{II} = 1$, implies $n_0 = t_0 b_0$ and yields the dimensionless PBE

$$\frac{\partial n^*}{\partial t^*} + Z_I \frac{\psi^*}{L^*}\left(\frac{\partial n^*}{\partial L^*} - \frac{n^*}{L^*}\right) = b^*. \tag{6.147}$$

This is the governing equation for all precipitation processes, in which only nucleation and diffusion-controlled growth occur. It describes the time-dependent particle size distribution $n(L,t)$ and can be used for a global simulation of such precipitation processes. The single particles are not simulated, only their distribution in size classes)

The continuous PBE can be transformed into a set of ODEs by semi-discretization of Eq. 6.147 with respect to the particle size L. Using a non-uniform grid for the particle size, the backward-differences approximation gives for $i \geq 2$,

$$\frac{dn_i^*}{dt^*} = \frac{Z_I}{r-1}\frac{\psi^*}{L_i^{*2}}(r\,n_{i-1}^* - n_i^*) \tag{6.148}$$

where the spacing between adjacent grid cells is given by the progression $L_i = rL_{i-1}$ with the progression ratio r and $L_1 = L_N$ as the size of the critical nucleus. The number density in the nucleation class n_1^*, i.e. the left boundary condition for Eq. 6.148, is given by

$$n_1 = n_0 n_1^* = \frac{B_{hom}}{G_N}, \tag{6.149}$$

the ratio of the nucleation rate, B_{hom}, to the growth rate of nuclei G_N, see Mahoney and Ramkrishna [448].

6.5.3 Precipitation in the interdiffusion zone

The next step is to incorporate mixing and diffusion between the fluid lamellae into the model. Both the number density and the species concentrations are dependent on the spatial coordinate x: $n = n(L,x,t)$ and $c = c(x,t)$. We assume that the solid particles are spatially fixed, i.e. neither convective nor diffusive motions of particles occur, or the particle growth is much faster than the change of the environment. Solving simultaneously the two integro-differential equations,

$$\frac{\partial n}{\partial t} = -\frac{\partial (Gn)}{\partial L} + b, \tag{6.150}$$

$$\frac{\partial c}{\partial t} = D\frac{\partial^2 c}{\partial x^2} - c_s \int_{L_N}^{L_{\max}} \pi L^2 Gn\,dL, \tag{6.151}$$

gives the accurate approach to couple one-dimensional diffusive mixing with the PSD evolution. The first equation is simply the PBE, Eq. 6.141, without particle death, and the second is the one-dimensional diffusion-reaction equation. The integral term accounts for the change in the species concentration due to precipitation. The term $\pi L^2 Gn\,dL$ is the increase in the volume of particles with size L over time.

Convective mixing devices like the T-shaped micromixer, see Figure 6.21 left side, combine short mixing times with excellent heat removal and low clogging potential. According to Engler [127], convective mixing is a laminar process, "where the mixing is mainly caused by convective flow and stretching due to secondary vortices perpendicular to the main flow direction". Figure 6.21 right side, shows the typical lamellae structure encountered in convective mixing. The two lamellae are rolling up into each other via two intertwining vortices (engulfment flow), which are induced by the redirection and collision of the feed streams within the T-junction. Comparing the engulfment flow structure on the right side of Figure 6.21 with an

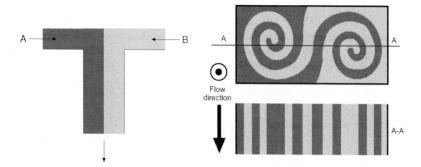

Fig. 6.21. Channel geometry and scheme of convective mixing in T-shaped microchannels, taken from [127]. Left: Straight laminar flow in a T-shaped micromixer with 1:1 feed rate (top view). Right: Convective mixing: thinning of fluid lamellae by roll-up of the two inlet flows (engulfment flow). Cross section of the mixing channel (top) and cut through the channel along intersecting line A-A (bottom).

unrolled lamellae shows that the interface length between the two distinct phases is twice the lamellae length λ. The velocity $w(t)$ is used only in comparison to the characteristic diffusion length $\delta = \sqrt{2Dt}$ to estimate the relative width of the so-called interdiffusion zone. For diffusion in a semi-infinite space, this characteristic is defined as the distance, over which the concentration drops to half of the initial value. The diffusion δ characterizes the distance, over which the concentration drops to zero.

The central notion to simplify Eqs. 6.150 and 6.151 is to replace the spatially variable quantities $\psi(x,t)$ and $B_{hom}(x,t)$ with their mean values $\langle \psi(t) \rangle$ and $\langle B_{hom}(t) \rangle$ in the interdiffusion zone:

$$\langle \psi \rangle = \frac{1}{\delta} \int_0^\delta \psi(\bar{c}_A(x), \bar{c}_B(x)) \, dx = \int_0^1 \psi(\bar{c}_A(x^*), \bar{c}_B(x^*)) \, dx^* \qquad (6.152)$$

$$\langle B_{hom} \rangle = \frac{1}{\delta} \int_0^\delta B_{hom}(\bar{c}_A(x), \bar{c}_B(x)) \, dx = \int_0^1 B_{hom}(\bar{c}_A(x^*), \bar{c}_B(x^*)) \, dx^* \quad (6.153)$$

wherein δ is the zone width, and $\bar{c}_A(x)$ and $\bar{c}_B(x)$ are the species concentrations along the diffusion zone, see Figure 6.22. It is assumed that the number density n does not vary within the diffusion zone and is thus taken as a mean value.

Utilizing the mean values eliminates the diffusion term in Eq. 6.151. The integral is replaced by the sum of the particle growth in each class. The following mass balance accounts for the concentration change Δc for each of the two species A and B due to the particle growth.

$$\Delta c_B = -2 \frac{\pi}{6} c_s n_0 L_N^4 \frac{r-1}{r} \sum_i (n_i^*(t^* + \Delta t^*) - n_i^*(t^*)) L_i^{*4} \qquad (6.154)$$

$$\Delta c_A = \Delta c_B \qquad (6.155)$$

For equi-molar precipitation, the concentration change in both species must be equal.

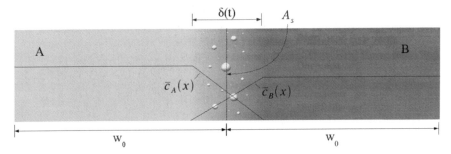

Fig. 6.22. Cross section through a two-phase system with concentration profiles according to the linear interdiffusion zone model. The left phase contains species A, the right phase species B. Both phases are w_0 wide and share a common interface, A_s. The zone width $\delta(t)$ increases with time by diffusion. For simplification in this model, it is assumed that the concentrations increase linearly within the interdiffusion zone and approach a constant value outside.

6.5.4 Binary compound precipitation in convective micromixers

A two-component system with equal phase width w_0 and planar interface area A_s, see Figure 6.22, contains the total solution volume $V_{sol} = 2w_0 A_s$. The volume where nucleation occurs is given by $V_{nuc} = \delta A_s$. Assuming that a single lamellae of width $w(t)$ and interface length $\lambda(t)$, which is continuously stretched in the mixing channel, decreases the phase width and increases the interface length. The lamellae depth h is constant and, therefore, cancels out from the ratio V_{nuc}/V_{sol}. The mixing model provides two simple correlations for the lamellae width and the interface length. This is valid for convective mixing in T-shaped micromixers over the range $350 \leq \mathrm{Re} \leq 1000$

$$w(t) = w_\infty + (w_0 - w_\infty) \exp\left(-\frac{t}{t_m}\right),$$

$$\lambda(t) = \frac{\lambda_0 w_0}{w(t)}.$$

The values w_0 and λ_0 are the initial lamellae parameter. The characteristic mixing time t_m is given by

$$t_m \approx \frac{w_\infty}{2D} \tag{6.156}$$

where w_∞ is the final lamellae width, after which the convective effects have subsided. There, the flow is relaminarized, and the width can be estimated as

$$w_\infty \approx \sqrt{\frac{8D}{v \mathrm{Re}} d_h}. \tag{6.157}$$

Since the whole phase is equally stretched, the width δ of the diffusion zone is not only affected by diffusion, but also by the decrease in the phase width $w(t)$. To keep the model simple, it is assumed that $\delta(t)$ is still determined by $\sqrt{2Dt}$. On that condition, the initial phase width w_0 has to be replaced with the time-dependent width $w(t)$ for calculating the nucleation volume ratio ρ_{nuc}:

$$\rho_{nuc} = \frac{V_{nuc}}{V_{sol}} = \frac{\delta(t)}{2w(t)} = \frac{1}{\sqrt{2}} \frac{\sqrt{Dt}}{w_\infty - (w_0 - w_\infty) \exp(-t/t_m)} \tag{6.158}$$

With this ratio, the boundary condition in Eq. 6.149, along with the reference values, gives the realistic nucleation density

$$n_1^* = \frac{\rho_{nuc} \langle B_{hom} \rangle}{Z_I b_0 L_N \langle \psi^* \rangle}. \tag{6.159}$$

Of course, the nucleation volume can be no greater than the solution volume. Hence, it has to be ensured that $\rho_{nuc} \leq 1$. This condition simply implies that the fluid is perfectly mixed as soon as $\rho_{nuc} = 1$. The second issue is a simple adjustment of the mass balances to compute the linear concentration profiles of the bulk concentrations within the respective phases. Thus, the mass balance in Eq. 6.155 must be multiplied with two. With these assumptions the algorithm can be generated to calculate the PSD:

1. Initialization: compute size classes L_i^*, reference values $(n_0, t_0, L_N, c_0, b_0)$, dimensionless growth coefficient Z_I and initialize vector of number densities $n_i^* = 0$; set c_A and c_B to initial values (based on the phase volume $1/2V_{sol}$):

2. while $c_B > 0$:
   a) determine current nucleation volume ratio ρ_{nuc} from Eq. 6.158 and ensure $\rho_{nuc} \leq 1$
   b) calculate $\langle \psi^*(c_A, c_B) \rangle$, using Eq. 6.152
   c) calculate $\langle B_{hom} \rangle$ with Eq. 6.153 and $n_1^*(t^* + \Delta t^*)$ through Eq. 6.159
   d) compute $n_i^*(t^* + \Delta t^*)$ from $n_i^*(t^*)$ by numerical integration of ODE 6.148, using $\langle \psi^* \rangle$
   e) calculate new concentrations $c_A(t^* + \Delta t^*)$ and $c_B(t^* + \Delta t^*)$ through mass balance 6.154 and 6.155

This algorithm enables the estimation of the PSD evolution for binary compound precipitation during convective mixing in T-shaped micromixer with a relatively simple, yet physically reasonable model. The procedures developed above allow the simulation of actual precipitation processes by coupling the PBE with various mixing models. These algorithms are applied to predict the PSD for barium sulfate precipitation, induced by convective mixing of barium chloride solution and sulfuric acid according to

$$BaCl_2 + H_2SO_4 \longrightarrow BaSO_4 \downarrow + 2HCl \tag{6.160}$$

or, considering only the active ion species:

$$Ba^{2+} + SO_4^{2-} \longrightarrow BaSO_4 \downarrow \tag{6.161}$$

The computations are carried out for initial feed concentrations of 0.33 molar sulfuric acid and 0.5 molar barium chloride solution with a feed rate ratio of 1:1, as used also by Schwarzer [444].

The equilibrium constants given by Schwarzer [444] are valid for 25 °C: $K_s = 1.2 \times 10^{-3}$ mol/l, $K_{ip} = 5.4 \times 10^{-3}$ mol/l, and $K_{sp} = 9.82 \times 10^{-11}$ mol^2/l^2. The rate of homogeneous nucleation B_{hom} was derived in terms of the supersaturation S_a and the interface energy γ_s between solid and liquid phase. Schwarzer gives a correlation based on more recent research [444], which which is used in the simulation to ensure comparability with his results

$$B_{hom} = 1.5 \cdot D \cdot \left(\sqrt{K_{sp}} \cdot S_a \cdot N_A \right)^{7/3} \cdot \sqrt{\frac{\gamma_s}{kT}} \cdot V_m \cdot$$
$$\cdot \exp \left[\frac{-16\pi}{3} \cdot \left(\frac{\gamma_s}{kT} \right)^3 \cdot \frac{V_m^2}{(v \ln S_a)^2} \right]. \tag{6.162}$$

with the Avogadro constant $N_A = 6.022 \cdot 10^{23}$ 1/mol and the Boltzmann constant k. For the test system of barium sulfate, the molar volume is $V_m = 8.64 \cdot 10^{-29}$ m^3, the stoichiometric coefficient $v = 2$, and the common diffusivity of the precipitating ion is $D = 1.67 \cdot 10^{-9}$ m^2/s, given in [449]. More details on thermodynamic modeling and simulation are given by Kastner [49] and Schwarzer [444].

The exponential term in Eq. 6.162 strongly depends on supersaturation and interface energy, both quantities, for which various estimating models exist. Aside from variations in supersaturation, the value of the interfacial energy may introduce errors in nucleation rates and, therefore, the PSD. For barium sulfate, quite different values for γ_s at $T = 25\ °C$ can be found in literature, for example, $0.136\,J/m^2$ [445] and $0.126\,J/m^2$ [449], or, including the effects of solved species concentrations, $0.1181\,J/m^2$ in case of $0.33\,M$ sulfuric acid and $0.5\,M$ barium chloride solution [61]. Here, the value of $0.1181\,J/m^2$ is applied as used by Schwarzer.

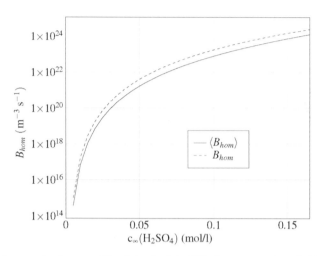

Fig. 6.23. Mean nucleation rate $\langle B_{hom} \rangle$ in the interdiffusion zone and nucleation rate B_{hom} for precipitation in complete and instantaneous mixing. The acid concentration $c_\infty\,(H_2SO_4)$ is based on the entire solution volume, i.e. twice the volume of the reactant feeds with $c_0\,(H_2SO_4) = 0.33\,mol/l$ and $c_0\,(BaCl_2) = 0.5\,mol/l$.

To calculate of the linear growth rate G, Eq. 6.140 is applied with the minimal Sherwood number Sh= 2 for spherical particles and the molar density $c_s = 19\,194.516\ mol/m^3$. Since barium sulfate is a binary compound, a model for the growth potential ψ is chosen from concentration difference or chemical activity. Using Eqs. 6.152 and 6.153, the mean values of growth potential and nucleation rate within the linear diffusion zone are calculated in terms of current acid bulk concentration $c_\infty\,(H_2SO_4)$. Figure 6.23 compares the mean values for nucleation rate and growth potential during barium sulfate precipitation in the interdiffusion zone with the corresponding values during precipitation in complete and instantaneous mixing. For diffusive mixing, both quantities are slightly below the values for instantaneous mixing.

To evaluate the designed mixing devices and the corresponding mixing model, two experimental series were executed. First, the operating performance for simple reactive precipitation was tested with barium sulfate. Secondly, the suitability of microchannel devices for particle-generating processes under industrial conditions in general, and for a competitive reaction-precipitation in particular, were investigated with the synthesis of a typical azo-pigment.

6.5.5 Experimental investigations of barium sulfate precipitation

In Figure 6.24, the experimental setup for the barium sulfate precipitation experiments is shown for typical test conditions. The micromixers are installed in the chip mount and connected with the periphery using polyethylene tubing and push-in connectors from Festo AG, Germany. The chip mount is placed beneath a microscope to allow optical observation of the precipitation within the mixing channel. The delivery of reactant flows is accomplished with syringe pumps ("Injectomat 2000", Fresenius HemoCare, Germany). These pumps deliver an accurate flow rate up to 400 ml/h per pump with almost no flow pulsation. Up to approx. 40 ml can be delivered per pass; hence, long-run experiments are not possible. To ensure symmetrical flow rates, the tubings must be of equal length after the junction. The product suspension flows through a short tube of approx. 15 cm and is collected in a beaker. At typical flow rates of some hundred ml/h, the residence time of the entire system is only a few seconds.

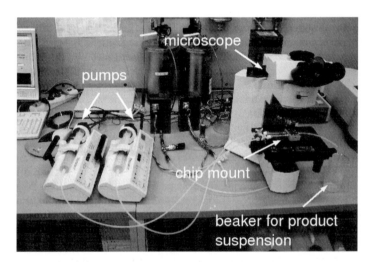

Fig. 6.24. The experimental setup for barium sulfate precipitation with two syringe pumps, chip mount under the microscope, and data acquisition system.

The experiments presented here have been carried out with 0.33 mol/l sulfuric acid and 0.5 mol/l barium chloride solution with equal flow rates in simple T-mixers, resulting in net concentrations $c_0(\mathrm{H_2SO_4}) = 0.165\,\mathrm{mol/l}$ and $c_0(\mathrm{BaCl_2}) = 0.25\,\mathrm{mol/l}$. Barium was supplied in higher concentration than necessary from stoichiometry to avoid particle agglomeration [444]. The reactant solutions were prepared using barium chloride di-hydrate, analytical grade, and 0.5 molar sulfuric acid from VWR, Germany. After the pumps have been started, a start-up time of at least three times of mean residence time was allowed before the product flow was sampled for PSD measurement. The course of mixing and precipitation within the outlet channel could be observed through the microscope and was documented by video recordings. Figure 6.25 shows images of a T-3-4x30 and a T-3-6x30 mixer. The nomenclature of the chip names (T-3-6x20) consists of the mixer type (T-mixer), the inlet channel width in 100 μm (300 μm), the mixing channel width in 100 μm (600 μm) and the mixing channel length in 100 μm (2 mm). The indicated Re numbers refer to the total flow rate through the mixing channel. The four mixer types T-3-4x10, T-3-4x30, T-3-6x10 and T-3-6x30 have been investigated for various Re numbers from 200 to 500 with symmetrical inlet conditions.

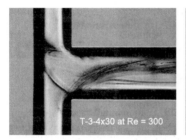

Fig. 6.25. Images of the T-junction during barium sulfate precipitation; flow direction: left to right. Precipitation and engulfment flow are indicated by the particle-laden streamlines (dark streaks).

The particle-laden streak lines in Figure 6.25 indicate the interface between the components where precipitation occurs, and also visualize the flow regime. Due to the moderate concentrations and small particles, clogging did not occur, however, barium sulfate particles are deposited on the channel walls, especially the Pyrex glass lid. This leads to the formation of larger aggregates, which are partially washed away by the flow itself. Such disengaged aggregates presumably do not break up completely and can, hence, affect the PSD if they enter the measured portion of the PSD. They showed up as secondary peaks at comparatively large particle sizes, and, for some samples, they disappeared in a second measurement of the PSD.

The laser light scattering instrument LS 230 from CoulterBeckmann has been used for measuring the PSD. This instrument uses laser Mie scattering for particles > 400 nm. For smaller particles, a tungsten halide lamp is used and the intensity ratio of vertically and horizontally polarized light is measured. The tungsten halide lamp covers the particle size range from 4 nm up to 400 nm. The entire device measures from 40 nm to 2 000 μm. The PSD measurements have a typical delay of 1 hour between sampling and measurement, which is somewhat problematic, since aggregation of the particles occurs, although the excess of barium ions slows this process notedly. Thorough investigations from Schwarzer show that for suspensions precipitated from 0.33 molar sulfuric acid and 0.5 molar barium chloride solution, the mean volume-averaged particle diameter ($\bar{L}_{Va}$) increases with a rate of approx. nm/min [444]. In addition, the reproducibility of the entire experimental procedure was checked; the deviations are acceptable, particularly the mean particle size changes only slightly with approx. 1 %.

Table 6.3. Comparison of number-averaged mean diameter and standard deviation for barium sulfate precipitation by 1:1 mixing of 0.33 molar sulfuric acid and 0.5 molar barium chloride solution in four different T-shaped mixers (T-3-4x10, T-3-4x30, T-3-6x10, and T-3-6x30)

| Re | $\bar{L}$ [nm] | | | | σ_L [nm] | | | |
	4x10	4x30	6x10	6x30	4x10	4x30	6x10	6x30
200	252.0	199.5	213.6	211.0	105.6	104.3	86.6	89.3
300	118.9	120.6	149.1	126.3	68.5	69.5	77.4	71.7
400	92.6	122.3	98.4	92.3	45.3	41.4	54.1	45.1
500	123.2	123.0	121.5	91.2	43.4	41.8	42.5	44.1
600	120.6	118.9	–	–	34.5	33.8	–	–

Considering the PSD measurements given in Table 6.3, the following systematic similarities can be noticed in all four mixer types:

- The mean diameter $\bar{L}$ falls at first with increasing Re number until a minimum is reached somewhere between Re = 300 and Re = 500. From there, the mean diameter increases again toward an upper limit of approximately 120 nm. The sole exception is the T-3-6x30, for which the mean diameter remains at 91 nm for Re = 500.
- The standard deviation, σ_L, decreases continuously for all mixer types and increasing Re numbers.
- A comparison of the different mixer types shows no systematic pattern in particle size and distribution for the same Re numbers.
- For low Re numbers, all mixers exhibit a secondary peak between 1 and 2 μm in the volume PSD. This peak diminishes with growing Re numbers.
- For medium Re numbers (between Re = 300 and 400), the number percentages show a non-zero axis intercept. Hence, a considerable amount of particles are smaller than the measurement limit of 40 nm, indicating that the applied PSD measurement method is only reliable to a certain degree.
- All PSDs are notably skewed, i.e. they ascend rapidly from small sizes and decline gradually toward larger sizes. It is not clear whether this effect is due to particle growth governed by mixing, particle aggregation, or the result of both phenomena.

An interesting relationship between PSD and Re number can be depicted in a quite descriptive way if the peak in the number PSD is considered: it moves clockwise, starting at 3.00 h for Re = 200 and moving along with increasing Re number, until finally 12.00 h is reached, indicated by the dashed arrow in Figure 6.26. This development of the PSD with the Re number is a hint for mixing quality development and will be included in further investigations.

Both mean size and standard deviation tend toward lower values with increasing Re number, indicating a mixing enhancement with the formation of engulfment flow for Re > 300. At high Re numbers, the mean size tends toward a limit value of approx. 100 nm, while the standard deviation continues to fall. Ultimately, the

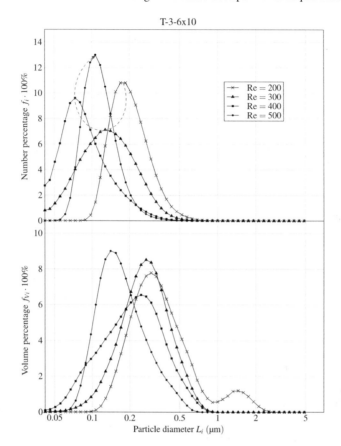

Fig. 6.26. Logarithmic particle size distribution (PSD) for barium sulfate precipitation in a T-3-6x10 for various Re numbers (1:1 mixing of 0.33 molar sulfuric acid and 0.5 molar barium chloride).

measured values are scattered over a comparatively broad range. The reason for this behavior is not clear, however, unstable flow conditions, due to the accelerated flow, are one reasonable explanation. The decline of both characteristics with increasing Re numbers is more distinctive and the limit for $\bar{L}$ seems to be below the corresponding value for the T-3-4 mixers. No influence from the mixing channel length is observed.

In Figure 6.27, the experimental data for the number density PSDs for barium sulfate precipitation in a T-3-6x30 are compared with the corresponding simulations for Re = 350 and Re = 500. While the peak of the experimental PSD changes only in size, and not position with increasing Re number, the simulations show a slight decrease of the peak particle size. Furthermore, the simulations give a narrow, almost symmetrical PSD, whereas the experimental PSD has a long "tail" toward larger particle sizes.

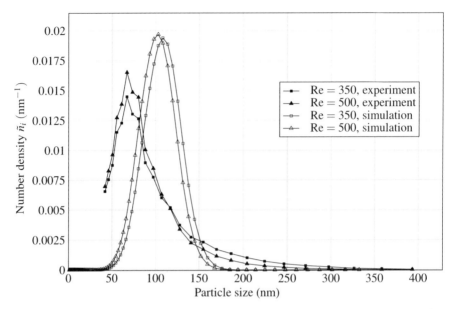

Fig. 6.27. Comparison of experimental with simulated number density PSD for barium sulfate precipitation from 0.33 molar sulfuric acid and 0.5 molar barium chloride solution in a T-3-6x30. The plots of experimental results are cut off at the lower measurement limit of the LS 230.

It can be seen from Figure 6.27 that both simulation and experiment show a common tendency, i.e. the characteristic properties decrease with growing Re numbers. On the other hand, there is a remarkable deviation between simulated and experimental PSD for Re < 350. The reason for this lies in the convective mixing model used. According to Engler [127], his model is only valid for Re > 350, where the engulfment flow has fully developed and the characteristic mixing length l_m can be considered as constant. Correspondingly, the mean values $\bar{L}$ from experiment and simulation are closer for high Re numbers.

Yet, there remains a major difference between the standard deviations, which can be explained as a result of the simplifications made in the mixing model. It is assumed that a single lamellae of homogeneous thickness is equally stretched during the mixing process, but, in fact, the flow splits up into many lamellae of non-uniform width. The difference between model and experiment diminishes with increasing Re numbers, since in both cases the mean lamellae width is stretched almost instantaneously for high Re numbers.

6.5.6 Experimental investigation of pigment synthesis

The synthesis of azo dyes is generally a two-step process, consisting of diazotization and azo-coupling [430]. The diazotization is a highly exothermic reaction and is, in most cases, also subject to a consecutive-competitive side reaction, which reduces the yield of the desired diazonium ion. Hence, the diazotization can benefit from microreactor technology, due to the excellent heat transfer and good mixing performance of microchannel devices, see also [450, 443]. The coupling step of azo compound pigments (i.e. dye particles) in convective micromixers is investigated with the pigment CLA1433 from Ciba, in terms of two primary questions:

- Are convective micromixers appropriate for particle-laden/particle-generating flow under typical industrial conditions?
- Can effectively operated micromixers overcome the drawbacks of conventional processes for azo pigment coupling?

While the first aspect is essential for a wide applicability of microreactor technology in the chemical industry, the second objective is necessitated by the particular properties of azo pigment synthesis. Due to economic reasons, the coupling is carried out in aqueous media. Conventional batch processes suffer from the complex solubility situation leading to process times of one or more days. Thus, the basic idea was to enhance conversion with convective micromixers due to their rapid mixing process. The diazonium compound solution was prepared according to the process regulations for batch operation, only with dilution by a factor of 10 with water to reduce the particle concentration in the product flow. The diazonium solution and the solved coupling compound are mixed within a T-shaped convective mixer and collected in a beaker, see Figure 6.28. The yielded product suspension is filtered with a suction strainer and washed with water. The moist pigment is analyzed with high performance liquid chromatography HPLC by the analytical department at Ciba and compared to their reference standard for this product. Figure 6.28 shows the experi-

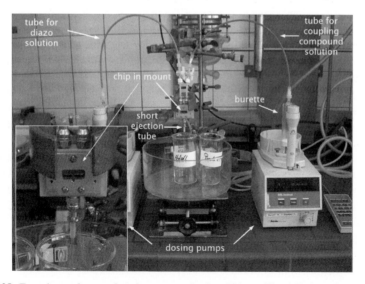

Fig. 6.28. Experimental setup for pigment synthesis within a silicon T-shaped micromixer with short mixing channel, see also Figure A.6.

mental setup with pumps (Dosimat665, Metrohm) and chip mount. The mixer chip is heated with a thermostat from underside to approximately 50 °C. The detail image in Figure 6.28 shows the chip mount during the process. A steady jet with coupled pigment (orange) leaves the short outlet nozzle and is collected in the beaker.

For the coupling experiments, three T-shaped micromixers were used: T-3-6x30, T-3-6x20, and T-3-6x10. The net flow rate was varied between 800 ml/h and 1 800 ml/h, corresponding to the range Re = 500 to Re = 1 100. As the results from barium sulfate precipitation show, there is no major difference in the mixing behavior of these three types, see Section 6.5. Blocking of the mixers occurred frequently after 2–3 minutes of operation, particularly at low flow rates and in mixers with long outlet channel. The fact that some experiments could be completed without blocking, i.e. the total pump capacity of 100 ml was processed, suggests that the blocking is not caused by gradual fouling, but rather by flow instabilities within the T-junction or flow pulsations into the opposite inlet channel. These phenomena can lead to backflow of one of the two reactants into the opposite inlet, where the precipitated particles block the channel and cannot be easily removed.

For successfully completed experiments, the chemical composition of the yielded product was analyzed by HPLC according to Ciba protocols. All pigment samples contain a residual of precipitated coupling compound: more than 30 area percent in the HPLC chromatogram in the worst case, but only 6.5 percent in the best case. Also, the batch process does not reach full conversion and approximately 2 to 3 area percent of unreacted coupling compound remain in the batch product.

In summary, it can be stated that convective micromixers with short outlet channels are in principle suitable for particle-laden flow handling. Optimization and redesign of the structures is required in relation to backmixing and flow pulsations. The desired full conversion to pigment CLA1433 was not accomplished, however, the production process for this pigment neither comes up to this goal. Whereas the batch process takes almost a day and requires additives to reach 97 % conversion, the continuous micromixer process has a reaction time of some milliseconds and uses no additives. The experiments were not conducted under ideal conditions concerning the chemical needs of diazo coupling reactions. In particular, the optimization of pH with respect to the reaction rate of coupling should increase the yield. Although the first results of the experimental investigation of competitive reaction and precipitation show some drawbacks, micromixers can feasibly handle particulate systems under industrial conditions, however, chemical recipes must be adjusted to the continuous operation of microstructured devices. If azo coupling steps could be operated directly with solved components, the process could be simplified, the product quality could be improved, and the process time would be reduced.

7

Coupled Transport Processes

The aim of this chapter is to introduce the thermodynamic theory of coupled irreversible processes and to illustrate these complex phenomena with examples of microstructured devices. Important coupled processes, such as thermoelectric energy conversion or thermodiffusion are introduced briefly with their actual application and potential for future microsystems and process equipment. The integration of microstructured devices or elements will be a key issue for future development.

The methodology of thermodynamics of irreversible processes is not very widespread in engineering thermodynamics, which deals more with equilibrium thermodynamics and linear efficiency calculations. Any real process is connected with irreversible effects and should be treated from the beginning with that emphasis [336]. For a short introduction and the implication on engineering application see also Müller [451]. For further reading on the general topics see the following textbooks of Kluge and Neugebauer [55] or Bejan [173], for more detailed information, refer to the following books [452, 453, 454, 455]. The matrix in Figure 7.1 gives an overview of the coupling of momentum, diffusion, thermal, and electrical currents and forces.

flux \ force	momentum	temperature	concentration	voltage
momentum	$\tau = -\eta \dfrac{du}{dx}$	thermo-osmosis Knudsen pump	concentration pressure	electroosmosis
heat	mech.-caloric effect thermo-mol. press.	$\dot{q} = -\lambda \dfrac{dT}{dx}$	diff.-thermo eff. **Dufour** effect	electrotherm. eff. **Peltier** effect
mass	pressure diffusion	thermodiffusion **Soret** effect	$\dot{n} = -D \dfrac{dc}{dx}$	electrophoresis
el. current	el. current by mass flow	thermoelectricity **Seebeck** effect	el. current by concentration diff.	$I = -\sigma \dfrac{dU}{dx}$

Fig. 7.1. Coupling matrix of transport processes.

7.1 Thermodynamics of Irreversible Processes

Generally, thermodynamics is divided into statistical and phenomenological description. The statistical description begins with initial events, like molecular encounters, and describes larger systems with statistical methods. An introduction and some results of statistical thermodynamics are given in Section 2.3.1. The phenomenological description of thermodynamic processes starts with experimental observations and continuum assumption, and follows a more technical perspective.

The phenomenological description of an irreversible process is given by the correlation between the generalized force X_i and the related flux J_i:

$$J_i = L_{ij} \cdot X_i. \tag{7.1}$$

The phenomenological coefficients L_{ij} are mostly of empirical nature and must be determined experimentally depending on the material characteristics. Examples for the linear correlations are found on the diagonal of the coefficient matrix $\mathbf{L}$, for example, Fourier's law of heat transfer

$$Q = -\lambda \frac{A}{s} \Delta T \quad \text{or} \quad \dot{q} = -\lambda \, \Delta T, \tag{7.2}$$

or the description of species flow by Fick's law $\dot{n} = -DA\Delta c$. Additionally, Newton's law for shear stress in fluid flow, and Ohm's law for electrical current are given in Figure 7.1. Further description of transport phenomena are given in Section 2.3, and can be found in Haase [452] and de Groot and Mazur [453].

The product of the generalized flux J_i with the generalized force X_i is called the local entropy production in a system, the dissipation function $\dot{\sigma}_i$.

$$\dot{\sigma}_i = J_i \cdot X_i \tag{7.3}$$

For real processes, entropy production is always positive due to the irreversible nature of transport processes. The entropy production is measured in [J/sK], hence, the generalized forces and fluxes must be adequately chosen. For example, the heat flux between two thermally coupled systems, see Figure 7.2, is treated to introduce the entropy production $\dot{\sigma}$.

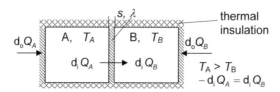

Fig. 7.2. Two thermally coupled systems A and B and heat exchange.

Both systems A and B do not exchange heat with their ambient, $d_oQ = 0$. The temperature in system A is higher than that in system B, hence, heat flows from system A to system B. The entropy dS_A in system A decreases due to the heat loss, while the entropy dS_B in system B increases. The entropy balance for the entire systems achieves the irreversible entropy production dS_{irr} during heat exchange.

$$dS_{irr} = \frac{|dQ_B|}{T_B} - \frac{|dQ_A|}{T_A} = dQ_i \left(\frac{1}{T_B} - \frac{1}{T_A} \right) \tag{7.4}$$

with the heat Q_i internally transferred. The temporal derivative of dS_{irr} is the entropy production rate or dissipation function

$$\dot\sigma = \frac{dS_{irr}}{dt} = \frac{dQ_i}{dt} \cdot \left(\frac{1}{T_B} - \frac{1}{T_A} \right). \tag{7.5}$$

According to Eq. 7.3, the first term on the right side is the generalized flux (heat flux), and the second term characterizes the generalized force (temperature gradient) of heat transfer.

$$J = \frac{dQ_i}{dt} = \dot Q_i \text{ and } X = \left(\frac{1}{T_B} - \frac{1}{T_A} \right) = \frac{T_A - T_B}{T_A \cdot T_B} \tag{7.6}$$

Comparison of the correlation in Eq. 7.1 with Fourier's first law in Eq. 7.2 and the above equation gives for the phenomenological coefficient L_q

$$L_q = \frac{\lambda A T_A T_B}{s}, \tag{7.7}$$

with the wall thickness s and the heat transfer area A. The entropy production rate $\dot\sigma$ is correlated with the temperature difference $\Delta T = T_A - T_B$ and the reference temperature T_A.

$$\dot\sigma = J \cdot X = \frac{\lambda A}{s} \cdot \frac{(T_A - T_B)^2}{T_A \cdot T_B}$$

$$= \frac{\lambda A}{s T_A^2 \cdot \left(1 + \frac{\Delta T}{T_A} \right)} \Delta T^2 \tag{7.8}$$

This expression is often too complex for heat transfer calculations, however, it facilitates the unified treatment of coupled transport phenomena, see Haase [452]. In microsystems engineering, coupled phenomena are gaining importance, which has been shown in Section 1.5 as a consequence of the gradient enhancement. Coupled transport phenomena can be expressed with an equation system or in matrices, here for two coupled transport processes in general.

$$\begin{aligned} J_1 &= L_{11} X_1 + L_{12} X_2 \\ J_2 &= L_{21} X_1 + L_{22} X_2 \end{aligned} \quad \text{or} \quad \begin{bmatrix} J_1 \\ J_2 \end{bmatrix} = \begin{bmatrix} L_{11} & L_{12} \\ L_{21} & L_{22} \end{bmatrix} \cdot \begin{bmatrix} X_1 \\ X_2 \end{bmatrix} \tag{7.9}$$

The relationship between the non-diagonal elements of the symmetrical matrix is given by the so-called Onsager reciprocity relation and is based on the microscopic reversibility of the fundamental kinetic processes [452].

$$L_{12} = L_{21} \quad \text{or} \quad \left.\frac{J_1}{X_2}\right|_{X_1} = \left.\frac{J_2}{X_1}\right|_{X_2} \tag{7.10}$$

The vertical line indicates that the displayed parameter vanishes and equals zero. The Onsager reciprocity relation reduces the number of parameters to three independent coefficients. With three experimental test series, the complete matrix in Eq. 7.9 can be determined. Additional to the first order correlations, there are another six combination possibilities for coupling of two transport processes.

$$\left.\frac{J_1}{J_2}\right|_{X_1} = -\left.\frac{X_2}{X_1}\right|_{J_2} = -\frac{L_{12}}{L_{22}}$$

$$\left.\frac{J_1}{J_2}\right|_{X_2} = -\left.\frac{X_2}{X_1}\right|_{J_1} = -\frac{L_{11}}{L_{12}} \tag{7.11}$$

$$\left.\frac{J_1}{X_2}\right|_{J_2} = -\left.\frac{J_2}{X_1}\right|_{J_1} = L_{12} - \frac{L_{11} L_{22}}{L_{12}}$$

For thermoelectric coupling of heat transfer ($J_1 = Q$, $X_1 = \Delta T$) with electrical currents ($J_2 = I_{el}$, $X_2 = \Delta U_{el}$), some application examples of the above equations are given. The direct heat conduction over a thermocouple and the electrical measurement gives the conductivity values for L_{11} and L_{22}. For no electrical current I_{el}, the voltage ΔU_{el} for a given temperature difference ΔT gives the Seebeck coefficient L_{12} in [V/K]. The Peltier coefficient L_{21} is determined from the temperature difference with given voltage ΔU_{el} and vanishing heat flux. The Peltier coefficient can also be determined from the Onsager relation in Eq. 7.10. Measuring the heat flux and the voltage for no electrical current gives additional experimental data for the coupling coefficients L_{12} and L_{21}. The relationship in Eq. 7.11 give six possibilities for determining the phenomenological coefficients of coupled thermodynamic processes. Some examples of electro-kinetic coupling are given by Hasselbrink [456].

As a further result from the entropy production in Eq. 7.3, the sign of the coefficients can be determined [457].

$$L_{12} > 0; \quad L_{11} L_{22} - L_{12}^2 > 0 \tag{7.12}$$

The second equation gives an estimation of the order-of-magnitude for the coefficients of the coupled processes. For example, the thermodiffusion coefficient in an aqueous system can be estimated with $O(L_{12}) = 10^{-9}$ m²/s using the order of the heat conductivity $O(L_{11}) = 0.6$ W/mK and the order of the diffusivity $O(L_{22}) = 10^{-9}$ m²/s, see [452] p. 406.

Furthermore, Curie's symmetry principle allows only the coupling of the same tensor systems with identical transformation behavior, according to rotation in the space, see [452] or [453]. This principle allows only combinations with the same tensor character: a scalar can be combined with a scalar process (chemical reactions, pressure diffusion), a vector with a vector (heat transfer, diffusion, electrical conduction) and a tensor with a tensor process (pressure and shear tensor in fluid flow of complex media) [453] p. 57.

Many micro electro-mechanical systems (MEMS) employ coupled processes for special purposes, such as sensing (magneto-electro coupling in a Hall sensor [54]) or cooling (thermo-electric coupling in a Peltier element [458]). Further applications with special interest for micro process engineering are given in the following.

7.2 Thermoelectric Energy Conversion

The thermoelectric effect describes the coupling of electrical and thermal currents, especially the occurrence of an electrical voltage due to a temperature difference between two material contacts, called the Seebeck effect. In reverse, an electrical current can produce a heat flux or cooling of a material contact, known as the Peltier effect. Both effects originate from different diffusivities of electrons and phonons in different materials [459], and are, therefore, related to thermodiffusion described in Section 7.4. A third effect also connected with thermoelectricity, the Thomson effect, occurs where an electric current flowing along a temperature gradient can absorb or release heat from or to the ambient surroundings [319, 460]. The Thomson effect depends on the Peltier coefficient and its derivation with the temperature. The relationship between the first two effects can be described by methods of irreversible thermodynamics and the linear transport theory of Onsager in vector form.

$$\begin{bmatrix} J_{el} \\ J_{th}/T \end{bmatrix} = \begin{bmatrix} \sigma_{el} & L_S \\ L_P & L_q \end{bmatrix} \begin{bmatrix} \Delta U \\ \Delta T \end{bmatrix} \tag{7.13}$$

The specific electric current density J_{el} and the thermal entropy flux J_{th}/T are linked to the driving forces, the temperature gradient ΔT and the potential gradient ΔU, with linear coefficients L_q for the heat transfer, L_S for the Seebeck effect, and L_P for the Peltier effect. This notation can not be used for calculation of entropy production $\dot{\sigma}$. The derivation of the coefficients here follows the notation of Nolas et al. [460], another good description is given by Kalitzin [461].

Solving the first equation for ΔU and substituting the result into the second equation yields

$$-\Delta U = \frac{J_{el}}{\sigma_{el}} + \frac{L_S}{\sigma_{el}} \Delta T, \tag{7.14}$$

$$J_{th} = T \frac{L_P}{\sigma_{el}} q_{el} + T \left(\frac{L_S L_P}{\sigma_{el}} - L_q \right) \Delta T. \tag{7.15}$$

The matrix in Eq. 7.13 is symmetrical according to the Onsager reciprocity law [319], which represents $L_S = L_P$. The linear coefficients can be identified with known material properties

$$\rho_{el} = \frac{1}{\sigma_{el}}, \tag{7.16}$$

$$\lambda_{th} = -T\left(\frac{L_S L_P}{\sigma_{el}} - L_q\right) = -T\left(\frac{L_S^2}{\sigma_{el}} - L_q\right), \tag{7.17}$$

$$\alpha_{ab} = \frac{L_S}{\sigma_{el}}, \quad \text{and} \tag{7.18}$$

$$\pi_{ab} = T\frac{L_P}{\sigma_{el}}. \tag{7.19}$$

The thermoelectric temperature measurement employs the Seebeck effect, the voltage between two material pair contacts on a different temperature level without the flow of electrical current ($Q_{el} = 0$). The Seebeck effect is also described with the correlation

$$\Delta U = (\alpha_a - \alpha_b)\Delta T = \alpha_{ab}\Delta T, \tag{7.20}$$

with the Seebeck coefficient α_{ab} with unit [V/K] for the material pair $a - b$. The Seebeck coefficient is dependent on the material pair. Certain applications, such as measurement or energy conversion, demand certain material properties, like heat or electrical conductivity or temperature linearity. For temperature measurement, standardized elements and material pairs are used with good linearity of the material properties, reproducibility, and long-term stability.

Table 7.1. Some standardized and commercially available thermocouples for temperature measurement (IC 584-1)

Material	Type	Seebeck coeff. [μV/K]	operating range [°C]
Fe/CuNi	J	51	$-200 \ldots 760$
NiCr/AlNi	K	41	$-200 \ldots 1260$
PtRh/Pt	R	6	$0 \ldots 1480$
Cu/CuNi	T	41	$-200 \ldots 370$

A summary of common standardized thermocouples is given in Table 7.1 according to IC 584-1. Besides the thermoelectric properties of the Seebeck coefficient, the thermal and electrical conductivity are important for the sensitivity and response time of the thermoelectric flow sensor.

The materials of standardized thermocouples are rarely used in microfabrication, which relies mostly on CMOS technology. Commonly used metals and semiconductors and their properties are given in Table 7.2. A comparison of the values gives remarkable differences between bulk material and deposited film material. Baltes et al. [318] give material properties of n- and p-doped silicon together with the fabrication process, which are widely differing. The doping concentration determines the thermoelectric properties of silicon in a wide range, see also [462, 463]. For practical applications, it is very important to characterize the fabrication process

Table 7.2. Thermoelectric properties of bulk and thin film semi-conductor materials [318]

Material	Seebeck coeff. [µV/K]	therm. conductivity [W/mK]	electrical resistivity [10^{-9} Ωm]
bulk materials			
p-$Bi_{0.5}Sb_{1.5}Te_3$	218	0.59	4.0
Bi_2Te_3	210	2.0	7.7
Sb_2Te_3	124	1.3	13.5
Bi	-65	8.3	83
Sb	36	24	238
PbTe	-200	3.5	13.0
Si (n-type)	-450	145	2.9
thin film materials			
p-$Bi_{0.5}Sb_{1.5}Te_3$	230	1.05	5.8
p-$Bi_{0.87}Sb_{0.13}$	-100	3.1	14.0
Bi	-65	5.2	28.5
Sb	35	13	100.0
PbTe	-170	2.5	5.0
Si (n-poly)	-108 ... -520	16 ... 29	0.2
Si (p-poly)	190 ... 330	19 ... 20	n.a.

and the relevant outcome and ensure the reproducibility of material properties. Further material properties are given by van Herwaarden [312], or a summary of micro thermocouple materials by Rowe [464].

Often, the thermoelectric material pairs are used for energy conversion from heat to electricity with the help of thermoelectric generators. Here, the so-called figure-of-merit Z gives an indication for good energy conversion and is given by

$$Z = \frac{\alpha_{ab}^2}{\sqrt{\lambda_a \rho_a} \cdot \sqrt{\lambda_b \rho_b}} = \frac{\alpha_{ab}^2 \sigma_{el,a}}{\lambda_a} \qquad (7.21)$$

The second part of Eq. 7.21 is often used to characterize the thermoelectric activity of a single material a. This parameter, with the unit [K^{-1}], may help to identify sensitive material pairs for thermoelectric temperature measurement. In summary, a high Seebeck coefficient and high electrical conductivity of the sensor, as well as low thermal conductivity of the sensor and the substrate provide a sensitive, low energy-consuming sensor with fast response time. Hence, the figure-of-merit Z of a thermoelectric device is a good quality indicator alongside other factors. The material combinations of the device must also guarantee low thermal stress, due to expansion through heating and cooling.

7.2.1 Microscale thermoelectric energy conversion

Besides sensing applications, the thermoelectric effect is often used for the generation of electrical energy from a given temperature difference. The energy supply from the direct vicinity is becoming more and more important for electronic devices, sensors, actuators, or other functional equipment. In addition to photoelectric supply via solar cells and conversion of mechanical energy from vibrations, thermoelectric generation is the most important possibility for supplying different applications. In thermoelectric generators, the Seebeck effect is employed for the direct conversion of a temperature difference into an electrical potential. At device level, high electrical conductivity together with good thermal insulation between the material contacts leads to a high temperature difference over the thermocouple and ensures enhanced power output for a given heat source. Deep space missions like "Voyager" are supplied by radioisotope thermoelectric generators RTG [465]. The enduring power supply of the "Voyager" RTG over the last 25 years indicates the high reliability and long-term performance of thermoelectric converters. Wrist watches are powered by heat released from the human body [466] where first application was reported from Bulowa in the early 1970s. The generator uses the temperature difference from the human skin to the ambient to generate a few µW of power, sufficient to drive the electronic clockwork. More information can be found on www.peltier.info.com.

The miniaturization of thermoelectric devices has great potential due to integration of a high number of material pairs in a small device, which results in high voltages and flexible applications of small energy converters [467]. The microfabrication technologies for thermoelectric devices can be classified into two groups: thick-film technology or bulk technology with mostly electrochemical deposited thermopiles, like the generator for wrist watches, or thin-film technology, which is based on the fabrication technologies and materials from microelectronics. The company "micropelt", www.micropelt.com, develops and manufactures thermoelectric generators and Peltier coolers on the base of BiTe alloys. Each of the two materials are structured on a wafer and then bonded together [467, 468]. One of the fabricated structures has an area of 1.12 mm^2 and delivers a power output of 67 µW at a temperature difference of 5 K. Electrochemically deposited BiTe alloys have also been developed and investigated by Jet Propulsion Laboratory at CalTec [469]. Further research work on thermoelectric generators in thin-film technology can be found in [470, 471]. Recent developments in thermoelectric material science are reviewed by Böttner et al. [472] for thin-film structures and by Nolas et al. [473] for bulk materials.

This section presents a modular, flexible, and multipurpose fabrication platform for electrical energy harvesting from environmental heat sources with low temperature difference by microstructured thermoelectric generators µTEG. The meandering thermocouples are fabricated on silicon wafer in CMOS-compatible thin-film technology to achieve a high "in-plane" integration density, however, there remains the opportunity using other thin-film processes and materials. The heat flux is guided perpendicular to the thermocouple plane to one contact junction by metal stripes with high thermal conductivity. The remainder of the thermocouple, especially the second contact junction, is thermally insulated by a polymer (SU-8) or photoresist layer, and DRIE opening in the silicon substrate. The metal stripes on the thermocouple junctions from the respective opposite side produce "cross-plane" heat flux with a high temperature gradient between the thermocouple junctions, see Figure 7.3. The heat flux from the ambient is introduced over the device foot-print area, guided through the generator to invoke electrical power, and also released over the opposite foot-print area. This guarantees maximum thermal coupling of the generator with the ambient and, additionally, maximum temperature gradient over the thermocouples.

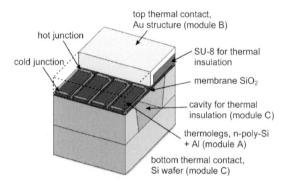

Fig. 7.3. Schematic setup of the investigated generators with thin-film thermoelectric structures and perpendicular heat flux, see also Figure A.28. Module A with SiO$_2$ membrane and thermocouples, module B with SU-8 insulation and gold layer, and module C with insulation cavity in the silicon wafer.

For modular fabrication, the structure was segmented in three modules, A, B, and C, according to their function and used materials. Test structures are fabricated for each module, optimized, and arranged together for a straightforward fabrication process. The properties of the thermoelectric materials were tested with special structures. The whole setup was fabricated and also tested in relation to their energy conversion characteristics.

7.2.2 Design and fabrication of the generator

The thermoelectric generator consists of planar thin-film thermocouples of n-doped poly-crystalline silicon (n-poly Si) and aluminum (Al) on a silicon substrate (module A) with perpendicular heat conducting structures. These structures allow the heat flux to flow from the planar side over electrochemically deposited gold Au (module B) to one junction, over the thermocouples, and from the second junction over silicon ridges (module C) to the opposite side, see Figure 7.3. This enables good thermal coupling to the ambient over a large area and maximum temperature gradient over the thermocouple for high performance and efficiency. Aside from the good heat transfer through gold and silicon ridges, optimized thermoelectric structures (750 nm n-poly thick Si and 250 nm thick Al films) on the thinnest possible membrane (300 nm thermal oxide SiO$_2$ and 300 nm LPCVD Si$_3$N$_4$) are also important for low thermal conductivity between the junctions. Additionally, the thermoelectric structures are electrically insulated from the gold structure with a 1.2 µm thick PECVD oxide layer. A second wafer is bonded to the back side of the chip to realize good thermal contact, to provide sufficient mechanical stability, and avoid a cavity contamination from the ambient. μTEG chips were designed with 10×10 mm² footprint and fabricated in the in-house clean room. The chips have integrated test structures with up to 13 460 thermocouples in module A.

The heart of the generator is the meandering thermocouple (n-poly Si and Al), in series switching with bond pads on the outside, see Figure 7.4. Each single thermocouple line can be measured on its own, which enlarges the yield of operable devices. The typical generator setup in Figure 7.4 illustrates the high integration density of the device as well as the complexity of the fabrication.

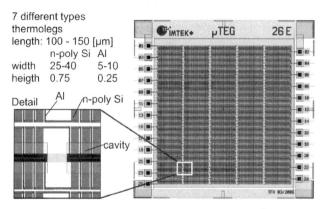

Fig. 7.4. CAD image of a complete generator chip, 10×10 mm² footprint, with typical meander-like thermocouple arrangement with 6 000 thermocouples, see also Figure A.29. The bond pads are on the left and right side of the chip for electrical connection.

Assisting the prototype fabrication, finite-element simulations (FEM) were performed with ANSYS for design optimization and estimation of the generator's theoretical performance. The simulation includes the two-dimensional temperature profile through the various material layers and an estimation of the thermoelectric power, see Figure 7.5. A key issue is the integration of material properties into the simulation code. In particular, the heat conductivity and electrical resistance of the thin-film structures are hard to measure, not to mention reliable Seebeck coefficients for the material pairs. The Seebeck coefficient of doped silicon depends not only on the doping concentration, but also on the fabrication equipment and procedure [318].

The two-dimensional temperature profile of a typical generator structure is given in Figure 7.5. The performance benchmarks were calculated for a prototype generator with 7 500 thermocouples made of n-poly Si and Al, each measuring 120 μm. Simulating a temperature difference of 5 K across the entire chip, an open-circuit voltage of 3.48 V and a theoretical power output of 7.94 μW was determined. The effective temperature difference across the thermocouples was 4.72 K, which is 94 % of the total temperature difference. An output voltage per thermocouple of 92.8 μV/K was simulated with these boundary conditions, giving a theoretical power factor of $3.17 \cdot 10^{-3}$ μW/mm²K² for 7 000 thermocouples on 1 cm². The thermoelectric simulation of 13 464 thermocouples gave a voltage of approx. 1.3 V/K from the chip with a footprint of 10×10 mm^2.

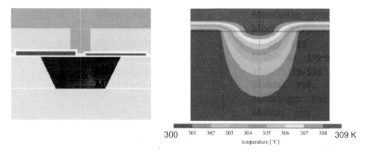

300   301   302   303   304   305   306   307   308   309 K
temperature [°C]

Fig. 7.5. Temperature profile within the µTEG for 9 K temperature difference, representing approx. 8.61 K effective temperature difference (95.7 %) over the thermocouple, see also Figure A.30.

In the first fabrication run, test structures are manufactured with n-poly Si and Al to characterize the fabrication procedure and the yielded material properties. In particular, the electrical conductivity, contact resistance, and Seebeck coefficient have been measured for typical geometrical arrangements. A test structure is shown in Figure 7.6 with a resistance heater on the center line together with temperature sensors beside the heater, as well as above and below the thermoelectric meander. The darker structure is the n-poly Si layer, the light gray structure is the Al layer. The thermoelectric properties of the middle structure are measured to neglect side effects and ensure uniform heating. It was difficult to measure thermal conductivity with this structure due to the parasitic side effect of parallel heat loss to the ambient.

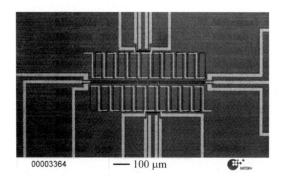

00003364   —— 100 µm

Fig. 7.6. Geometrical setup of the test structure to measure the Seebeck coefficient.

The modular arrangement of the generator and the related fabrication process is displayed in Figure 7.7. A 300 µm thick silicon wafer serves as the substrate and is covered with an SiO$_2$ and Si$_3$N$_4$ electrical insulation layer to hold the thermocouple structures. The n-poly Si structures are deposited by LPCVD processing, structured with photolithography, and covered by a sputtered Al layer. This layer is structured by wet-chemical etching. Finally, the thermocouples are covered with PECVD silicon oxide to electrically insulate the thermocouples from the gold layer.

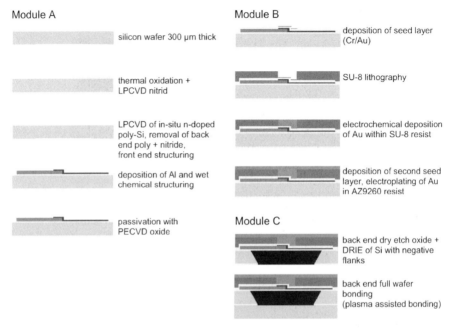

Fig. 7.7. Setup of the µTEG structure and related fabrication process steps of the thermocouple structure (module A), the thermal connector structure (module B), and the silicon cavity (module C), see also Figure A.31.

For electro deposition, a chromium/gold start layer is deposited and structured on the thermocouple junctions. An SU-8 layer is spun on the start layer and structured with photolithography. The light-exposed areas are developed and form trenches on one thermocouple junction, which is electrochemically filled with gold. After filling the trenches, a second start layer is deposited on the entire wafer, structured, and serves as a base for the second gold layer, which covers the complete area of the thermocouples to provide good thermal coupling with the ambient.

Following the electrochemical process, the back side of the 300 µm thick silicon wafer is structured. The opposite junction to the gold contact and most of the thermocouple length are opened via DRIE process to thermally insulate the structure. The second thermocouple junction has thermal contact with the silicon substrate to provide a high temperature difference over the thermocouple length between the junctions. The DRIE process is operated with increasing power to yield walls with negative angle and a small contact area at the thermocouple junction, a large contact area at the outside, and high mechanical strength for the silicon grid. The cross section of a fabricated generator chip is displayed in Figure 7.8. The notch from DRIE etching near the dielectric layer in the vicinity of the thermocouple membrane gives additional thermal insulation, but must be considered during process and geometry design. The gold layer is smeared slightly over the thermocouple membrane from cutting. The membrane setup is not visible due to the small layers.

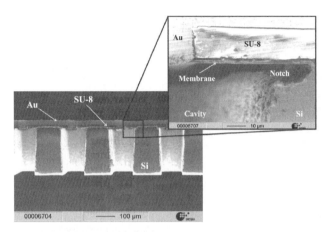

Fig. 7.8. Cross section of a fabricated thermoelectric generator with detailed membrane view.

Figure 7.9 shows complete generator chips with gold layer, type number, and bond pads at the left and right side. These devices are characterized according to their electrical and thermoelectrical properties. Due to the complex device structure with approx. 40 process steps, only a small number of operable chips have been yielded.

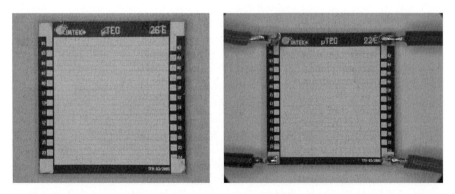

Fig. 7.9. Total view of the thermoelectric generator chip, right side with the electrical contacts.

7.2.3 Experimental characterization and discussion

For device characterization, a measurement setup was developed where a defined temperature difference is applied across the chip and the gained output voltage is measured. In particular, the electrical resistance, output voltage, and electrical power are measured with the fabricated generator chips. The test structures displayed in Figure 7.6 are characterized with electrical and thermoelectrical properties and compared with data from literature [474], see Table 7.3.

Table 7.3. Electrical resistivity of material pair in micro thermoelectric generators

property	n-poly Si	Al
ρ own measurement	6.294 $\mu\Omega$ m	$4.303 \cdot 10^{-8}$ Ω m
ρ from [474]	6.61 ± 0.17 $\mu\Omega$ m	$2.733 \cdot 10^{-8}$ Ω m

The difference between measured values of electrical resistance and literature data can be explained from additional contact resistance between the material pair. The Al film is partially porous from the wet-etching process. The Seebeck coefficient of the material pair n-poly silicon and aluminum is $\alpha_{n-poly-Al} = 106.18 \pm 4.63 \mu V/K$, which is in the range of literature data (gate poly CAE: $\alpha_{n-poly} = -111.3 \pm 1.5$ $\mu V/K$, aluminum: $\alpha_{Al} = -1.7$ $\mu V/K$), see Table 7.2. The measurement of heat conductivity is not possible with the test setup, due to high parasitic heat loss to the ambient.

From the fabricated structures, some operable thermoelectric generators are chosen to measure their thermoelectric characteristics and performance. In Figure 7.10, the measured output voltage of the thermoelectric generator chip with 125 active thermocouples is displayed over the applied temperature difference. The incline of the linear fit curve gives the Seebeck coefficient, which is determined to 1.33 mV/K. The ratio of the experimental value to the simulated value of 11.63 mV/K is approx. 11.4 %, which is quite large. The experimental temperature measurement error alone accounts for approx. 68 %. Other reasons for the deviation between simulation values and experimental data include contact resistance at the thermoelectrical junction. Major problems in fabrication have been the porosity of the Al structures and corresponding over-etchings (mouse-bites) during wet etching, resulting in uncertainties in the material properties. Additionally, misalignment of the SU-8/gold structures to the thermocouples leads to an offset of the heat conducting structures. This can be seen in Figure 7.8, where the gold structures are not placed over the middle of the silicon trench. A simulation of the misaligned structures results in a 70 % decrease of the effective temperature difference between the thermocouple junctions.

The comparison with similar thermoelectric generators from literature illustrates the performance of the developed structures. The presented microstructured thermoelectric generators with 7500 thermocouples on a footprint of 1 cm^2 exhibit a measured power factor of $1.37 \cdot 10^{-5}$ $\mu W/mm^2$ K^2, which is the power earned from

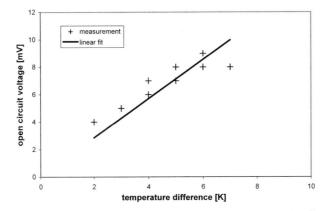

Fig. 7.10. The idle voltage of the thermoelectric generator section with 125 thermocouples depending on the applied temperature difference. The linear fitting curve gives a Seebeck coefficient of 11.63 mV/K.

a generator with a certain footprint area under a certain temperature difference. The power factor is a better indicator of the generator effectiveness than the material parameter of figure-of-merit Z, see Section 7.2. The internal resistance was measured to 5.04 MΩ compared to a simulated value of 0.383 MΩ, which originates from the contact resistance between the n-poly Si and the Al film. By neglecting the misalignment, a thermoelectric generator chip with 7 500 thermocouples will deliver a power of 4.11 μW under a temperature difference of 5 K or has a power factor of $8.22 \cdot 10^{-4}$ μW/mm^2 K^2.

Infineon [458, 475] has developed thermoelectric generators with comparable integration density of 15 782 thermocouples on an area of 7 mm^2. The generators have a high internal resistance with 2.1 MΩ, a power factor of $4.26 \cdot 10^{-4}$ μW/mm^2 K^2, and deliver 0.112 μW of power at a temperature difference of 5 K. Thermoelectric generators developed by HSG-IMIT and Kundo consist of aluminum (p-material) and doped silicon (n-material) and have 1 000 thermocouples on 16.5 mm^2 chip area [476, 470]. The devices have a power factor $9.1 \cdot 10^{-4}$ μW/mm^2 K^2 and deliver 0.75 μW of power at a temperature difference of 5 K. Thermoelectric generators on the basis of BiTe thermocouples [466] have a power factor up to 0.5 μW/mm^2 K^2. In conclusion, heat flux optimized microstructured thermoelectric generators with modular fabrication offer many opportunities for the high integration of thermoelectric structures with good yield of the given temperature gradient and heat flux. The fabricated prototypes achieve power factors of similar commercial devices and exhibit an immense development potential.

An interesting new application field for miniaturized thermoelectric generators is their integration into the wall of compact heat exchangers between the two media to generate electrical energy from process heat exchangers. The small size and compact structure of μTEGs allows thin wall structures and a high temperature difference over thin-film thermocouples. A rough performance estimation of such an integrated compact heat exchanger yields an efficiency of 2 to 5 % conversion rate from heat to electrical power. Still, many issues must be clarified, such as efficient thermoelectrical material pair, wall integration, as well as safe operation and long-term reliability.

7.2.4 Thermocouples and microfluidic chips

Another interesting application is the skillful combination of a microreactor with the temperature control of Peltier elements [477] to yield a micro calorimeter. A microreactor array on a chip is covered on both sides with a Peltier element and submerged in a heat bath to control the ambient temperature. The heat release from an exothermic reaction in the microreactor is adsorbed by one Peltier element, while the other element controls the temperature of the calorimeter. The device exhibits a time constant of approx. 3 s for a volume of 50 µl. Two example reactions of highly energetic reactions are given by Schifferdecker et al. [478], which match quite well with data from literature. With microstructured thermocouples and Peltier elements, temperature and released reaction heat can be determined with high local resolution, as well as reaction progress and kinetics in a continuous-flow reactor.

The temperature measurement with thermoelectric microdevices is an established technology [54, 318], which also includes thermal flow sensors [314], see Section 4.1.5. Rencz et al. [479] propose a heat-flux sensor based on the Seebeck effect of p-doped Si and Al thin film elements. Shen and Gau [480] developed an thermal Si chip for flow rate and heat transfer measurements based on poly Si and heavily Boron-doped Si elements. Unfortunately, no details on the thermoelectric structures were given. To investigate biochemical processes, Baier et al. [481] developed a thermopile heat sensor based on SU-8 technology. The device consists of BiSb/Sb thin-film thermocouples and are combined with thin-film NiCr heaters on a silicon wafer. The heat power sensitivity of the calorimeter is approx. 5 V/W, and a detection limit of less than 100 nW can be obtained for flow rates lower than 5 µl/min. Heat amounts of lower than $4 \cdot 10^{-6}$ J can be detected with a differential method.

To determine the heat transfer in microchannels, the accurate temperature measurement is essential. Figure 7.11 shows the lithography mask for a mixer chip (20×20 mm^2 footprint) with the arrangement of 30 thermocouples. The thermocouples are located on the back side of the chip with a Venturi mixer and residence time channel structures, see Section 6.4. The second thermoelectric contact is located at the lower edge of the chip together with a silicon heater and a resistance temperature sensor. The second contact is formed as bond pad to contact the thermocouple to an external electronic circuit board with operational amplifiers. For thermal insulation, a channel with ligaments for mechanical strength is located between the mixing channel and the second contact with heater and sensor.

The fabrication process of the thermoelectric structures was described in Section 7.2.2, but the geometry was adapted to the sensing function. A row with six thermoelectric couplings is displayed in Figure 7.11, right side. The gray structure represents n-poly Si with a thickness of 400 nm, while the light gray structure represents the sputtered Al layer, which is 300 nm thick.

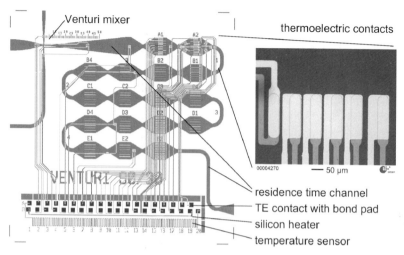

Fig. 7.11. Left: Arrangement of temperature sensors on a mixer chip with 20×20 mm^2 footprint; Right: Row of thermoelectric contacts for temperature measurement.

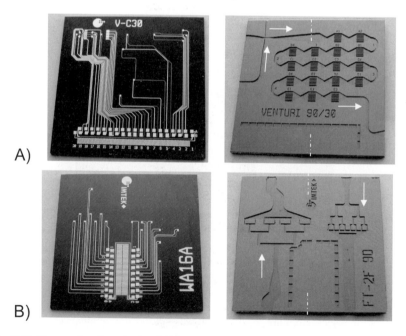

Fig. 7.12. Thermoelectric and microfluidic structures on a silicon chip. The axis of rotation is marked for both chips with the dotted line. A) Arrangement of temperature sensors on a mixer chip similar to Figure 7.11. B) Heat exchanger fluidic network with thermocouples.

Local temperature measurement on the backside of microreactors with high spatial resolution can give information about the development of the reaction and its kinetics. In Figure 7.12, two fabricated microfluidic chips (20×20 mm^2) are displayed with thermoelectric temperature sensors on the backside. The flow directions are indicates by arrows, the mirror axis is given by the dotted line. The channels have a depth of 300 μm and are covered by Pyrex glass. The high heat conductivity of silicon with $\lambda = 150$ W/mK enables an accurate measurement of the channel wall temperature. The second thermoelectric contact is clearly visible in the lower region of the chip with bond pads for electrical contact, electrical heater, and resistance temperature sensor. The additional separation channel for thermal insulation is clearly visible on the microfluidic side.

For the Venturi mixer (A), the sensors are located directly at the mixing element and the first two distribution chambers to study the temperature and flow distribution. The second chip (B) has two flow manifolds to distribute fluid over an area for cooling purposes, see Section 4.2.4. Between both channel systems, the area with the second TE contact, electrical heater and resistance temperature sensor is located. The sensors measure the wall temperature of the inlet flow and at each junction of the manifold.

The complete measurement setup is displayed in Figure 7.13 with the chip integrated in the fluidic mount and electronic amplifiers on printed circuit board. Unfortunately, temperature measurements could not be performed up to now, due

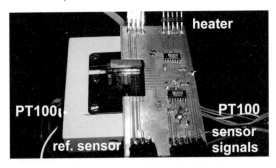

Fig. 7.13. Complete arrangement of cooling chip, fluidic mount and electronic board for signal processing. Two reference PT100 sensors are integrated to measure the temperature of the hot plate underneath the cooling chip.

to fabrication issues. Bonding of aluminum was not feasible in most cases, the wires broke or do not attach to the bond pad. In the next fabrication run, copper Cu will be used as thermoelectric material instead of Al. It has similar thermoelectric properties than Al with $\alpha_{Cu} = 1.84$ μV/K at 27 °C [482], however, wire bonding of Cu is an established process.

7.3 Electro-osmotic and Electro-kinetic Effects

The correlation between the electrical current and the fluid dynamic flow is called the electro-kinetic flow. A related type of the electro-induced flow is the electrophoretic flow of species through a matrix, a gel, or a resting fluid. This effect is often used for separation and detection of larger molecules and biological reagents on microfluidic chips [483]. Pfeiffer et al. give an overview and some heuristic and numerical optimization design techniques [484]. Beginning with geometric channel elements, such as straight channels, 180° turns, meandering channels, and spirals, complete separation chips are designed and compared to optimize the channel arrangement.

Kirby and Hasselbrink [485] describe in a review article the typical surface characteristics and the Zeta potential of silicon, glass, and polymers for microfluidic applications. They recommend precisely controlling the temperature and the pH value of the solution to achieve reliable experimental data. Measuring the streaming potential and the electro-osmotic mobility are the most effective methods to determine the Zeta potential. Thermodynamics of irreversible processes and the Onsager reciprocity relations are very helpful for minimizing experimental effort [486]. In a system with two coupled effects, here the electric and hydrodynamic flux, only three measurements are necessary to determine all coefficients [456]. Thermodynamics can also help finding possible applications, for example, the energy conversion to produce an electric current from hydrodynamic motion.

Electro-osmotic pumps, also known as electro-kinetic pumps, employ the negative charge of the solid wall and the corresponding positive charge of ions in the fluid [105]. An external electrical field generates the motion of the counter ions and, therefore, the entire fluid in the capillary. The electro-osmotic effect decreases with increasing channel diameter, which have characteristic dimensions of approx. 1 μm or less. High pressure differences of up to 2 MPa for water have been realized for low flow rates of 3.6 μl/min. High flow rates up to 33 ml/min have been measured, which indicates that the main applications of electro-osmotic pumps are for dosing and analytical applications [105]. The efficiency of electro-kinetic pumps is mainly determined by the channel width compared to the EDL length. An optimum ratio of 2 to 5 was found by Min et al. [487] in an analytical and numerical study for silica and a pH-value larger than 8, which gives an EDL thickness of approx. 10 nm.

Electro-kinetic devices for energy conversion and power generation are described by van der Heyden et al. [488]. Hydrostatic energy is converted into pressure-driven flow of aqueous solution, inducing a charge separation at the wall of small channels. For a glass device with 500 nm channels, an estimation gives an energy conversion efficiency of 12 % and a Watt-level electrical power from an applied pressure difference of 5 bar. Design and fabrication of the entire device is one of the future tasks in this field. Further physical principles with their technical realization are presented by Schmidt [489] in his book on unconventional energy converters. Aside from thermoelectric and electro-kinetic generators, other energy sources are used to generate electrical energy, such as radiation, chemical sources, or thermionic converters. These device only play a major role in niche applications.

7.4 Thermodiffusion

The physical effect of thermodiffusion characterizes the different migration velocities of molecules with different size or mass in a thermal gradient field generating of
a concentration gradient. Assisted by natural convection, thermodiffusion was employed for isotope separation in the Clusius-Dickel column, see London [99]. Pelster
et al. [459] describe simple thermodiffusion experiment to demonstrate different diffusion velocities in a hot and cold body. Brodkey and Hershey [178] gave a detailed
energy and mass balance and described the relevance of different effects and phenomena. While Fourier's law describes the coupling between heat flux and temperature gradient, Fick's first law gives the linear relationship between mass or species
flow and concentration gradient. The cross coupling between heat flux and concentration gradient (Dufour effect L_{qD}), as well as the cross coupling between species
flux and temperature gradient (Soret effect L_{Dq}), takes the following form.

$$\begin{aligned} J_q &= -\lambda \, \Delta T + L_{qD} \, \Delta c \\ J_n &= L_{Dq} \, \Delta T - D \, \Delta c \end{aligned} \tag{7.22}$$

The coefficient of the Soret effect L_{Dq} is 3 to 5 orders-of-magnitude smaller than
the first-order effects of heat conduction $L_{qq} = \lambda$ or diffusion $L_{DD} = D$ [56].
Thermodiffusion can be employed as single-phase separation process for gaseous
and liquid mixtures without unwanted interfacial forces present in distillation or extraction. The fluid mixture flows through a number of parallel microchannels with a
superimposed temperature gradient. Smaller molecules tend to diffuse to the hotter
side, larger remain at the colder side. At the end of the channel, the flow is split up
and rearranged into the next channels. By cascading the parallel channels and appropriate split-and-recombine, the concentration of the smaller molecules is gradually
enriched at the hot side, see Figure 7.14.

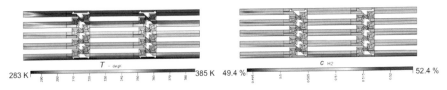

Fig. 7.14. Temperature profile (left) and concentration profile of H_2 (right) in a single channel, 1 500 µm long, with temperature gradient (100 K), CO_2-H_2 system, 4 000 Pa pressure
difference, 120 K temperature difference, Soret coefficient 10^{-9} m²/sK, see also Figure A.32.

7.4.1 Design and fabrication of thermodiffusion devices

Kockmann et al. [70] investigated the separation of a gaseous mixture (gas, 50 % H_2
and 50 % CO_2) in a microchannel array with a given temperature profile at the walls

and resulting temperature profile over the channels. Numerical simulations of an array of 5×3 channels (200 µm wide and 400 µm deep) were performed with the commercial FVM-code CFD-ACE+ by ESI group. The thermodiffusion coefficients were calculated according to the kinetic theory of gases (Chapman-Enskog model [153]) implemented in the code, see CFD-ACE documentation [490]. The simulation results indicate a trend and help explain the technical realization of physical effects.

The simulation assists in designing the geometry layout for an equal flow distribution in the channel network and gives an indication of the separation performance of the process. The separation effect is dependent on the temperature gradient, as well as the cascaded separation and collection of the species. The number of elements includes almost 800 000 grid points, shown in Figure 7.14, and limits the simulated geometry. Based on the simulations and the available fabrication techniques, the structures were designed and fabricated in silicon to demonstrate the separation in liquid mixtures, see Figure 7.15.

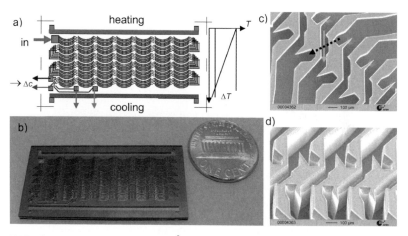

Fig. 7.15. a) Chip layout of a 40×20 mm² separation chip with 35 blocks of 5 curved channels each. The inlet and the four outlet connections are arranged on the left side, the heating and cooling channel on the top and bottom side of the chip, respectively. b) Fabricated silicon chip with glass lids. c) and d) SEM images of the RIE fabricated crossing structure, 400 µm deep. The narrow gaps in the RIE crossing channel compensate the pressure loss in the KOH crossing (not displayed) on the bottom side. The arrows in image c) indicate the fluid crossing direction between the channel blocks.

The mask design consists of different basic elements, which were successively arranged together on the final chip layout. The software L-Edit from Tanner Research was used to design the chips and the wafer layout. A silicon wafer was structured with successive DRIE and KOH etching and finally covered with two Pyrex lids from the top and bottom side. The mask layout and details of fabricated devices at the critical channel crossing area are shown in Figure 7.16. In the left bottom area,

KOH etching lines are displayed on the compensation structures, which are described in detail by Wacogne et al. [491].

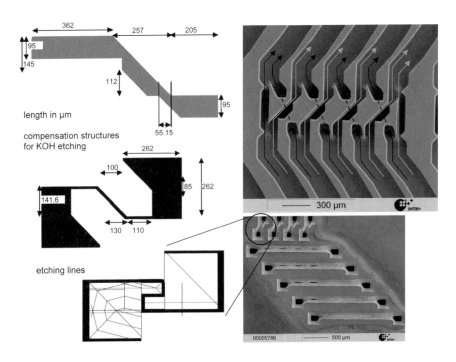

Fig. 7.16. Mask layout of the crossing, RIE channel, and KOH etched bottom crossing with compensation structures; top right) SEM image of the crossing structure and the flow directions; bottom right) SEM picture of the KOH etched bottom side of a chip. The through etch of V-trenches to the front side channels is visible. On the top left, four crossings are visible connecting five channels in two different channel blocks.

Four 40×20 mm^2 chips and four 20×20 mm^2 chips with different geometric structures were arranged and fabricated on a 4-inch (1 0 0) silicon wafer. The mechanical strength of the DRIE processed wafer was weakened by the 400 μm deep channels, which, in addition, were oriented in crystal plane direction. After DRIE structuring of the separation channels, the crossing channels are KOH etched on the backside. Finally, both wafer sides are cleaned, carefully covered with Pyrex wafer, and diced. However, the combination of the front side DRIE process and the back side KOH-etching process creates difficulties. Etching the bottom crossing, see Figure 7.16 bottom right, and the V trenches was a difficult step because of the relationship between the V trenches' depth and the inlet structures' size. For the inlet structures, a through-etch must be performed to connect the front and back sides. Due to depth variations of the DRIE fabricated front side channels over the whole wafer, the through-etch by KOH-etching was realized at different times, and the long

KOH-etching process time would connect the V trenches with the front side channels, which is depicted in Figure 7.16 bottom right. These undesirable openings and channel connections disturb the separation effect in the previous channels.

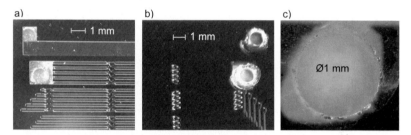

Fig. 7.17. a) Inlet heating channel and separation channels on the top left side, RIE fabrication; b) bottom view of the KOH etched crossing and transfer structures; the drilled holes in the bottom Pyrex lid are clearly visible. c) Fluid contacting hole in a Pyrex wafer fabricated by drilling with a 0.8 mm mounted point rotating with 50 000 rpm.

With the two glass lids, the entire chip is mechanically stiff enough to be integrated into the fluidic mount. Another difficult point was the drilling or grinding of the fluid connection with 1 mm diameter into the bottom glass lid. In Figure 7.17, some results are shown where cracks and splits are clearly visible in the glass. An optimization of the grinding speed, as well as a stable and vibration-free tool holder produced acceptable results for the drilled holes.

7.4.2 Experimental investigations with thermodiffusion chips

To test the function and the dimension of the physical effect, an experimental arrangement is designed. In Figure 7.18 right side, the experimental setup with the complete device is displayed with syringe pump, flow meters, and thermostat. The fluidic mount in Figure 7.18 left side is made of poly-etheretherketone (PEEK) for an optimal thermal insulation. Two pressure sensors (RS Components, 100 psi, 235-5863) are installed, one at the inlet and one at the outlet. The fluid connections are placed around the tool holder and on its back side. The whole system is mounted on a perforated aluminum plate. The warm and cold fluid streams are generated by two thermostats with internal rotating pumps, heating and refrigeration system, see Figure 7.18. The fluid to be separated is driven by an injection pump (Injectomat 2000, Fresenius Vial, Germany) with a given flow rate. The temperature is measured with a mobile thermometer from Testo (Testo 925, Germany), and with two PT100 sensors integrated into the fluidic mount. The electric signal is processed by a Keithley Integra 2700 multimeter. The chip sealing is casted with the silicone Dublisil-15 from Drewe Dentamid, Germany. The fluidic sealing of the chips was not tight for high pressures and difficult to handle. The inlet hole within the fluidic mount were too small; when the lid of the fluidic mount was tightened, the sealing shrinks under the applied pressure and blocks the passages. To prevent such blocking, the sealing passages were widened with a special cutting tool.

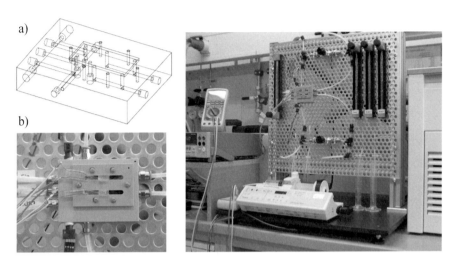

Fig. 7.18. a) Technical drawing of the fluidic mount for separation silicon chips 40×20 mm^2. b) Image of the experimental setup for liquids. c) Entire experimental setup for liquids with injection pump in the middle, digital multimeter, pressure and temperature measurement. The heating thermostat (Jumo, Germany) is visible at right. See also Figure A.7.

The treatment of pressurized air by an applied magnetic field, concerning the paramagnetic properties of oxygen in this system, is measured with an oxygen analyzer (Oxybaby V, Witt Gastechnik, Germany). The measurement results clearly indicate that no separation effect could be observed within a magnetic field lower than 1 Tesla and the detection limit of the oxygen analyzer.

The separation of a liquid mixture, water with 0.1 M BaCl$_2$ salt, is described in the following. The volume flow rate through the system is measured at four different outlet channels with a marked cylinder and the time. The inlet flow is varied from 50 ml/h up to 250 ml/h. Except for one chip, no outlet flow at 50 ml/h is achieved at measurement point F21 and F22. The remainder of the chips show a homogeneously distributed outlet flow.

To achieve a temperature gradient across the chip, two thermostat pumps are connected with the chip to pump deionized water through the heating and cooling channels. The heating channel is fed with warm water (90 °C) and the cooling channel is supplied with cold water (10 °C), measured at the thermostat. Due to insufficient thermal tube insulation, a discrepancy exists between the chip temperature and the adjusted temperature at the circulating pump.

For the pressure measurement, a sensor is positioned at the inlet and outlet channel, respectively. The location is given in Figure 7.19 with the measurement results for different chips. A square fitting function clearly represents the pressure loss over the chip indicating, straight channel flow (linear part) and convective flow in bends and junctions (square term), see also Section 3.1.2.

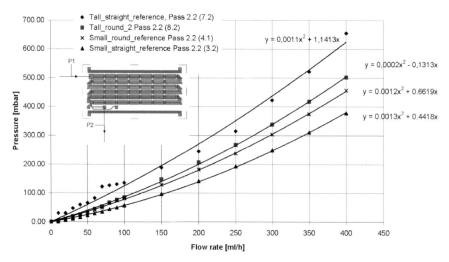

Fig. 7.19. Pressure measurement for the thermodiffusion chips with the schematic drawing of the measuring points for the pressure measurement.

To evaluate the separation effect of thermodiffusion, a 0.1 mol/l barium chloride ($BaCl_2$) solution is pumped through the system. The concentration is indirectly measured over the conductivity of the aqueous solution. At both the inlet and the outlet channels, the conductivity of the solution is measured with a conductivity meter (WTW InoLab Cond 730, Germany) and a standard-conductivity cell (WTW Tetra-Con 325, Germany). The differences between inlet concentration C11 and the outlet concentration (C21 to C24) range from 0.016 mmol/l to 6.04 mmol/l. The concentration differences are quite small with less than 6 %, which is due to the small effect of thermodiffusivity. Control measurements indicate some difficulties in measuring the concentrations, however, the major hurdle is temperature control on the separation chip, which must be overcome in future chip improvements. The little effect of thermodiffusion was reproducible, although further measurement are necessary to improve the separation effect.

7.4.3 Improvement of thermodiffusion devices

The next steps will be to optimize the chip fabrication for an equal flow distribution in the channel network and a proper KOH crossing at the bottom. The temperature gradient of the chip from one side to the other causes severe problems with thermal losses and temperature measurements. The next chip generation will have a temperature gradient from one chip face to the other, comparable to the heat flux path in the thermoelectric generator, see Section 7.2.2.

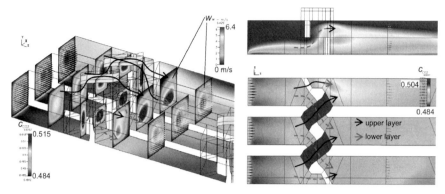

Fig. 7.20. Flow, temperature, and concentration simulation of 50/50 % mixture of H_2 and CO_2 in three channels. The top wall of the channels has a temperature of 300 K, the bottom a temperature of 360 K. At the end of the channels, the flow is horizontally split and rearragend according their concentration into the following channels.

The geometrical setup is displayed in Figure 7.20 with velocity and concentration profiles in three parallel channels and connecting channels. Simulations were performed with a 50/50 % gas mixture of H_2 and CO_2 flowing under a temperature difference of 60 K between bottom (360 K) and top (300 K) channel wall. The flow is horizontally split at the end of the channel and rearranged according the concentration. The H_2 enriched, lower hot stream is guided to the right side, and the CO_2 enriched, upper cold part is guided over the lower stream to the left side. Hence, CO_2 is enriched in the left side channels at the cold wall up to a concentration of approx. 51.5 % after four cascade steps. On the other side, H_2 is enriched up to a concentration of 51.6 %. These values are in the range of the first simulations depicted in Figure 7.14.

The main advantage of horizontally splitting the channel flow originates from facilitated fabrication procedure with two DRIE structured wafers covered with a glass wafer. The temperature gradient is applied from the top to the bottom side, which give higher gradients for the single channels and simplified experimental setup. Fabrication and experimental investigation are the next steps to evaluate this separation technique. With thermodiffusion, a single-phase separation process was presented without unwanted interfacial forces, which may complicate or disturb the fluid flow and separation.

7.4.4 Knudsen pump

A similar coupled process with a thermal gradient is used in vacuum technology to pump rarefied gases, the so-called Knudsen effect or thermal transpiration, see Figure 7.1. The effect is based on the correlation between the pressure and temperature of two connected systems in the free molecular flow regime (Kn > 10)

$$\frac{p_1}{p_2} = \sqrt{\frac{T_1}{T_2}}. \tag{7.23}$$

For a silicon micro pump, McNamara and Gianchandani [492] give a short explanation of the underlying theory and discuss the microfabrication of the 1.5 x 2.0 mm silicon chip. The pump consists of five large channels, 10 μm deep and 30 μm wide, connected with narrow channels, 10 μm wide and 100 nm deep. Two of the large channels are heated up to 600 °C with a poly-silicon heater. With an on-chip pressure sensor, a pressure difference of approx. 3 kPa was measured for a power input of 60 mW. Multiple stages may be cascaded to get a pump with a higher pressure difference for applications in gas sampling, pneumatic actuation, or vacuum encapsulation. Alexeenko et al. [493] describe the theoretical modeling of the flow in a Knudsen pump with the Monte-Carlo method (DSMC).

Further interesting effects, which are not covered here, are connected with rarefied gas flow, surface effects, and thermal gradients. Sone [154] reports a "ghost effect", which describes a flow phenomena of rarefied gases under heat transfer, however, its application is still awaiting.

7.5 Pressure Diffusion

As advanced process of the micro cyclone, see Section 1.5.3, the pressure diffusion within a flow causes a species flow and concentration gradient by a pressure gradient. Within a separation nozzle, see Figure 7.21, pressure diffusion is used to separate uranium isotopes assisted by centrifugal forces. An early application for isotope separation in chemically-etched microstructures is described by Schütte [56].

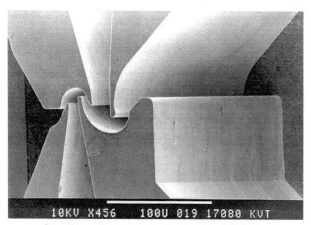

Separation nozzle as an example of arbitrary lateral shaping.

Fig. 7.21. Separation nozzle for pressure diffusion, LIGA fabricated, courtesy of Research Center Karlsruhe, Germany.

A few years later, Becker et al. [58] reported upon experimental investigations with separation nozzle devices, which show only slightly higher energy consumption than contemporary gas diffusion plants. For microstructured devices fabricated with the help of synchrotron radiation, lithography, and galvanoplastics, Becker et al.[59] reported an increase by a factor of three for the gas pressure and "considerable savings" for the enrichment of uranium isotopes . The nozzle channels were fabricated with a channel height of approx. 400 µm and structural details of approx. 0.1 µm. The successive development of more appropriate and advanced geometries led to the development of the LIGA technology, see [494] and [495], Chapter 7, which was one of the birthplaces for microsystems technology and micro process engineering in Germany.

The high energy requirements compared to ultracentrifuges was the limiting factor for the application of separation nozzles. Modern processes are working with ultracentrifuges, which cause a higher centrifugal force, easier staging, and higher separation factors [56].

Enhanced transport processes and equipment show excellent promise for successful application in various fields where high transfer rates, intelligent and prolific combinations of microstructures in macro devices, and new process routes are needed. With the fundamentals and some basic applications of heat and mass transfer, the benefits of miniaturization have been shown in Chapters 3 and 4. High gradients and high specific surface area in devices with various construction materials lead to fast equilibrium. The characteristic dimensions of microstructure internals are in the range of boundary layers, where high gradients enhance the transfer processes. Coupled transport phenomena gain importance, however, they are not yet exploited by far. Additionally, the cascading of effects and the integration of various elements enhance and guide the entire process. This encompasses new possibilities for difficult processes and opens new opportunities for the future.

8

Conclusion and Final Remarks

Micro process engineering at the interface of (at least) chemistry, process engineering, and microsystems engineering provides good opportunities to introduce new possibilities and options in many application fields. The basic features of integrated and enhanced transfer phenomena and small dimensions are explained and illustrated in this work. It reinforces ideas and concepts from many areas with the goal of collecting and introducing elegant solutions and methods into new research and development fields. Their superior transport characteristics are the main motivation to use microstructured devices in mixing, heat exchange, and/or chemical reaction applications. Micro process engineering can work with controlled fluid dynamics, short residence time, and its narrow distribution in continuous processes, as well as short diffusion paths. The accurate control of diffusion length and time, convective transport, and reaction kinetics allow short reaction times for high yield and selectivity in complex chemical reactions. The typical ranges can lead to transient effects, such as flow frequencies and temperature cycling, which are used for new process windows and routes with reactant concentrations or temperature. Chemical reactions can be processed at their kinetic limits.

Aside from these process engineering issues, high integration density of channels with internal and external numbering-up following the equal-up design method gives the opportunity of devices with high mass flow rates. The integration of sensors and actuators serves for precise parameter and process control in microchannels, which may also be integrated in multi-purpose plants. The high surface-to-volume ratio leads to dominating surface effects, which enhance catalytic reactions, sensing, or adsorption processes, but are detrimental due to particle deposition or corrosion. Small volumes in microstructured devices require less amount of reactants leading to low risk for hazardous media, low solvent consumption, and less waste, however, small production volumes have large influence on product logistics. More information about the process in shorter time leads to better process understanding and low risk to scale-up a process from laboratory to production scale. This enables short time to market from laboratory development to production.

Multiple-channel devices demand a new design methodology to increase the channel number and providing safe operating conditions. Design methodology must also consider materials, surface effects, and their related fabrication technologies, device integration, and experimental characterization and testing to yield a successful device for new process conditions or product properties. A lead role is taken by correct and reasonable control and manipulation of transport processes. The designer must consider multi-scale physical and chemical processes, often combined with multiphase flow and heterogeneous reactions including dominant surface effects. Biological examples may indicate new solutions and give new insights into complex, interwoven process networks – the rhizome of technology.

The already wide acceptance is illustrated by applications and research projects in Europe, United States and the Far East, however, there are still many educational needs for chemical process engineers, microsystems engineers, mechanical engineers, physicists, or chemists. All groups are involved in successful selection of auspicious applications, in device design, fabrication, and systems integration, as well as in economical production. Often, potential users have to be convinced of the benefits, which is also an educational task. With the successful introduction of microstructured devices, some drawbacks and further challenges are also combined, see Section 1.7. Microstructured devices enable to operate in enlarged process windows, but there are still limits, such as highly exothermic reactions where the Damköhler numbers Da must be considered, uniform flow distributions over all channel elements, highly viscous media and polymerization, or multiphase flow and phase separation. Particle processing has been studied, however, particle attachment, precipitation, fouling, and channel blocking are still major issues for successful operation. Fouling is often connected with corrosion, to which microstructured devices are very sensitive. Cost-effective mass production is essential for wider application. Inherent safety of microstructured devices must be checked for the actual case. The major issue remains the education and training on microstructured devices. In chemical education, continuous processing should be taught from the beginning, with process engineering and chemical engineering methods. In mechanical engineering, microchannels give the opportunity to grasp the essentials of fluid flow, heat transfer, boiling, evaporation, and condensation within continuously operating equipment.

The apparent limits of microstructured equipment invite to work on these issues in future research projects. For this, suitable protocols, descriptions, and equipment are required, however, this will lead to a paradigm change many industrial applications. Some applications are mentioned in this book from energy harvesting for autonomous microsystems, through mixing, heat transfer, and chemical reactions for process intensification and selectivity engineering, to appropriate product preparation, such as nanoparticles in defined size and quality. Here, the appropriate integration of microstructures into conventional equipment gives a very promising route to combine the benefits of both areas. Aside from process engineering, energy technology can benefit from integration of microstructured equipment to enhance transfer processes, such as micromixers in premixed combustion or isothermal gas turbines. This leads to increased efficiency of conventional devices and opens new routes for energy conversion. Yet, experience and successful references are still missing to guide the appropriate application of microstructured devices, however, this book should contribute a little step for wider application.

A

Color Figures

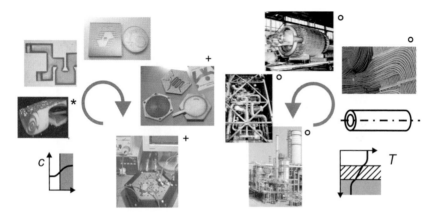

Fig. A.1. Setup of microstructured and conventional equipment from single microstructures over combined elements to an entire plant, see also Fig. 2.2.

Fig. A.2. Micro gas chromatograph; left: separation column on a silicon chip, covered with a glass plate; right: IC plate and analytical equipment of the micro-GC, with courtesy from SLS micro technology [81], see also Fig. 1.8.

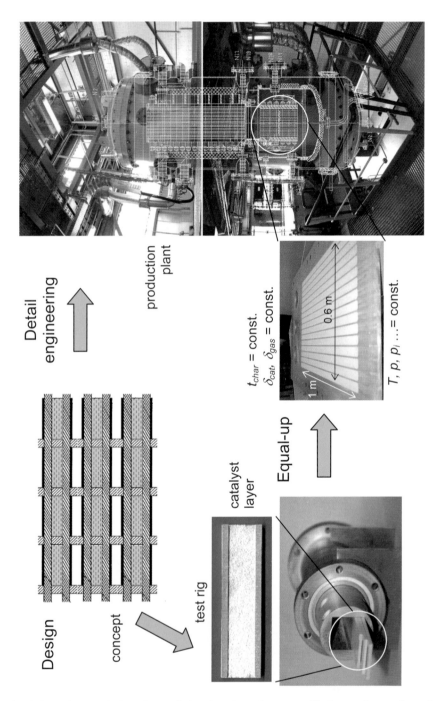

Fig. A.3. Equipment design of a catalytic microchannel reactor with the equal-up and equal-down method acc. to [34, 33]. The modified residence time, the catalyst thickness, the channel height, and the process conditions are kept constant. Courtesy of Degussa AG, Hanau and Uhde GmbH, Dortmund, see also Fig. 2.3.

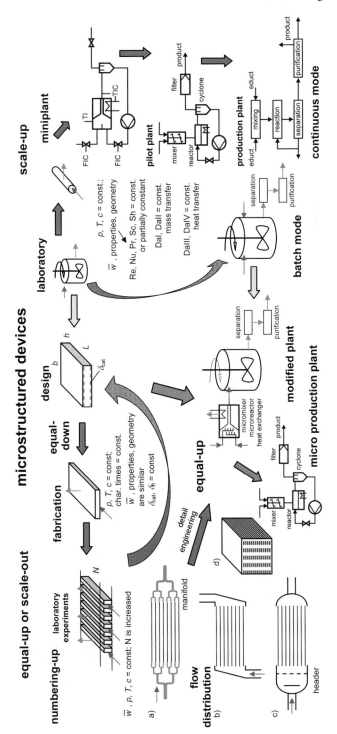

Fig. A.4. Equal-up, scale-out, and scale-up procedures for chemical engineering purposes with microstructured devices, see also Figures 1.3 and 2.3. Development of microstructured devices, partial integration of microstructured devices, see also Fig. 6.10.

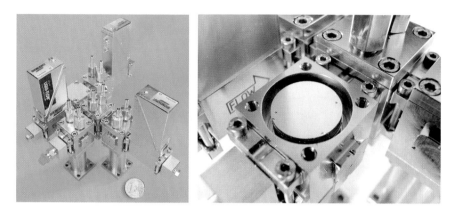

Fig. A.5. Left: Syntics plant layout with a modular setup of flow sensors, mixers and heat exchangers. Right: Detail of a turbulence mixer with cooling chamber and housing. See also Fig. 1.12.

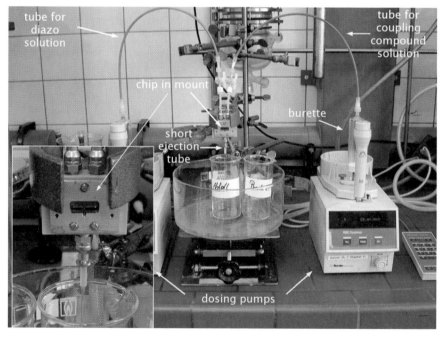

Fig. A.6. Pigment synthesis within a silicon T-shaped micromixer with short mixing channel indicating a typical experimental setup for laboratory microstructured device testing, see also Fig. 6.28.

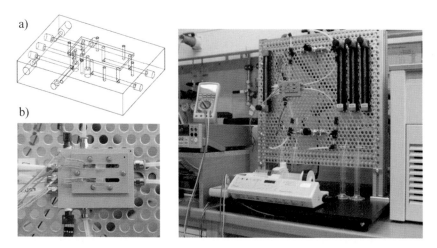

Fig. A.7. a) Technical drawing of the fluidic mount for separation silicon chips 40×20 mm^2. b) Picture of the experimental setup for liquids. c) Entire experimental setup for liquids with injection pump in the middle, digital multimeter, pressure and temperature measurement. On the left side the heating thermostat is visible, see also Fig. 7.18.

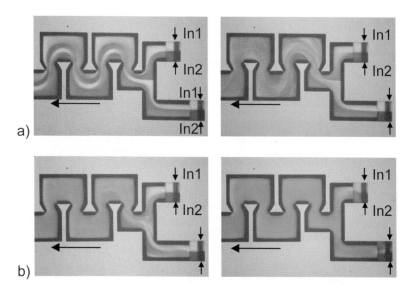

Fig. A.8. Flow and mixing visualization in a single TTree mixer with four different mass flow rates: a) In = 11 g/h, Re$_{in}$ = 11.9; In = 86 g/h, Re$_{in}$ = 93.1; b) In = 250 g/h, Re$_{in}$ = 270.7; In = 755 g/h, Re$_{in}$ = 817.6, see also Fig. 5.41.

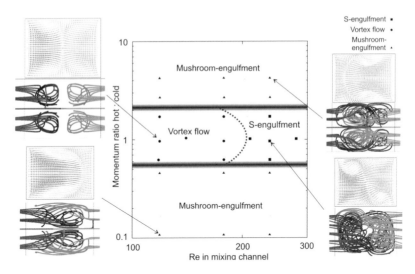

Fig. A.9. Simulations of flow regimes in T-mixers: flow regimes found in a T-mixer (300×300 μm² inlets, 300×400 μm² mixing channel) with stream lines and velocities with color coded temperature in a lateral cut through the mixing channel; mixing of nitrogen at 20 °C and 70 °C, see also Fig. 6.14.

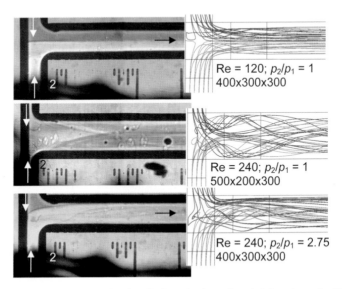

Fig. A.10. pH neutralization reaction in a T-shaped micromixer (mixing channel width × inlet channel width × channel depth in μm) together with the corresponding simulations: vortex flow (top), S-engulfment (middle) and mushroom-engulfment (bottom), see also Fig. 6.15.

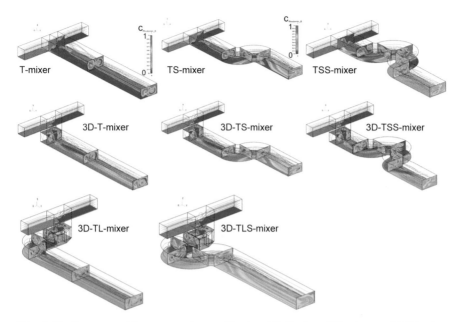

Fig. A.11. Geometry with concentration profiles in a 3D-T-mixer, TS-mixer, and TSS-mixer for Re = 200 in the mixing channel, see also Fig. 5.17.

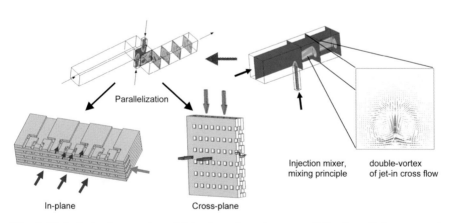

Fig. A.12. Left: implementation of injection micromixers, parallelization of elements and stack setup; Right: mixing principle of one channel injection mixer, see also Fig. 5.45.

Simulation

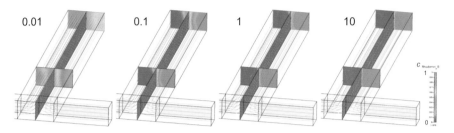

Fig. A.13. Regime I, straight laminar flow in T600×300×300 mixer; Re numbers from 0.01 to 10; Sc = 3 700, see also Fig. 3.13.

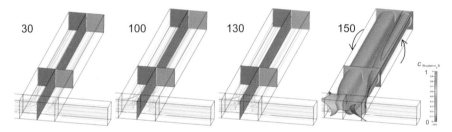

Fig. A.14. Regime II, curved laminar flow in T600×300×300 mixer, Dean vortices for 30 < Re < 130 (approx.) and asymmetrical flow for Re = 150 (engulfment flow regime); Re numbers from 30 to 150; Sc = 3 700, see also Fig. 3.14.

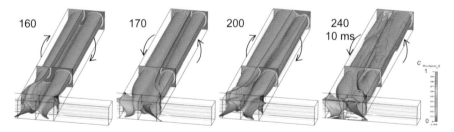

Fig. A.15. Regime III, asymmetrical, laminar flow in T600×300×300 mixer, engulfment flow; Re numbers from 160 to 200 and Re = 240 at t = 10 ms due to periodic oscillating flow; Sc = 3 700, see also Fig. 3.15.

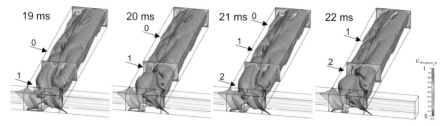

Fig. A.16. Regime IV, periodic pulsating laminar flow in T600×300×300 mixer; Re number Re = 300, observation times are 1.9 ms, 2.0 ms, 2.1 ms, and 2.2 ms, see Figure 3.22; Sc = 3700, see also Fig. 3.21.

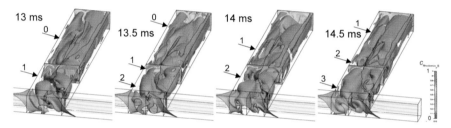

Fig. A.17. Regime V, quasi-periodic pulsating, laminar flow in T600×300×300 mixer; Re = 500, observation times are t = 1.3 to 1.45 ms, see Figure 3.22; Sc = 3700, see also Fig. 3.26.

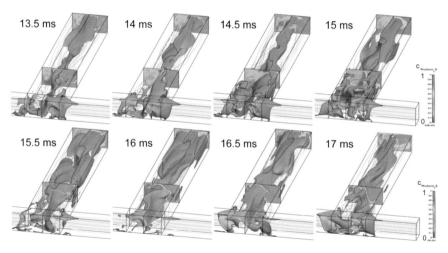

Fig. A.18. Regime VI, chaotic pulsating, quasi-laminar flow and quasi-vortex flow in T600×300×300 mixer; Re = 1000, observation times are t = 1.35 to 1.5 ms in the top row, t = 1.55 to 1.7 ms in the bottom row; Sc = 3700, see also Fig. 3.27.

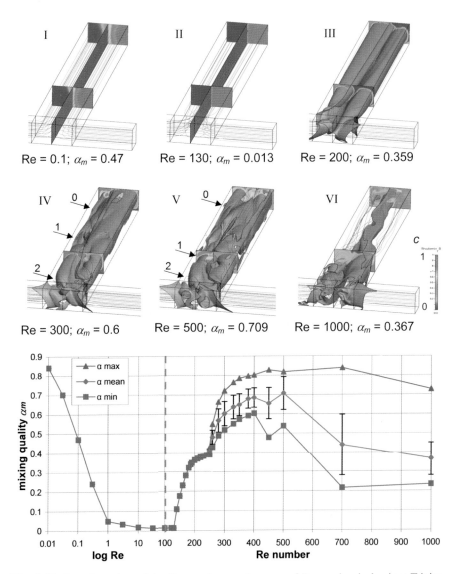

Fig. A.19. Top: Overview of the flow regimes and ranges of Re number in laminar T-joint flow with 1:1 mixing ratio, Sc = 3700, T600×300×300. Bottom: The mixing quality α_m is determined at a constant length $l = 5 \cdot d_h$ of the mixing channel, here $l = 2000$ µm, see also Figures 3.12 and 5.12.

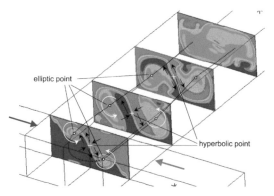

Fig. A.20. Typical points of the concentration field in a T-mixer (T600×300×300), Re = 300, see also Fig. 5.25.

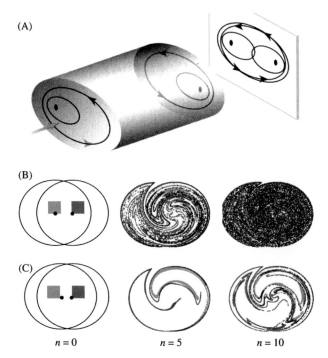

$n = 0$ $n = 5$ $n = 10$

Fig. A.21. Schematic representation of a channel type micromixer constructed from the concatenation of basic mixing elements. The linked twist map (LTM) mechanism causes the flow to mix completely after passing through five periodic elements of the mixer. Acc. to Sturman et al. [354], see also Fig. 5.23.

Heat transfer

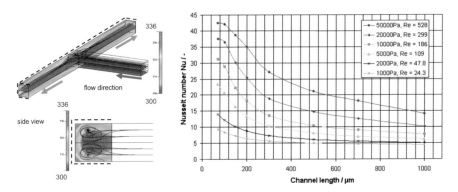

Fig. A.22. Simulation of the heat transfer in a microchannel with square cross section $d_h = 100$ μm, constant heat flux of 150 W/cm^2; Left: Wall temperature distribution and stream lines in the side view; Right: Development of the local Nu number in the curved flow with various pressure losses and Re numbers, see also Fig. 4.6.

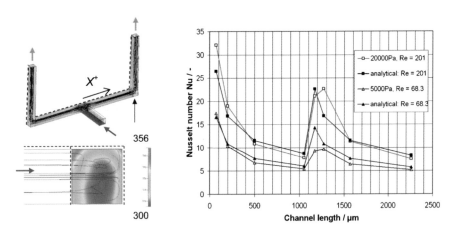

Fig. A.23. Heat transfer simulation of a fork-shaped microchannel setup with a cross section of 100×100 μm^2; left: Temperature distribution and stream lines; right: Local Nu number over the channel length, see also Fig. 4.7.

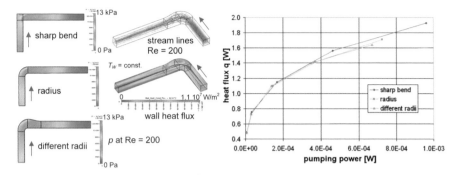

Fig. A.24. Pressure loss in bends and curves with corresponding heat flux for water, 300 K inlet temperature. The microchannel with square cross section has a width of 100 μm, an inlet length of 600 μm, and an outlet length of 1 400 μm. The boundary condition at the inlet is a constant laminar velocity profile, see also Fig. 4.8.

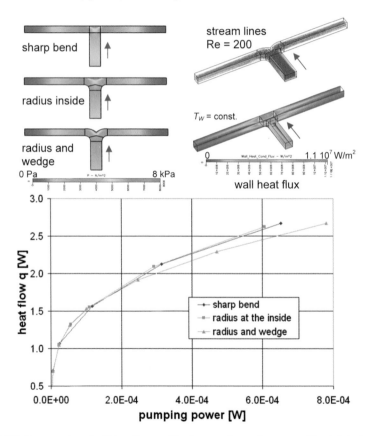

Fig. A.25. Pressure loss in T-junctions with sharp and round corners. The inlet channel is 170 μm wide, the outlet channel is 100 μm wide, and all channels are 100 μm deep, see also Fig. 4.9.

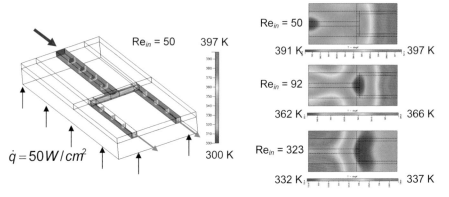

Fig. A.26. Temperature distribution in a fluidic element with fork-shaped microchannels and surrounding material for different Re numbers in the outlet, see also Fig. 4.10.

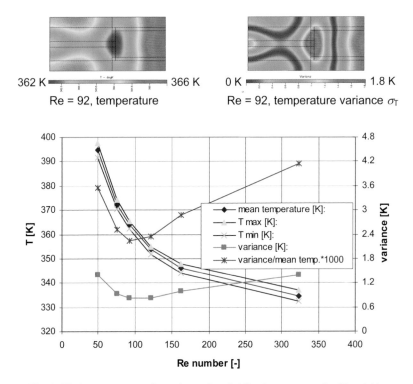

Fig. A.27. Temperature variance in a microfluidic element, see also Fig. 4.11.

Coupled transport phenomena: Thermoelectrics and Thermodiffusion

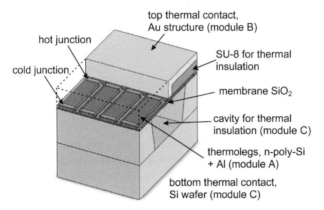

Fig. A.28. Schematic setup of the investigated generators with thin-film thermoelectric structures and perpendicular heat fl ux. Module A with SiO₂ membrane and thermolegs, module B with SU-8 insulation and gold layer, and module C with insulation cavity in the silicon wafer, see also Fig. 7.3.

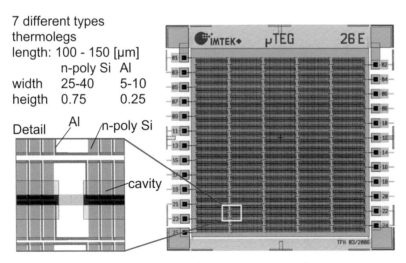

Fig. A.29. CAD image of a complete generator chip, 10×10 mm² footprint, with typical meander-like thermocouple arrangement with 6 000 thermocouples. The bond pads are on the left and right side of the chip to connect the generator, see also Fig. 7.4.

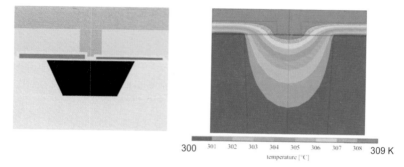

300 `301` `302` `303` `304` `305` `306` `307` `308` 309 K

temperature [°C]

Fig. A.30. Temperature profile within the µTEG for 9 K temperature difference, which gives approx. 8.61 K effective temperature difference (95.7 %) over the thermocouple, see also Fig. 7.5.

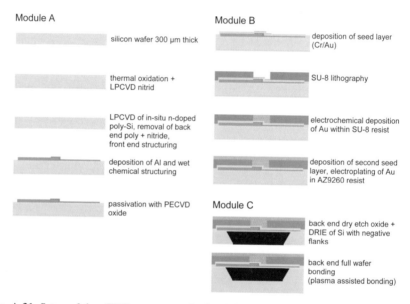

Module A

silicon wafer 300 µm thick

thermal oxidation + LPCVD nitrid

LPCVD of in-situ n-doped poly-Si, removal of back end poly + nitride, front end structuring

deposition of Al and wet chemical structuring

passivation with PECVD oxide

Module B

deposition of seed layer (Cr/Au)

SU-8 lithography

electrochemical deposition of Au within SU-8 resist

deposition of second seed layer, electroplating of Au in AZ9260 resist

Module C

back end dry etch oxide + DRIE of Si with negative flanks

back end full wafer bonding (plasma assisted bonding)

Fig. A.31. Setup of the µTEG structure and related fabrication process steps of the thermocouple structure (Module A), of the thermal connector structure (Module B), and of the silicon cavity (Module C), see also Fig. 7.7.

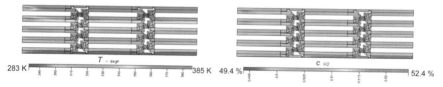

283 K T - degK 385 K   49.4 % c H2 52.4 %

Fig. A.32. Temperature profile (left) and concentration profile of H_2 (right) in a single channel, 1500 µm long, with temperature gradient (100 K), CO_2-H_2 system, 4000 Pa pressure difference, 120 K temperature difference, Soret coefficient 10^{-9} m^2/sK, see also Fig. 7.14.

References

1. Bird RB, Stuart WE, and Lightfoot EN. *Transport Phenomena*. John Wiley, New York, 1960.
2. Bird RB, Stuart WE, and Lightfoot EN. *Transport Phenomena*. John Wiley, New York, 2002.
3. Kockmann N, editor. *Micro Process Engineering*. Wiley-VCH, Weinheim, 2006.
4. Ehrfeld W, editor. *Microreaction Technology - Proc. First Int. Conf. Microreaction Technology*. Springer, Berlin, 1998.
5. Löhder W and Bergmann L. Verfahrenstechnische Mikroapparaturen und Verfahren zu ihrer Herstellung. German patent DD 246 247 A1, Akademie der Wissenschaften der DDR, 1987.
6. Behrens H. *Mikrochemische Technik*. Voss, Hamburg, 1895.
7. Behrens H. *Mikrochemische Technik*. Voss, Hamburg, 1900.
8. Boricky E. *Elemente einer neuen chemisch-mikroskopischen Mineral- und Gesteinsanalyse. Archiv der naturwissenschaftlichen Landesdurchforschung von Böhmen*. Rivnac, Prag, 1877.
9. Jäger C. *Gilles Deleuze - Eine Einführung*. UTB für Wissenschaft, Wilhelm Fink, München, 1997.
10. Deleuze G and Guattari F. *Rhizom*. Suhrkamp, Berlin, 1980.
11. Kornwachs K. *Technikbilder und Technikkonzepte im Wandel - eine technikphilosophische und allgemeintechnische Analyse*, chapter Neue Bilder?, pages 69–74. Forschungszentrum Karlsruhe GmbH, FZKA 6697, 2002.
12. Kockmann N. *Micro Process Engineering*, chapter Transport Processes and Exchange Equipment, Chapter 3. Wiley-VCH, Weinheim, 2006.
13. Hessel V, Hardt S, and Löwe H. *Chemical Micro Process Engineering: Fundamentals, Modelling and Reactions*. Wiley-VCH, Weinheim, 2004.
14. Arno R. Was Studenten wissen müssen - Schlüsselqualifikationen. *Die Zeit*, 27:72, 2006.
15. Vauck WRA and Müller HA. *Grundoperationen chemischer Verfahrenstechnik*. Deutscher Verlag für Grundstoffindustrie, Stuttgart, 11. ed. edition, 2000.
16. Winde B and Heim J. *Technik en miniature - Stippvisite bei der Mikrotechnik*. Urania-Verlag Leipzig, 1988.
17. Gad-el-Hak M, editor. *The MEMS Handbook, 2nd Ed.* CRC Press, Boca Raton, 2006.
18. Quinn DJ, Spearing SM, Ashby MF, and Fleck NA. A Systematic Approach to Process Selection in MEMS. *Journal of MEMS*, 15:1039–1050, 2006.

19. Fujita H, editor. *Micromachines as Tools for Nanotechnology.* Springer, Berlin, 2002.

20. Young DJ, Zormann CA, and Mehregany M. *Springer Handbook of Nanotechnology,* chapter MEMS/NEMS Devices and Applications, Chapter 8. Springer, Berlin, 2004.

21. Mijatovic D, Eijkel JCT, and van den Berg A. Technologies for nanofluidic systems: top-down vs. bottom-up - a review. *Lab on a Chip,* 5:492–500, 2005.

22. Hessel V, Löwe H, Müller A, and Kolb G. *Chemical Micro Process Engineering: Processing and Plants.* Wiley-VCH, Weinheim, 2005.

23. Wang Y and Holladay JD, editors. *Microreactor Technology and Process Intensification.* ACS Symposium Series 914, Oxford University Press, 2005.

24. Pais A. *Raffiniert ist der Herrgott... Albert Einstein - Eine wissenschaftliche Biographie.* Spektrum akademischer Verlag, Heidelberg, 2000.

25. Knösche CM. Wärmeabfuhr und Rückvermischung in mikrostrukturierten Apparaten. *Chemie Ingenieur Technik,* 77:1715–1722, 2005.

26. Kockmann N, Engler M, Haller D, and Woias P. Fluid dynamics and transfer processes in bended microchannels. *Heat Transfer Engineering,* 26:71 – 78, 2005.

27. Kockmann N, Engler M, and Woias P. Fouling processes in micro structured devices. In *ECI Heat Exchanger Fouling and Cleaning: Challenges and Opportunities,* http://services.bepress.com/eci/ heatexchanger2005/28, 2005.

28. Kobayashi I and Nakajima M. *Micro Process Engineering,* chapter Generation and Multiphase Flow of Emulsions in Microchannels, Chapter 5. Wiley-VCH, Weinheim, 2006.

29. GroßA. *Theoretical Surface Science – A Microscopic Perspective.* Springer, Berlin, 2003.

30. Renken A and Kiwi-Minsker L. *Micro Process Engineering,* chapter Chemical Reactions in Continuous-flow Microstructured Reactors, Chapter 6. Wiley-VCH, Weinheim, 2006.

31. Duduković MP, Larachi F, and Mills PL. Multiphase catalytic reactors: a perspective on current knowledge and future trends. *Catalysis Reviews,* 44:123–246, 2002.

32. Zlokarnik M. *Scale-up - Modellübertragung in der Verfahrenstechnik, 2nd ed.; english translation: Scale-up in Process Engineering.* Wiley-VCH, Weinheim, 2002.

33. Emig G and Klemm E. *Technische Chemie, Einführung in die chemische Reaktionstechnik, Chapter 16.* Springer, Berlin, 2005.

34. Becker F, Albrecht J, Markowz G, Schütte R, Schirrmeister S, Caspary KJ, Döring H, Schwarz T, Dietzsch E, Klemm E, Kruppa T, and Schüth F. DEMiS: Results from the development and operation of a pilot-scale micro reactor on the basis of laboratory measurements. In *IMRET 8,* number TK131f, 2005.

35. Bayer T and Kinzl M. Werkzeug für Verfahrensentwickler. Mikroverfahrenstechnik - Beispiele aus der industriellen Praxis. *Verfahrenstechnik,* 38:7–8, 2004.

36. Roberge DM, Ducry L, Bieler N, Cretton P, and Zimmermann B. Microreactor Technology: A Revolution for Fine Chemical and Pharmaceutical Industries? *Chem. Eng. & Techn.,* 28:318–323, 2005.

37. Roberge DM. An Integrated Approach Combining Reaction Engineering and Design of Experiments for Optimizing Reactions. *Org. Proc. Res. & Dev.,* 8:1049–1053, 2004.

38. MicroChemTec. URL: www.microchemtec.de, 2006.

39. Stief T, Langer OU, and Schubert K. Numerische Untersuchungen zur optimalen Wärmeleitfähigkeit in Mikrowärmeübertragerstrukturen. *Chemie Ingenieur Technik,* 70:1539–1544, also in Chem.–Eng.–Tech. 21 (1999) 297–302, 1998.

40. Truckenbrodt E. *Fluidmechanik, Bd. 1.* Springer, Berlin, 1996.

41. Perry RH, Green DW, and Maloney JO, editors. *Perry's Chemical Engineers Handbook.* McGraw-Hill, New York, 1998.

42. Yue J, Chen G, and Yuan Q. Pressure drops of single and two-phase flows through T-type microchannel mixers. *Chemical Engineering Journal*, 102:11–24, 2004.

43. Kockmann N, Kiefer T, Engler M, and Woias P. Channel networks for optimal heat transfer and high throughput mixers. In *ECI Int. Conf. Heat Transfer and Fluid Flow in Microscale, Il Ciocco, Tuscany*, 2005.

44. Hessel V, Serra C, Löwe H, and Hadziioannou G. Polymerisation in mikrostrukturierten Reaktoren: Ein Überblick. *Chemie Ingenieur Technik*, 77:1693–1714, 2005.

45. Herwig H. *Micro Process Engineering*, chapter Momentum and Heat Transfer in Microsized Devices, Chapter 2. Wiley-VCH, Weinheim, 2006.

46. Kandlikar SG, Schmitt D, Carrano AL, and Taylor JB. Characterization of surface roughness effects on pressure drop in single-phase flow in minichannels. *Physics of Fluids*, 17:100606, 2005.

47. Perry JL and Kandlikar SG. Investigation of Fouling in Microchannels. In *ASME 4th Int. Conf. Nano Micro Mini Channels, Limerick, Ireland*, number ICNMM2006-96248, 2006.

48. Kockmann N, Kastner J, and Woias P. Numerical and experimental study of nanoparticle precipitation in microreactors. In *ECI 2nd Int. Conf. Transport Phenomena Micro Nano, Il Ciocco, Tuscany*, 2006.

49. Kastner J. Mixing and reactive precipitation in microchannels. Master's thesis, University of Freiburg, IMTEK, 2006.

50. Batchelor GK. *An Introduction to Fluid Dynamic, 2nd ed.* Cambridge University Press, Cambridge, U.K., 2000.

51. Arpentier P, Cavani F, and Trifiro F. *Technology of Catalytic Oxidations, Vol. 2: Safety Aspects*. Editions TECHNIP, Paris, 2001.

52. Veser G. Experimental and theoretical investigation of H_2 oxidation in a high-temperature catalytic microreactor. *Chemical Engineering Science*, 56:1265–1273, 2001.

53. Miesse CM, Jensen CJ, Masel RI, Shannon MA, and Short M. Sub-millimeter Scale Combustion. *AIChE Journal*, 50:3206, 2004.

54. Middelhoek S and Audet SA. *Silicon Sensors*. TU Delft, Dep. of Electrical Engineering, 1994.

55. Kluge G and Neugebauer G. *Grundlagen der Thermodynamik*. Spektrum Akademischer Verlag, Heidelberg, 1994.

56. Schütte R. *Ullmanns Encyklopädie der technischen Chemie*, volume Vol. 2, Verfahrenstechnik I: (Grundoperationen), chapter Diffusionstrennverfahren. 1972.

57. Becker EW, Ehrfeld W, Hagmann P, Maner A, and Münchmeyer D. Fabrication of microstructures with high aspect ratios and great structural heights by synchrotron radiation lithography, galvanoforming, and plastic moulding (LIGA process). *Microelectronic Engineering*, 4:35–56, 1986.

58. Becker EW, Bier W, Ehrfeld W, Schubert K, Schütte R, and Seidel D. Uranium enrichment by the separation-nozzle process. *Naturwissenschaften*, 63:407 – 411, 1976.

59. Becker EW, Ehrfeld W, Münchmeyer D, Betz H, Heuberger A, Pongratz S, Glashauser W, Michel HJ, and von Siemens R. Production of separation-nozzle systems for uranium enrichment by a combination of X-ray lithography and galvanoplastics. *Naturwissenschaften*, 69:520 – 523, 1982.

60. Heckele M and Schomburg WK. Review on micro molding of thermoplastic polymers. *J. Micromech. Microeng.*, 14:R1–R14, 2004.

61. Schwarzer HC and Peukert W. Combined Experimental/Numerical Study on the Precipitation of Nanoparticles. *AIChE J.*, 50:3234–3247, 2004.

62. Verfahren zur Herstellung von Diketopyrrolopyrrol-Pigmenten. European patent application, EP 1 162 240 A2.

63. Verfahren zur Konditionierung von organischen Pigmenten. European patent application, EP 1 167 461 A2.

64. Blass E. *Entwicklung verfahrenstechnischer Prozesse - Methoden, Zielsuche, Lösungssuche, Lösungsauswahl.* Springer-Verlag, Berlin, 1997.

65. Schönbucher A. *Thermische Verfahrenstechnik.* Springer, Berlin, 2002.

66. Geankoplis CJ. *Transport Processes and Separation Process Principles.* Prentice Hall, Upper Saddle River, 2003.

67. Sowata KI and Kusakabe K. Design of microchannels for use in distillation devices. In *IMRET7*, 2003.

68. Fink H and Hampe MJ. Designing and Constructing Microplants. In *IMRET3, 664-673*, 1999.

69. Klemm E, Rudek M, Markowz G, and Schütte R. *Winnacker-Küchler: Chemische Technik - Prozesse und Produkte, Vol. 2*, chapter Mikroverfahrenstechnik, Chap. 8. Wiley-VCH, Weinheim, 2004.

70. Kockmann N and Woias P. Separation Principles in Micro Process Engineering. In *IMRET8, TK129a*, 2005.

71. Schlünder EU and Thurner F. *Destillation,Absorption, Extraktion.* Thieme, Stuttgart, 1986.

72. Schweitzer PA, editor. *Handbook of Separation Techniques for Chemical Engineers.* McGrawHill, New York, 1988.

73. Häberle S, Schlosser HP, Zengerle R, and Ducree J. A Centrifuge-Based Microreactor. In *IMRET8, 129f*, 2005.

74. Tonkovich AL, Jarosch K, Arora R, McDaniel J, Silva L, Daly F, and Litt B. Methanol production FPSO plant concept using multiple microchannel unit operations. In *IMRET9*, 2006.

75. Iwatsubo Y, Yamada M, Yasuda M, and Seki M. Microfluidic device for single distillation system. In *IMRET9, P9*, 2006.

76. Knoche KF and Bošnjaković F. *Technische Thermodynamik*, volume Teil II. Steinkopff, Darmstadt, 1997.

77. Cypes SH and Engstrom JR. Analysis of a toluene stripping process: a comparison between a microfabricated stripping column and a conventional packed tower. *Chem. Eng. J.*, 101:49–56, 2004.

78. Ameel TA, Papautsky I, Warrington RO, Wegeng RS, and Drost MK. Miniaturization Technologies for Advanced Energy Conversion and Transfer Systems. *J. Propulsion Power*, 16:577–582, 2000.

79. Stenkamp VS and teGrotenhuis W. Microchannel Absorption for Portable Heat Pumps. In *IMRET8, 129e*, 2005.

80. Bockhardt HD, Güntzschel P, and Poetschukat A. *Grundlagen der Verfahrenstechnik für Ingenieure.* Deutscher Verlag für Grundstoffindustrie, Stuttgart, 1997.

81. SLS Micro Technology GmbH, 2005.

82. Kang Q, Golubovic NC, Pinto NG, and Henderson HT. An integrated micro ion-exchange separator and detector on a silicon wafer. *Chemical Engineering Science*, 56:3409–3420, 2001.

83. Tice JD, Song H, Lyon AD, and Ismagilov RF. Formation of Droplets and Mixing in Multiphase Microfluidics at Low Values of the Reynolds and the Capillary Numbers. *Langmuir*, 19:9127–9133, 2003.

84. Günther A, Khan SA, Thalmann M, Trachsel F, and Jensen KF. Transport and reaction in microscale segmented gas-liquid flow. *Lab-on-a-Chip*, 4:278–286, 2004.

85. Ehrfeld W, Hessel V, and Löwe H. *Microreactors*. Wiley-VCH, Weinheim, 2000.

86. Okubo Y, Toma M, Ueda H, Maki T, and Mae K. Microchannel devices for the coalescence of dispersed droplets produced in rapid extraction process. *Chem. Eng. J.*, 101:39–48, 2004.

87. teGrotenhuis WE, Cameron R, Butcher MG, Martin PM, and Wegeng RS. Micro Channel Devices for Efficient Contacting of Liquids in Solvent Extraction. In *Separation Science and Technology, PNNL-SA-28743*, 1998.

88. Kusakabe K, Sotowa KI, and Katsuragi H. Internal flow in a droplet formed in a microchannel contactor for solvent extraction. In *IMRET7*, 2003.

89. Baehr HD and Stephan K. *Wärme- und Stoffübertragung*. Springer, Berlin, 4. aufl. edition, 2004.

90. Bejan A. *Convection Heat Transfer*. John Wiley, New York, 2004.

91. Billat S, Glosch H, Kunze M, Hedrich F, Frech J, Auber J, Sandmaier H, Wimmer W, and Lang W. Micromachined inclinometer with high sensitivity and very good stability. *Sensors & Actuators A*, 97-98:125–130, 2002.

92. Barron R. *Cryogenic Systems*. McGraw-Hill, New York, 1966.

93. Frey H, Haefer RA, and Eder FX, editors. *Tieftemperaturtechnologie*. VDI-Verlag, Düsseldorf, 1981.

94. Peterson RB. Size Limits for Regenerative Heat Engines. *Microscale Thermophysical Eng.*, 2:121–131, 1998.

95. Schmidt R. Challenges in Electronic Cooling. In *ICMM2003-1001*, pages 951–959, 2003.

96. Melin T and Rautenbach R. *Membranverfahren – Grundlagen der Modul- und Anlagenauslegung*. Springer, Berlin, 2004.

97. Jaeckel R and Oetjen GW. Molekulardestillation. *Chemie-Ingenieur-Technik*, 21:169–208, 1949.

98. Jorisch W. *Vakuumtechnik in der chemischen Industrie*. Wiley-VCH, Weinheim, 1999.

99. London E, editor. *Separation of Isotopes*. Georges Newnes Ltd., 1961.

100. Albring W. *Elementarvorgänge fluider Wirbelbewegungen*. Akademie-Verlag, Berlin, 1981.

101. Ookawara S, Street D, and Ogawa K. Quantitative Prediction of Separation Efficiency of a Micro-Separator/Classifier by an Euler-Granular Model. In *IMRET8, TK130c*, 2005.

102. Ookawara S, Oozeki N, and Ogawa K. Experimental Benchmark of a Metallic Micro-Separator/Classifier with Representative Hydrocyclone. In *IMRET8, TK129g*, 2005.

103. *Product catalogue*. Little Things Factory, 2005.

104. Woias P. Micropumps. *Sens. Act. B*, 105:28–38, 2005.

105. Laser DJ and Santiago JG. A review of micropumps. *J. Micromech. Microeng.*, 14:R35–R64, 2004.

106. HNP Mikrosysteme GmbH, 2005.

107. Tuchbreiter A, Marquardt J, Kappler B, Honerkamp J, Kristen MO, and Mülhaupt R. High-Output Polymer Screening: Exploiting Combinatorial Chemistry and Data Mining Tools in Catalyst and Polymer Development. *Macromol. Rapid Commun.*, 24:48–62, 2003.

108. IMRET8: 8th Int. Conf. on Microreaction Technology. In *AIChE National Spring Meeting, Atlanta,*, 2005.

109. IMRET9: 9th Int. Conf. on Microreaction Technology. In *DECHEMA, Potsdam, Germany*, 2006.

110. ICNMM2006: 4th Int. Conf. on Nanochannels, Microchannels, and Minichannels, Limerick, Ireland. In *S.G. Kandlikar, M. Davies, ASME*, 2006.

111. ACHEMA2006: DECHEMA World Forum for the Process Industries, 28. International Exhibition – Congress on Chemical Engineering, Environmental Protection and Biotechnology, Frankfurt am Main, Germany. 2006.

112. Bayer T and Kinzl M. *Micro Process Engineering*, chapter Industrial Applications in Europe, Chapter 14. Wiley-VCH, Weinheim 2006, 2006.

113. Matlosz M. Mikroverfahrenstechnik: Neue Herausforderungen für die Prozessintensivierung. *Chemie Ingenieur Technik*, 77:1393–1398, 2005.

114. Henke L and Winterbauer H. Modularer Mikroreaktor zur Nitrierung mit Mischsäure. *Chemie Ingenieur Technik*, 76:1783–1790, 2004.

115. Kirschneck D and Marr R. Anlagenkonzepte in der Mikroverfahrenstechnik. *Chemie Ingenieur Technik*, 78:29 – 38, 2006.

116. Palo DR, Stenkamp VS, Dagle RA, and Jovanovics GN. *Micro Process Engineering*, chapter Industrial Applications of Microchannel Process Technology in the United States, Chapter 13. Wiley-VCH, Weinheim, 2006.

117. Yoshida J and Okamoto H. *Micro Process Engineering*, chapter Industrial Production Plants in Japan and Future Developments, Chapter 15. Wiley-VCH, Weinheim, 2006.

118. Yoshida J, Nagaki A, Iwasaki T, and Suga S. Enhancement of Chemical Selectivity by Microreactors. *Chem. Eng. Technol.*, 28:259–266, 2005.

119. Mae K. Advanced chemical processing using microspace. *Chemical Engineering Science*, 2007.

120. Newsletter IMMage January 2006. Technical report, IMM Mainz, 2006.

121. Jeon MK, Kim JH, Noh J, Kim SH, Park HG, and Woo SI. Design and characterization of a passive recycle micromixer. *J. Micromech. Microeng*, 15:346–350, 2005.

122. Hessel V. Novel Tools - Novel Chemistry. *Chemical Engineering & Technology*, 30:289, 2007.

123. Klemm W, Ondruschka B, Köhler M, and Günther M. *Micro Process Engineering*, chapter Laboratory applications of microstructured devices in student education, Chapter 16. Wiley-VCH, Weinheim, 2006.

124. Bourne JR. Mixing and the Selectivity of Chemical Reactions. *Organic Process Research & Development*, 7:471–508, 2003.

125. Ottino JM. *The kinematics of mixing: stretching, chaos, and transport.* Cambridge University Press, 1997 reprint.

126. Wünsch O. *Strömungsmechanik des laminaren Mischens*. Springer, Berlin, 2001.

127. Engler M. *Simulation, Design, and Analytical Modelling of Passive Convective Micromixers for Chemical Production Purposes*. PhD thesis, PhD thesis, University of Freiburg, IMTEK, Shaker Verlag, Aachen, 2006.

128. Kockmann N, Kiefer T, Engler M, and Woias P. Convective Mixing and Chemical Reactions in Microchannels with High Flow Rates. *Sensors & Actuators B*, 117:495–508, 2006.

129. Tonomura O. *Micro Process Engineering*, chapter Simulation and Analytical Modeling for Microreactor Design, Chapter 8. Wiley-VCH, Weinheim, 2006.

130. Thoma J and Ould Bouamama B. *Modelling and Simulation in Thermal and Chemical Engineering*. Springer, Berlin, 2000.

131. Löbbecke S. *Micro Process Engineering*, chapter Integration of Sensors and Process-analytical Techniques, Chapter 9. Wiley-VCH, Weinheim, 2006.

132. Schirrmeister S, Brandner JJ, and Kockmann N. *Micro Process Engineering*, chapter Design Process and Project Management , Chapter 7. Wiley-VCH, Weinheim, 2006.

133. Kockmann N, Kiefer T, Engler M, and Woias P. Silicon microstructures for high throughput mixing devices. *Microfluidics and Nanofluidics*, 2:327–335, 2006.

134. Emich F. *Lehrbuch der Mikrochemie*. Bergmann, Wiesbaden, 1911.

135. Baerns M, Hofmann H, and Renken A. *Chemische Reaktionstechnik*. Wiley-VCH, Weinheim, 3rd edition, 2002.

136. Hoffmann M, Schlüter M, and Räbiger N. Experimental Investigation of Mixing in a Rectangular Cross-section Micromixer. In *AIChE Annual Meeting, paper 330b*, 2004.

137. Seider WD, Seader JD, and Lewin DR. *Product and Process Design Principles: Synthesis, Analysis and evaluation*. Wiley, New York, 2004.

138. Lewin DR, Seider WD, Seader JD, Dassau J, Golbert J, Goldberg DN, Fucci MJ, and Nathanson RB. *Using Process Simulators in Chemical Engineering (CD-ROM)*. Wiley, New York, 2003.

139. Löwe A. *Chemische Reaktionstechnik*. Wiley-VCH, Weinheim, 2001.

140. Wilkes JO. *Fluid Mechanics for Chemical Engineers*. Prentice Hall, New York, 2006.

141. Hagen J. *Chemiereaktoren - Auslegung und Simulation*. Wiley-VCH, Weinheim, 2004.

142. Himmelblau DM and Riggs JB. *Basic Principles and Calculation in Chemical Engineering*. Prentice Hall, Upper Saddle River, 2004.

143. Benitez J. *Principles and Modern Applications of Mass Transfer Operations*. John Wiley, New York, 2002.

144. Kutter JP and Fintschenko Y, editors. *Separation Methods in Microanalytical Systems*. CRC Press, Boca Raton, 2006.

145. Pahl G and Beitz W. *Konstruktionslehre – Methoden und Anwendung*. Springer, Berlin, 1997.

146. Schlünder, E.U. *Einführung in die Stoffübertragung*. Vieweg, Braunschweig, 1997.

147. Bird GA. *Molecular Gas Dynamics and the Direct Simulation of Gas Flows*. Clarendon Press, Oxford, 1994.

148. Sommerfeld A. *Vorlesungen über Theoretische Physik*, volume Band V, Thermodynamik und Statistik. Verlag Harry Deutsch, Thun, 1988.

149. Rowlinson JS. The Maxwell-Boltzmann distribution. *Molecular Physics*, 103:2821–2828, 2005.

150. Reif F. *Fundamentals of Statistical and Thermal Physics*. McGraw-Hill Intern'l Ed., 1985.

151. Barber RW and Emerson DR. Challenges in Modeling Gas-Phase Flow in Microchannels: From Slip to Transition. In *ICNMM2005-75074*, 2005.

152. Gad-el-Hak M. The Fluid Mechanics of Microdevices - The Freeman Scholar Lecture. *J. Fluids Engineering*, 121:5–33, 1999.

153. Chapman S, Cowling TG, and Burnett D. *The Mathematical Theory of Non-Uniform Gases*. University Press, Cambridge, 1970.

154. Sone Y. *Kinetic Theory and Fluid Dynamics*. Birkhäuser, Boston, 2002.

155. Vasudevaiah M and Balamurugan K. Heat transfer of rarefied gases in a corrugated microchannel. *Int. J. Therm. Sci*, 40:454–468, 2001.

156. Jones D. Crookes flies high. *Nature*, 392:337, 1998.

157. Jones D. High radiation levels. *Nature*, 392:443, 1998.

158. Jones D. Cruise micromissiles. *Nature*, 395:120, 1998.

159. Goodman FO and Wachman HY. *Dynamics of Gas-Surface Scattering*. Academic Press, New York, 1976.

160. Nedea SV, Frijns AJH, Steenhoven AA, Jansen APJ, Markvoort AJ, and Hilbers PAJ. Density distribution for a dense hard-sphere gas in micro/nano-channels: Analytical and simulation results. *J. Comp. Phys.*, 219:532–552, 2006.

161. Passian A, Warmack RJ, Ferrell TL, and Thundat T. Thermal Transpiration at the Microscale: A Crookes Cantilever. *Physical Review Letters*, 90:124503, 2003.

162. Landau LD and Lifschitz EM. *Lehrbuch der theoretischen Physik*, volume Band X, Physikalische Kinetik. Akademie-Verlag, Berlin, 1983.

163. Cercignani C. *The Boltzmann Equation and its Applications*. Springer, New York, 1987.

164. Ferziger JH and Kaper HG. *Mathematical theory of transport processes in gases*. Elsevier, Amsterdam, 1972.

165. Brush SG. Boltzmann's "Eta Theorem": Where's the Evidence? *Am. J. Physics*, 35:892, 1967.

166. Boltzmann L. *Entropie und Wahrscheinlichkeit (1872 - 1905)*. Ostwalds Klassiker der exakten Wissenschaften, Band 286, Verlag Harri Deutsch, Frankfurt, 2000.

167. Cercignani C. *Rarefied Gas Dynamics*. Cambridge University Press, Cambridge, 2000.

168. Krafzyk M. *Gitter-Boltzmann-Methoden: Von der Theorie zur Anwendung*. Habilitationsschrift, Technical University of Munich, 2001.

169. Sukop MC and Thorne DT. *Lattice Boltzmann Modeling*. Springer, Berlin, 2006.

170. Babovsky H. *Die Boltzmann-Gleichung: Modellbildung - Numerik - Anwendungen*. Teubner, Stuttgart, 1998.

171. Herwig H. Die irreführende Verwendung der thermodynamischen Größe Enthalpie – ein didaktischer Sündenfall. *Forschung im Ingenieurwesen*, 71:107–112, 2007.

172. Volz S, editor. *Microscale and Nanoscale Heat Transfer*. Springer, Berlin, 2007.

173. Bejan A. *Advanced Engineering Thermodynamics*. John Wiley, New York, 1997.

174. Szargut J, Morris DR, and Steward FR. *Exergy analysis of thermal, chemical, and metallurgical processes*. Hemisphere Publ. Group, Springer, Berlin, 1988.

175. Bejan A. *Entropy Generation Minimization*. CRC PRess, Boca Raton FL, 1996.

176. Shinskey FG. *Energy Conservation through Control*. Academic Press, New York, 1978.

177. Linnhoff B. Pinch Analysis - A State-of-the-Art Overview. *Chemical Engineering Research and Design*, 71a:503–522, 1993.

178. Brodkey RS and Hershey HC. *Transport Phenomena – A Unified Approach, Vol. 1 Basic Concepts*. Brodkey Publishing, Columbus, OH, 2001.

179. Brodkey RS and Hershey HC. *Transport Phenomena – A Unified Approach, Vol. 2 Applications, Transport Property*. Brodkey Publishing, Columbus, OH, 2001.

180. Deen WM. *Analysis of Transport Phenomena*. Oxford University Press, New York, 1998.

181. Kraume M. *Transportvorgänge in der Verfahrenstechnik: Grundlagen und apparative Umsetzungen*. Springer, Berlin, 2004.

182. Haase R. *Transportvorgänge*. Steinkopff, Darmstadt, 1973.

183. Wothe D. *Messen des Stofftransports durch die Phasengrenze zweier nicht mischbarer Flüssigkeiten*. PhD thesis, University of Hannover, 2006.

184. Boon JP and Yip S. *Molecular Hydrodynamics*. McGraw-Hill, New York, 1980.

185. Kittel C. *Physik der Wärme*. Oldenbourg, München, 1973.

186. Peters MH. *Molecular Thermodynamics and Transport Phenomena*. McGraw Hill, New York, 2005.

187. Frenkel J. *Kinetic theory of liquids*. Dover, New York, 1955.

188. Gad-el-Hak M, editor. *The MEMS Handbook*. CRC Press, Boca Raton, 2003.

189. Narayanan R and Schwage D, editors. *Interfacial Fluid Dynamics and Transport Processes*. Springer, Berlin, 2003.

190. Mersmann A. *Thermische Verfahrenstechnik – Grundlagen und Methoden*. Springer, Berlin, 1980.

191. Mitrovic J. On the equilibrium conditions of curved interfaces. *International Journal of Heat and Mass Transfer*, 47:809–818, 2004.

192. Joseph DD and Preziosi L. Heat Waves. *Rev. Modern Physics*, 61:41–73, 1989.

193. Chang CW, Okawa D, Majumdar A, and Zettl A. Solid-State Thermal Rectifier. *Science*, 314:1121–1124, 2006.

194. Müller I and Ruggeri T. *Rational Extended Thermodynamics*. Springer, New York, 1998.

195. Jou D, Casas-Vazquez J, and Lebon G. *Extended Irreversible Thermodynamics*. Springer, Berlin, 2001.

196. Zierep J. *Änlichkeitsgesetze und Modellregeln der Strömungslehre*. Braun, Karlsruhe, 1972.

197. Wetzler H. *Mathematisches System für die universelle Ableitung der Kennzahlen*. Hüthig, Heidelberg, 1987.

198. Kaspar M. *Mikrosystementwurf, Entwurf und Simulation von Mikrosystemen*. Springer, Berlin, 2000.

199. Senturia SD. *Microsystem Design*. Kluwer, Boston, 2000.

200. Schaedel HM. *Fluidische Bauelemente und Netzwerke*. Vieweg, Braunschweig, 1979.

201. Gerlach G and Dötzel W. *Grundlagen der Mikrosystemtechnik*. Hanser, München, 1997.

202. Nguyen NT. *Mikrofluidik - Entwurf, Herstellung und Charakterisierung*. Teubner, Stuttgart, 2004.

203. Nguyen NT and Werely ST. *Fundamentals and Applications of Microfluidics*. Artech House, Boston, 2002.

204. Karniadakis GE and Beskok A. *Microflows - Fundamentals and Simulations*. Springer, New York, 2002.

205. Faghri M and Sunden B, editors. *Heat and Fluid Flow in Microscale and Nanoscale Structures*. WIT press, Southampton, 2004.

206. Succi, S. *The Lattice Boltzmann Equation for Fluid Dynamics and Beyond*. Oxford University Press, New York, 2001.

207. Doraiswamy LK and Kulkarni BD. *The Analysis of Chemically Reacting Systems: A Stochastic Approach*. Gordon and Breach Science Publ., New York, 1987.

208. Gardiner CW. *Handbook of Stochastic Methods*. Springer, Berlin, 2004.

209. Versteeg HK and Malalasekera W. *An Introduction to Computational Fluid Dynamics - The Finite Volume Method*. Prentice Hall, Harlow, 1995.

210. Stephenson G. *Partial Differential Equations for Scientists and Engineers*. Imperial College Press, London, 1985.

211. Braun M. *Differential Equations and Their Applications*. Springer, New York, 1993.

212. Shah RK and London AL. *Laminar forced convection in ducts*. Academic Press, New York, 1978.

213. Weigand B. *Analytical Methods for Heat Transfer and Fluid Flow Problems*. Springer, Berlin, 2004.

214. Kays WM, Crawford ME, and Weigand B. *Convective Heat and Mass Transfer*. McGrawHill, Boston, 2005.

215. Li BQ. *Discontinuous Finite Elements in Fluid Dynamics and Heat Transfer*. Springer, London, 2006.

216. Ferziger JH and Peric M. *Computational Methods for Fluid Dynamics*. Springer, Berlin, 1999.

217. Polifke W and Kopitz J. *Wärmeübertragung - Grundlagen, analytische und numerische Methoden*. Pearson Studium, München, 2005.

218. Schubert J. *Dictionary of Effects and Phenomena in Physics Descriptions, Applications, Tables*. VCH, Weinheim, 1987.

219. Koller R. *Konstruktionslehre für den Maschinenbau*. Springer, Berlin, 1998.

220. nbm. Creativity techniques for engineers, 2004.

348   References

221. Buzan T. *Mind maps at work*. HarperCollins, 2004.

222. Otto A. Kreativität. *Psychologie heute*, 32:8, 2005.

223. Bernecker G. *Planung und Bau verfahrenstechnischer Anlagen*. VDI-Verlag, Düsseldorf, 1980.

224. Claussen U and Rodenacker WG. *Maschinensystematik und Konstruktionsmethodik - Grundlagen und Entwicklung moderner Methoden*. Springer, Berlin, 1998.

225. Dym CL and Little P. *Engineering Design - A project-based introduction*. John Wiley, New York, 1999.

226. Orloff MA. *Inventive Thinking through TRIZ - A Practical Guide*. Springer, Berlin, 2003.

227. Retseptor G. 40 Inventive Principles in Microelectronics. *TRIZ Journal*, 8, 2002.

228. Hipple J. 40 Inventive Principles with Examples for Chemical Engineering. *TRIZ Journal*, 6, 2005.

229. Marz J. *Mikrospezifischer Produktentwicklungsprozess (μPEP) für werkzeuggebundene Mikrotechniken*. PhD thesis, Institut für Produktentwicklung, Universität Karlsruhe, 2005.

230. Herwig H. *Strömungsmechanik*. Springer, Berlin, 2002.

231. Landau LD and Lifschitz EM. *Lehrbuch der theoretischen Physik*, volume Band VI, Hydrodynamik. Akademie-Verlag, Berlin, 1981.

232. Oertel H. *Strömungsmechanik - Methoden und Phänomene*. Springer, Berlin, 1995.

233. Schlichting H. *Boundary-Layer Theory*. McGraw-Hill, New York, 1968.

234. Foias C, Manley O, Rosa R, and Temam R. *Navier-Stokes Equations and Turbulence*. Cambridge University Press, Cambridge, 2001.

235. Hoyer U. Über das Turbulenzproblem. *philosophia naturalis*, 20:536–537, 1983.

236. Geier M, Greiner A, and Korvink JG. Cascaded digital lattice Boltzmann automata for high Reynolds number flow. *Phys. Rev. E*, 73:066705, 2006.

237. Gorban AN and Karlin IV. *Invariant Manifolds for Physical and Chemical Kinetics*. Springer, Berlin, 2005.

238. Gersten K and Herwig H. *Strömungsmechanik*. Vieweg, Braunschweig, 1992.

239. Beitz W and Grote KH. *Dubbel - Taschenbuch für den Maschinenbau*. Springer, Berlin, 2001.

240. Kockmann N, Föll C, and Woias P. Flow regimes and mass transfer characteristics in static micro mixers. In *SPIE Photonics West, Micromachining and Microfabrication, 4982-38, San Jose, USA*, 2003.

241. Petersen S, Costaschuk D, Elsnab J, Klewicki J, and Ameel T. Microtube fluid dynamic characterisitics with MTV qualified flow conditions. In *ICMM2005-75096*, 2005.

242. Taylor JB, Carrano AL, and Kandlikar SG. Characteriszation of the effect of surface roughness and texture on fluid flow - past, present, and future. In *ICNMM2005-75075*, 2005.

243. Gersten K. *Einführung in die Strömungsmechanik*. Vieweg, Braunschweig, 1989.

244. Brodkey RS. *The Phenomena of Fluid Motions*. Dover, New York, 1995.

245. Massey BS. *Mechanics of Fluids, 6th ed.* Chapman and Hall, London, 1989.

246. Becker E. *Gasdynamik*. Teubner, Stuttgart, 1966.

247. Iancu F and Müller N. Efficiency of shock wave compression in a microchannel. *Microfluid Nanofluid*, 2:50–63, 2006.

248. Eckert ERF. *Wärme- und Stoffübertragung*. Springer, Berlin, 1966.

249. Hetsroni G, Mosyak A, Pogrebnyak E, and Yarin LP. Fluid flow in micro-channels. *Int. J. Heat Mass Transfer*, 48:1982–1998, 2005.

250. Erbay LB, Yalcin MM, and Ercan MS. Entropy generation in parallel plate microchannels. *Heat Mass Transfer*, 43:729–739, 2007.

251. Kock F. *Bestimmung der lokalen Entropieproduktion in turbulenten Strömungen und deren Nutzung zur Bewertung konvektiver Transportprozesse*. Dissertation, TU Hamburg-Harburg, 2004.

252. Knoeck H. *Fundamentals of Laminar Fluid Flow*. BJJS & Associates P/L, Kerrimuir, Australia, 2000.

253. Natrajan VK and Christensen KT. Microscopic particle image velocimetry measurements of transition to turbulence in microscale capillaries. *Exp. Fluids*, 43:1–16, 2007.

254. Li H and Olsen MG. MicroPIV measurements of turbulent flow in square microchannels with hydraulic diameters from 200 μm to 640 μm. *International Journal of Heat and Fluid Flow*, 27:123–134, 2006.

255. Hof B, Westerweel J, Schneider TM, and Eckhardt B. Finite lifetime of turbulence in shear flows. *Nature*, 443:59–62, 2006.

256. Dean WR. Note on the motion of a fluid in a curved pipe. *Philosophical Magazine*, 4:208–223, 1927.

257. Schönfeld F and Hardt S. Simulation of Helical Flows in Microchannels. *AIChE Journal*, 50:771–778, 2004.

258. Jiang F, Drese KS, Hardt S, Küpper M, and Schönfeld F. Helical Flows and Chaotic Mixing in Curved Micro Channels. *AIChE Journal*, 50:2297–2305, 2004.

259. Thomson DL, Bayazitoglu Y, and Meade AJ. Series solution of low Dean and Germano number flows in helical rectangular ducts. *Int. J. Therm. Sci.*, 40:937–948, 2001.

260. Sudarsan AP and Ugaz VM. Fluid mixing in planar spiral microchannels. *Lab on a Chip*, 6:74–82, 2006.

261. Kumar V, Aggarwal M, and Nigam KDP. Mixing in curved tubes. *Chemical Engineering Science*, 61:5742–5753, 2006.

262. Pathak JA, Ross D, and Migler KB. Elastic flow instability, curved streamlines, and mixing in microfluidic flows. *Physics of Fluids*, 16:4028–40334, 2004.

263. Herwig H, Hölling M, and Eisfeld T. Sind Sekundärströmungen noch zeitgemäß? *Foschung im Ingenieurwesen*, 69:115–119, 2005.

264. Adler M. Strömung in gekrümmten Rohren. *ZAMM*, 14:257–275, 1934.

265. Nippert H. Über den Strömungsverlust in gekrümmten Kanälen. *Forsch. Ing. Wes., VDI-Verlag Berlin*, 320, 1929.

266. Berger SA, Talbot L, and Yao LS. Flow in curved pipes. *Ann. Rev. Fluid Mech.*, 15:461–512, 1983.

267. Kuchling H. *Physik*. VEB Fachbuchverlag Leipzig, 1985.

268. Brauer H. *Grundlagen der Einphasen- und Mehrphasenströmungen*. Sauerländer AG, Aarau, 1971.

269. Kiefer T. *Design and Fabrication of Highly Efficient Micromixers*. Diploma thesis, University of Freiburg, IMTEK, 2005.

270. Kockmann N, Engler M, Föll C, and Woias P. Liquid Mixing in Static Micro Mixers with Various Cross Sections. In *ASME, 1st Int. Conf. Micro and Minichannels [ICMM-1121], 911-918*, 2003.

271. Kockmann N, Engler M, and Woias P. Theoretische und experimentelle Untersuchungen der Mischvorgänge in T-förmigen Mikroreaktoren – Teil 3: Konvektives Mischen und chemische Reaktionen. *Chemie Ingenieur Technik*, 76:1777–1783, 2004.

272. Roache PJ. Quantification of uncertainty in computational fluid dynamics. *Annu. Rev. Fluid Mech.*, 29:123–160, 1997.

273. Hoffmann M, Schlüter M, and Räbiger N. Experimental investigation of liquid-liquid mixing in T-shaped micro-mixers using μ-LIF and μ-PIV. *Chemical Engineering Science*, 61:2968–2976, 2006.

274. Gobby D, Angeli P, and Gavriilidis A. Mixing characteristics of T-type microfluidic mixers. *J. Micromech. Microeng.*, 11:126–132, 2001.

275. Schlüter M, Hoffmann M, and Räbiger N. Theoretische und experimentelle Untersuchungen der Mischvorgänge in T-förmigen Mikroreaktoren – Teil 2: Experimentelle Untersuchung des Strömungsmischens. *Chemie Ingenieur Technik*, 76:1682–1688, 2004.

276. Herrmann M. A Eulerian level set/vortex sheet method for two-phase interface dynamics. *Journal of Computational Physics*, 203:539–571, 2005.

277. Williamson CHK and Govardhan R. Vortex-Induced Vibrations. *Annu. Rev. Fluid Mech.*, 36:413–455, 2004.

278. Deshmukh SR and Vlachos DG. Novel Micromixers Driven by Flow Instabilities: Application to Post-Reactors. *AIChE Journal*, 51:3193–3204, 2005.

279. wikipedia.de. http://de.wikipedia.org/wiki/Strouhal-Zahl, 2007.

280. Tritton DJ. *Physical Fluid Dynamics*. Oxford University Press, 1988.

281. Günther A and Jensen KF. Multiphase microfluidics: from flow characteristics to chemical and materials synthesis. *Lab on a Chip*, 6:1487–1503, 2006.

282. Cheng L and Mewes D. Review of two-phase flow and flow boiling of mixtures in small and mini channels. *Int. J. Multiphase Flow*, 32:183–207, 2006.

283. Jensen MK, Peles Y, Borca-Tasciuc T, and Kandlikar SG. *Micro Process Engineering*, chapter Chapter 4: Multiphase Flow, Evaporation, and Condensation at the Microscale, pages 115–147. Wiley-VCH, Weinheim, 2006, 2006.

284. Kawaji M and Chung PMY. Adiabatic Gas-Liquid Flow in Microchannels. *Microscale Thermophysical Engineering*, 8:239–257, 2004.

285. Kawahara A, Chung PMY, and Kawaji M. Investigation of two-phase flow pattern, void fraction and pressure drop in a microchannel. *Int. J. Multiphase Flow*, 28:1411–1435, 2002.

286. Bar-Cohen A and Rahim E. Modeling and Prediction of Two-Phase Refrigerant Flow Regimes and Heat Transfer Characteristics in Microgap Channels. In *ICNMM2007-30216*, 2007.

287. Yang Y and Fujita Y. Flow Boiling Heat Transfer and Flow Pattern in Rectangular Channel of Mini-Gap. In *ICMM2004-2383*, 2004.

288. Coleman JW and Garimella S. Two-Phase Flow Regimes in Round, Square and Rectangular Tubes During Condensation of Refrigerant R134a. *Int. J. Refrig.*, 26:118–129, 2003.

289. Wu HY, Cheng P, Siu BCP, and Lee YK. Alternating condensation flow patterns in microchannels. In *ICMM2004-2394*, 2004.

290. Qu W and Mudawar I. Measurement and correlation of critical heat flux in two-phase micro-channel heat sinks. *International Journal of Heat and Mass Transfer*, 47:2045–2059, 2004.

291. Wayner PC. Effects of Interfacial Phenomena and Conduction on an Evaporating Meniscus. In *ICNMM2007-30021*, 2007.

292. Gokhale SJ, Plawsky JL, and Wayner PC. Spreading, Evaporation, and Contact Line Dynamics of Surfactant-Laden Microdrops. *Langmuir*, 21:8188–8197, 2005.

293. Garimella S, Killion JD, and Coleman JW. An Experimentally Validated Model for Two-Phase Pressure Drop in the Intermittent Flow Regime for Circular Microchannels. *J. Fluids Eng.*, 124:205–214, 2002.

294. Friedlander SK. *Smoke, Dust and Haze – Fundamentals of Aerosol Dynamics*. Oxford University Press, Oxford, 2000.

295. Schuchmann HP and Danner T. Emulgieren: Mehr als nur Zerkleinern. *Chemie Ingenieur Technik*, 76:364–375, 2004.

296. Bergveld P and Sibbald A. *Analytical and Biomedical Applications of Ion-Selective Field-Effect Transistors*, volume Band 23 in Comprehensive Analytical Chemistry. Elsevier, Amsterdam, 1988.

297. Noy A, Vezenov DV, and Lieber CM. Chemical Force Microscopy. *Annu. Rev. Mater. Sci*, 27:381–421, 1997.

298. Verein Deutscher Ingenieure and GVC VDI-Gesellschaft Verfahrenstechnik und Chemieingenieurwesen, editors. *VDI-Wärmeatlas*. Springer, Berlin, VDI-Verlag, Düsseldorf, Berlin, 8. aufl. edition, 1997.

299. Lienhardt IV JH and Lienhardt V JH. *A Heat Transfer Textbook*. Phlogiston Press, Cambridge, 2003.

300. Hewitt GF, editor. *Heat Exchanger Design Handbook HEDH*. Begell House, New York, 2002.

301. Karniadakis GE, Beskok A, and Aluru N. *Microflows - Fundamentals and Simulations*. Springer, New York, 2006.

302. Zohar Y. *Heat Convection in Micro Ducts*. Kluwer, Boston, 2003.

303. Kandlikar SG, Garimella S, Li D, Colin S, and King MR. *Heat Transfer and Fluid Flow in Minichannels and Microchannels*. Elsevier, Amsterdam, 2005.

304. Garimella SV and Singhal V. Single-Phase Flow and Heat Transport in Microchannel Heat Sinks. In *ICMM03-1018*, pages 159–169, 2004.

305. Palm B. Heat transfer in microchannels. *Microscale Thermophys. Eng.*, 5:155–175, 2001.

306. Celata GP. Single-Phase Heat Transfer and Fluid Flow in Micropipes. In *ICMM2003-1019*, pages 171–179, 2003.

307. Koo J and Kleinstreuer C. Analysis of surface roughness effects on heat transfer in micro-conduits. *Int. J. Heat Mass Transfer*, 48:2625–2634, 2005.

308. Knudsen M. *Kinetic Theory of Gases*. Wiley, London, 1950.

309. Frohn A. *Einführung in die Kinetische Gastheorie*. Aula-Verlag, Wiesbaden, 1988.

310. Hadjiconstantinou NG and Simek O. Constant-Wall-Temperature Nusselt Number in Micro and Nano Channels. *J. Heat Transfer*, 124:356–364, 2002.

311. van Oudheusden BW. Silicon thermal flow sensors. *Sensors and Actuators A*, 30:5–26, 1992.

312. van Herwaarden AW and Sarro PM. Thermal sensors based on the Seebeck effect. *Sensors and Actuators*, 10:321–346, 1986.

313. Ashauer M, Glosch H, Hedrich F, Hey N, Sandmaier H, and Lang W. Thermal flow sensors for liquids and gases based on combinations of two principles. *Sensors and Actuators A*, 73:7–13, 1999.

314. Buchner R, Sosna C, Maiwald M, Benecke W, and Lang W. A high-temperature thermopile fabrication process for thermal flow sensors. *Sensors and Actuators A*, 130:262–266, 2006.

315. Al khalfioui M, Michez A, Giani A, Boyer A, and Foucaran A. Anemometer based on Seebeck effect. *Sensors and Actuators A*, 107:36–41, 2003.

316. van Oudheusden BW. The thermal modelling of a flow sensor based on differential convective heat transfer. *Sensors and Actuators A*, 29:93–106, 1991.

317. Welty JR, Wicks CE, and Wilson RE. *Fundamentals of Momentum, Heat, and Mass Transfer 3rd ed.* Wiley, 1984.

318. Baltes H, Paul O, and Brand O. Micromachined Thermally Based CMOS Microsensors. *Proc. IEEE*, 86, 8:1660–1678, 1998.

319. Rowe DM. *Thermoelectrics Handbook - Macro to Nano*, chapter General Principles and Theoretical Considerations. CRC, Tayler&Francis, Boca Raton, 2006.

320. Stroock, A.D., Dertinger, S.K.W., Ajdari, A., Mesic, I., Stone, H.A., and Whitesides, G.M. Chaotic Mixer for Microchannels. *Science*, 295:647–651, 2002.

321. Schenk R, Hessel V, Hofmann C, Kiss J, Löwe H, Schönfeld F, and Ziogas A. Numbering up von Mikroreaktoren: Ein neues Flüssigkeitsverteilsystem. *Chemie Ingenieur Technik*, 76:584–597, 2004.

322. Markowz G, Schirrmeister S, Albrecht J, Becker F, Schütte R, Caspary KJ, and Klemm E. Mikrostrukturierte Reaktoren für heterogen katalysierte Gasphasenreaktionen im industriellen Maßstab. *Chemie Ingenieur Technik*, 76:620–625, 2004.

323. Tonomura O, Kano M, Hasebe S, and Noda M. Optimal design approach for microreactors with desired temperature, concentration and residence time distributions. In *Proc. of IMRET8, Atlanta, paper TK132u*, 2005.

324. Tonomura O, Noda M, Kano M, and Hasebe S. Thermo-fluid Design Approach to Microreactors with Uniform Temperature and Residence Time Distribution. In *Proc. of AIChE Annual Meeting 2004, Austin, paper 430d*, 2004.

325. Schlünder EU. *Einführung in die Wärmeübertragung*. Vieweg, Braunschweig, 1991.

326. Kandlikar SG. Heat Flux Removal with Microchannels - A Roadmap of Challenges and Opportunities. In *Proc. of ICNMM2005-75086, Toronto*, 2005.

327. Peles Y, Kosar A, Mishra C, Kuo CJ, and Schneider B. Pressure drop over MEMS-based micro scale fin heat sinks. In *ICNMM2005-75156, Toronto*, 2005.

328. Cheong BCY, Ireland PT, and Siebert AW. High Performance Single Phase Liquid Coolers for Power Electronics. In *ICMM2005-75027, Toronto*, 2005.

329. Overholt MR, McCandless A, Kelly KW, Becnel CJ, and Motakef S. Micro-Jet Arrays for Cooling Electronic Equipment. In *ICMM2005-75250, Toronto*, 2005.

330. Bejan A. *Shape and Structure, From Engineering to Nature*. Cambridge University Press, 2000.

331. Sherman TF. On connecting large vessels to small. The meaning of Murray's law. *J. Gen. Physiology*, 78:431–453, 1981.

332. Emerson DR, Cieslicki K, and Barber RW. Hydrodynamic scaling approaches for the biomimetic design of microfluidic channels. In *ICNMM2006-96196*, 2006.

333. Emerson DR, Cieslicki K, Gu X, and Barber RW. Biomimetic design of microfluidic manifolds based on a generalised Murray's law. *Lab on a Chip*, 6:447–454, 2006.

334. Rosaguti NR, Fletcher DF, and Haynes BS. Low-Reynolds number heat transfer enhancement in sinusoidal channels. *Chemical Engineering Science*, 62:694–702, 2007.

335. Lewins J. Bejan's constructal theory of equal potential distribution. *Int. J. Heat Mass Transfer*, 46:1541–1543, 2003.

336. Favre-Marinet M, Le Person S, and Bejan A. Maximum Heat Transfer Rate Density in Two-Dimensional Minichannels and Microchannels. *Microscale Thermophys. Eng.*, 8:225–237, 2004.

337. Wezler K and Sinn W. *Das Strömungsgesetz des Blutkreislaufes*. Editio Cantor, Aulendorf, 1953.

338. Roth-Nebelsick A, Uhl D, Mosbrugger V, and Kerp H. Evolution and Function of Leaf Venation Architecture: A Review. *Annals of Botany*, 87:553–566, 2001.

339. Roth-Nebelsick A. Die Prinzipien der pflanzlichen Wasserleitung. *Biologie unserer Zeit*, 36:110–118, 2006.

340. Konrad W and Roth-Nebelsick A. The Significance of Pit Shape for Hydraulic Isolation of Embolized Conduits of Vascular Plants During Novel Refilling. *Journal of Biological Physics*, 31:57–71, 2005.

341. West GB, Brown JH, and Enquist BJ. A General Model for the Origin of Allometric Scaling Laws in Biology. *Science*, 276:122–126, 1997.

342. Sack L, Streeter CM, and Holbrook NM. Hydraulic Analysis of Water Flow through Leaves of Sugar Maple and Red Oak. *Plant Physiology*, 134:1824–1833, 2004.

343. Nachtigall W and Blüchel KG. *Das große Buch der Bionik, Neue Technologien nach dem Vorbild der Natur*. Deutsche Verlags-Anstalt, Stuttgart, 2000.

344. Harris C, Despa M, and Kelly K. Design and fabrication of a cross flow micro heat exchanger. *J. MEMS*, 9:502–508, 2000.

345. Maranzana G, Perry I, and Maillet D. Mini- and micro-channels: influence of axial conduction in the walls. *Int. J. Heat Mass Transfer*, 47:3993–4004, 2004.

346. Weigand B and Gassner G. The effect of wall conduction for the extended Graetz problem for laminar and turbulent channel flows. *Int. J. Heat Mass Transfer*, 50:1097–1105, 2007.

347. Foli K, Okabe T, Olhofer M, Jin Y, and Sendhoff B. Optimization of micro heat exchanger: CFD, analytical approach and multi-objective evolutionary algorithms. *Int. J. Heat Mass Transfer*, 49:1090–1099, 2006.

348. *Process Technology of Tomorrow – The Catalogue*. IMM Institut für Mikrotechnik Mainz GmbH, 2003.

349. Wengeler R, Heim M, Wild M, Nirschl H, Kaspar G, Kockmann N, Engler M, and Woias P. Aerosol particle deposition in a T-shaped micro mixer. In *ICMM2005-75199*, 2005.

350. Heinzel V, Jianu A, and Sauter H. Stategies against particle fouling in the channels of a micro heat exchanger subject to µPIV flow pattern measurements. In *ECI Heat Echanger Fouling and Cleaning, Kloster Irsee*, 2005.

351. Wu D, Fang N, Sun C, and Zhang X. Stiction problems in releasing of 3D microstructures and its solution. *Sensors and Actuators A*, 128:109–115, 2006.

352. Heim M, Wengeler R, Nirschl H, and Kasper G. Particle deposition from aerosol flow inside a T-shaped micro-mixer. *J Micromech. Microeng.*, 16:70–76, 2006.

353. Ottino JM . Mixing, Chaotic Advection, and Turbulence. *Ann. Rev. Fluid Mech.*, 22:207–253, 1990.

354. Sturman R, Ottino JM, and Wiggins S. *The Mathematical Foundations of Mixing*. Cambridge University Press, 2006.

355. Ehrfeld W, Golbig K, Hessel V, Löwe H, and Richter T. Characterization of Mixing in Micromixers by a Test Reaction: Single Mixing Units and Mixer Arrays. *Ind. Eng. Chem. Res.*, 38:1075–1082, 1999.

356. Lu LH, Ryu KS, and Liu C. A Novel Microstirrer and Arrays for Microfluidic Mixing. In *Micro Total Analysis Systems, pp. 28-30*, 2001.

357. Nguyen NT and Wu Z. Micromixers – a review. *J. Micromech. Microeng*, 15:R1–R16, 2005.

358. Karlsruhe Research Center, for DSM Linz. Presseinformation 13/2005.

359. Mae K. Advanced Chemical Processing using Micro Space. In *IMRET9, Potsdam*, 2006.

360. Jost W. *Diffusion – Methoden der Messung und Auswertung*. Steinkopff, Darmstadt, 1957.

361. Cussler EL. *Diffusion - Mass Transfer in Fluid Systems*. Cambridge University Press, Cambridge, 1997.

362. Stachel J. *Einsteins Annus mirabilis – Fünf Schriften, die die Welt der Physik revolutionierten, Zweiter Teil*. Rowohlt, Reinbek, 2001.

363. Kraume M, editor. *Mischen und Rühren*. Wiley-VCH, Weinheim, 2003.

364. Zlokarnik M. *Rührtechnik – Theorie und Praxis*. Springer, Berlin, 1999.

365. Bothe D, Stemich C, and Warnecke HJ. Theoretische und experimentelle Untersuchungen der Mischvorgänge in T-förmigen Mikroreaktoren – Teil 1: Numerische Simulation und Beurteilung des Strömungsmischens. *Chemie Ingenieur Technik*, 76:1480–1484, 2004.

366. Bothe D and Warnecke HJ. Berechnung und Beurteilung strömungsbasierter komplex-laminarer Mischprozesse. *Chemie Ingenieur Technik*, 79:1001–1014, 2007.

367. Chao SH, Holl MR, Koschwanez JH, Seriburi P, and Meldrum DR. Scaling for microfluidic mixing. *Proceedings of ICMM2005*, 2005.

368. Schönfeld F, Hessel V, and Hofmann C. An optimised split-and-recombine micro-mixer with uniform 'chaotic' mixing. *Lab Chip*, 4:65–69, 2004.

369. Fulford GF and Broadbridge P. *Industrial Mathematics - Case studies in the diffusion of heat and matter*. Cambridge University Press, 2002.

370. Cybulski A, Moulijn JA, Sharma MM, and Sheldon RA. *Fine Chemicals Manufacture - Technology and Engineering*. Elsevier, Amsterdam, 2001.

371. Baldyga J and Bourne JR. *Turbulent Mixing and Chemical Reactions*. John Wiley, Chichester, 1999.

372. Aoki N, Hasebe S, and Mae K. Design of Fluid Segments in Microreactors: Influences of Design Factors on Mixing by Diffusion and Product Compositions. In *IMRET8, TK132a*, 2005.

373. Engler M, Kockmann N, Kiefer T, and Woias P. Numerical and Experimental Investigations on Liquid Mixing in Static Micromixers. *Chem. Eng. J.*, 101:315–322, 2004.

374. Engler M, Kiefer T, Kockmann N, and Woias P. Effective Mixing by the Use of Convective Micro Mixers. In *IMRET8, TK128d*, 2005.

375. Kockmann N, Kiefer T, Engler M, and Woias P. Silicon microstructures for high throughput mixing devices. In *ICNMM05-75125*, 2005.

376. Kockmann N, Engler M, Haller D, and Woias P. Fluid dynamics and transfer processes in bended micro channels. In *ICMM04-2331*, pages 165–171, 2004.

377. Bökenkamp D, Desai A, Yang X, Tai YC, Marzluff EM, and Mayo SL. Microfabricated Silicon Mixers for Submillisecond Quench-Flow Analysis. *Anal. Chem.*, 70:232–236, 1998.

378. Wong SH, Ward MCL, and Wharton CW. Micro T-Mixer as a Rapid Mixing Micromixer. *Sens. Act. B*, 100:359–379, 2004.

379. Engler M, Föll C, Kockmann N, and Woias P. Investigations of Liquid Mixing in Static Micro Mixers. In *11th European Conference on Mixing, Bamberg, 14-17 October*, pages 277–284, 2003.

380. Hoffmann M, Räbiger N, Schlüter M, Blazy S, Bothe D, Stemich C, and Warnecke A. Experimental and numerical investigations of T-shaped micromixers. In *11th Europ. Conf. on Mixing, Bamberg, Germany*, pages 269–276, 2003.

381. Gobert C, Schwertfirm F, and Manhart M. Lagrangian scalar tracking for laminar micromixing at high Schmidt numbers. In *FEDSM2006-98035*, 2006.

382. Engler M, Kockmann N, Kiefer T, and Woias P. Convective mixing and its application to micro reactors. In *ICMM04-2412*, pages 781–788, 2004.

383. Zlokarnik M. *Stirring, Theory and Practice*. Wiley-VCH, Weinheim, 2001.

384. Ottino JM. Mixing and chemical reactions - A tutorial. *Chemical Engineering Science*, 49:4005–4027, 1994.

385. Aref H and El Naschie MS, editors. *Chaos applied to fluid mixing*. Pergamon, Oxford, 1995.

386. Hessel V and Löwe H. Organic Synthesis with Microstructured Reactors. *Chem. Eng. Technol.*, 28:267–284, 2005.

387. Ottino JM and Wiggins S. Designing Optimal Micromixers. *Science*, 305:485–486, 2004.

388. Stremler MA, Haselton FR, and Aref H. Designing for chaos: applications of chaotic advection at the microscale. *Phil. Trans. R. Soc. Lond. A*, 362:1019–1036, 2004.

389. Tice JD, Song H, Lyon AD, and Ismagilov RF. Formation of Droplets and Mixing in Multiphase Microfluidics at Low Values of the Reynolds and the Capillary Numbers. *Langmuir*, 19:9127–9133, 2003.

390. Gavriliidis A, Agneli P, and Salman W. A CFD model for characterising axial mixing in microreactors operating under Taylor flow. In *IMRET8, TK128c*, 2005.

391. Khan SA, Günther A, Schmidt MA, and Jensen KF. Microfluidic Synthesis of Colloidal Silica. *Langmuir*, 20:8604–8611, 2004.

392. Madou M. *Fundamentals of Microfabrication*. CRC Press, Boca Raton, 1997.

393. Menz W, Mohr J, and Paul O. *Microsystem Technology*. Wiley-VCH, Weinheim, 2001.

394. Lawes RA. Manufacturing costs for microsystems/MEMS using high aspect ratio microfabrication techniques. *Microsyst Technol*, 13:85–95, 2007.

395. Engler M, Kiefer T, Kockmann N, and Woias P. Efficient High-Throughput Micromixers for Production Purposes. In *Int. Conf. Commercialization Micro Nano Systems (COMS)*, 2005.

396. Bayer T, Matlosz M, and Jenck J. IMPULSE - Ein neuartiger Ansatz für die Prozessentwicklung. *Chemie Ingenieur Technik*, 76:528–533, 2004.

397. Fournier MC, Falk L, and Villermaux J. A new parallel competing reaction system for assessing micromixing efficiency - Determination of micromixing time by a simple mixing model. *Chemical Engineering Science*, 51(23):5187–5192, 1996.

398. Guichardon P and Falk L. Characterization of Micromixing Efficiency by the Iodide-Iodate Reaction System. Part I: Experimental Procedure. *Chemical Engineering Science*, 55:4233–4243, 2000.

399. Guichardon P, Falk L, and Villermaux J. Characterization of Micromixing Efficiency by the Iodide-Iodate Reaction System. Part II: Kinetic Study. *Chemical Engineering Science*, 55:4245–4253, 2000.

400. Panić S, Löbbecke S, Türcke T, Antes J, and Bošković D. Experimental Approaches to a Better Understanding of Mixing Performance of Microfluidic Devices. *Chem. Eng. J.*, 101:409–419, 2004.

401. Trippa G and Jachuk RJJ. Characterization of mixing efficiency in narrow channels by using the iodide-iodate reaction system. In *ICMM2004-2413*, pages 789–793, 2004.

402. Kockmann N, Engler M, and Woias P. Convective Mixing and Chemical Reactions in T-shaped Micro Reactors. In *AIChE Annual Meeting*, 2004. Session [330]: Mixing in Microdevices and Microreactors II.

403. Hardt S and Schönfeld F. Laminar mixing in different interdigital micromixers: II: Numerical simulations. *AIChE Journal*, 49:578–584, 2003.

404. Fournier MC, Falk L, and Villermaux J. A new parallel competing reaction system for assessing micromixing efficiency - Experimental approach. *Chemical Engineering Science*, 51(22):5053–5064, 1996.

405. Clicq D, Vervoort N, Ranson W, De Tandt C, Ottevaere H, Barona GV, and Desmet G. Axial dispersion measurements of Bodenstein > 100,000 flows through nano-channels etched on flat surfaces. *Chemical Engineering Science*, 59:2783–2790, 2004.

406. Aris R. *Elementary Chemical Reactor Analysis*. Dover, New York, 1999.

407. Homola J, Yee SS, and Gauglitz G. Surface plasmon resonance sensors: review. *Sensors and Actuators B*, 54:3–15, 1999.

408. Xinglong Y, Dongsheng W, Dingxin W, Hua OYJ, Zibo Y, Yonggui D, Wei L, and Xinsheng Z. Micro-array detection system for gene expression products based on surface plasmon resonance imaging. *Sensors and Actuators B*, 91:133–137, 2003.

409. Biesalski M, Rühe J, Kügler R, and Knoll W. *Handbook of Polyelectrolytes and Their Application, Vol. 1, Chap. 2*, chapter Polyelectrolytes at Solid Surfaces: Multilayers and Brushes. American Scientific Publishers, 2002.

410. Yoon SK, Fichtl GW, and Kenis PJA. Active control of the depletion boundary layers in microfluidic electrochemical reactors. *Lab on a Chip*, 6:1516–1524, 2006.

411. Gervais T and Jensen KF. Mass transport and surface reactions in microfluidic systems. *Chemical Engineering Science*, 61:1102–1121, 2006.

412. Chattopadhyay Sand Veser G. Heterogeneous-Homogeneous Interactions in Catalytic Microchannel Reactors. *AIChE Journal*, 52:2217–2229, 2006.

413. Abramovitz M and Stegun IA. *Handbook of Mathematical Functions*. Dover Publications, New York, 1970.

414. Musiol G and Mühlig H. *Bronstein, Semendjajew - Taschenbuch der Mathematik*. Harry Deutsch, Frankfurt, 1996.

415. Caygill G, Zanfir M, and Gavriilidis A. Scalable Reactor Design for Pharmaceuticals and Fine Chemicals Production. 1: Potential Scale-up Obstacles. *Organic Process Research & Development*, 10:539–552, 2006.

416. Levenspiel O. *Chemical Reaction Engineering*. Wiley, New York, 3rd edition, 1999.

417. Smith R. *Chemical Process - Design and Integration*. Wiley, Chichester, 2005.

418. Hartmann K and Kaplick K. *Analyse und Entwurf chemisch-technologischer Verfahren*. Akademie-Verlag, Berlin, 1985.

419. Felder RM and Rousseau RW. *Elementary Processes of Chemical Processes*. Wiley, Hoboken, 2000.

420. Belfiore LA. *Transport Phenomena for Chemical Reactor Design*. John Wiley, New York, 2003.

421. Rosner DE. *Transport Processes in Chemically Reacting Flow Systems*. Butterworths, Boston, 1986.

422. Brötz W. *Grundriß der chemischen Reaktionstechnik*. Verlag Chemie, Weinheim, 1958.

423. Henning T, Brandner JJ, Eichhorn L, Schubert K, Schreiber M, Güngerich M, Günther M, Klar PJ, Rebbin V, and Fröba M. Selective Adsorption of Solvents in a Multiscale Device. In *ICNMM2006-96190*, 2006.

424. Damköhler G. *Einfluss von Diffusion, Strömung und Wärmetransport auf die Ausbeute bei chemisch-technischen Reaktionen*. Sonderabdruck aus: Der Chemie-Ingenieur, Band 3, 1. Teil, Akademische Verlagsgesellschaft, Leipzig, 1937, 1957.

425. Baerns M, Behr A, Brehm A, Gmehling J, Hofmann H, Onken U, and Renken A. *Technische Chemie*. Wiley-VCH, Weinheim, 2006.

426. Wada Y, Schmidt MA, and Jensen KF. Flow Distribution and Ozonolysis in Gas-Liquid Multichannel Microreactors. *Ind. Eng. Chem. Res.*, 45:8036–8042, 2006.

427. Commenge JM, Falk L, Corriou JP, and Matlosz M. Optimal Design for Flow Uniformity in Microchannel Reactors. *AIChE Journal*, 48:345–358, 2002.

428. Rebrov EV, Ismagilov IZ, Ekatpure RP, de Croon MHJM, and Schouten JC. Header Design for Flow Equalization in Microstructured Reactors. *AIChE Journal*, 53:28–38, 2007.

429. Bourne JR, Kut OM, Lenzner J, and Maire H. Kinetics of the Diazo Coupling between 1-Naphthol and diazotized Sulfanilic Acid. *Industrial Engineering & Chemical Research*, 29(9):1761–1765, 1990.

430. Zollinger H. *Diazo chemistry*. VCH, Weinheim, 1994.

431. Kumar C, editor. *Biological and Pharmaceutical Nanomaterials*. Wiley-VCH, Weinheim, 2006.

432. Wegner K, Stark WJ, and Pratsinis SE. Flame-nozzle synthesis of nanoparticles with controlled size, morphology and crystallinity. *Materials Letters*, 55:318–321, 2002.

433. Borra JP. Nucleation and aerosol processing in atmospheric pressure electrical discharges: powders production, coatings and filtration. *J. Phys. D: Appl. Phys.*, 39:R19–R54, 2006.

434. Sue K, Kimura K, and Arai K. Hydrothermal synthesis of ZnO nanocrystals using microreactor. *Materials Letters*, 58:3229–3231, 2004.

435. Penth B. Nanopartikel aus dem MicroJetReactor. 2005.

436. Wegner K. *Nanoparticle Synthesis in Gas-Phase Systems: Process Design and Scale-up for Metals and Metal Oxides*. PhD thesis, Diss. ETH Zürich, No. 14568, 2002.

437. Schwarzer HC and Peukert W. Experimental Investigation into the Influence of Mixing on Nanoparticle Precipitation. *Chemical Engineering & Technology*, 25(6):657–661, 2002.

438. Schwarzer HC, Schwertfirm F, Manhart M, Schmid HJ, and Peukert W. Predictive simulation of nanoparticle precipitation based on the population balance equation. *Chemical Engineering Science*, 61(1):167–181, 2006.

439. Choi YJ, Chung ST, Oh M, and Kim HS. Investigation of Crystallization in a Jet Y-Mixer by a Hybrid Computational Fluid Dynamics and Process Simulation Approach. *Crystal Growth & Design*, 5(3):959–968, 2005.

440. Shirure VS, Pore AS, and Pangarkar VG. Intensification of Precipitation Using Narrow Channel Reactors: Magnesium Hydroxide Precipitation. *Industrial Engineering & Chemical Research*, 44(15):5500–5507, 2005.

441. Schur M, Bems B, Dassenoy A, Kassatkine I, Urban J, Wilmes H, Hinrichsen O, Muhler M, and Schlögl R. Kontinuierliche Cofällung von Katalysatoren in einem Mikromischer: nanostrukturierte Cu/ZnO-Komposite für die Methanolsynthese. *Angewandte Chemie*, 115(32):3945–3947, 2003.

442. Su YF, Kim H, Qiu H, Halder R, Koven S, and Lee WY. Synthesis of Functionalized Nanoparticles using Microreactor. In *IMRET9, P44*, 2006.

443. Pennemann H, Forster S, Kinkel J, Hessel V, Löwe H, and Wu L. Improvement of Dye Properties of the Azo Pigment Yellow 12 Using a Micromixer-Based Process. *Organic Process Research & Development*, 9(2):188–192, 2005.

444. Schwarzer HC. *Nanoparticle Precipitation – An Experimental and Numerical Investigation Including Mixing*. Dissertation, Universität Erlangen-Nürnberg, 2005. Logos Verlag, Berlin.

445. Söhnel O and Garside J. *Precipitation: Basic principles and industrial applications*. Butterworth-Heinemann, Oxford, 1992.

446. Ghez R. *A primer of diffusion problems*. John Wiley & Sons, New York, 1988.

447. Hounslow MJ, Ryall RL, and Marshall VR. A Discretized Population Balance for Nucleation, Growth, and Aggregation. *AIChE*, 34(11):1821–1832, 1988.

448. Mahoney AW and Ramkrishna D. Efficient solution of population balance equations with discontinuities by finite elements. *Chemical Engineering Science*, 57:1107–1119, 2002.

449. Judat B. *Über die Fällung von Bariumsulfat – Vermischungseinfluss und Partikelbildung*. Shaker Verlag, Aachen, 2003.

450. Schwalbe T, Kursawe A, and Sommer J. Application Report on Operating Cellular Process Chemistry Plants in Fine Chemical and Contract Manufacturing Industries. *Chemical Engineering & Technology*, 28(4):408–419, 2005.

451. Müller I. Zur Thermodynamik irreversibler Prozesse. *Chemie Ingenieur Technik*, 72:194–202, 2000.

452. Haase R. *Thermodynamik der irreversiblen Prozesse*. Steinkopff, Darmstadt, 1963.

453. de Groot SR and Mazur P. *Non-Equilibrium Thermodynamics*. Dover Publications, New York, 1984.

454. Kondepudi D and Prigogine I. *Modern Thermodynamics - From Heat Engines to Dissipative Structures*. Wiley, Chichester, 1998.

455. Demirel Y. *Nonequilibrium Thermodynamics - Transport and Rate Processes in Physical and Biological Systems.* Elsevier, Amsterdam, 2002.

456. Hasselbrink EF. Electrokinetic Pumping and Power Generation: Onsager Reciprocal Processes. In *ICMM2004, panel session*, 2004.

457. Demirel Y and Sandler SI. Linear-nonequilibrium thermodynamics theory for coupled heat and mass transport. *Int. J. Heat Mass Transfer*, 44:2439–2451, 2001.

458. Strasser, M. *Entwicklung und Charakterisierung mikrostrukturierter thermoelektrischer Generatoren in Silizium-Halbleitertechnologie.* Shaker, Aachen, 2004.

459. Pelster R, Pieper R, and Hüttl I. Thermospannungen – viele genutzt und fast immer falsch erklärt! *Physik und Didaktik in Schule und Hochschule*, 1/4:10–22, 2005.

460. Nolas GS, Sharp J, and Goldsmid HJ. *Thermoelectrics - Basic Principles and New Material Developments.* Springer, Berlin, 2001.

461. Kalitzin G. *Thermodynamik irreversibler Prozesse.* VEB Deutscher Verlag der Grundstoffindustrie, Leipzig, 1968.

462. Menz W, Mohr J, and Paul O. *Mikrosystemtechnik für Ingenieure.* Wiley-VCH, Weinheim, 2005.

463. von Arx M. *Thermal Properties of CMOS Thin Films.* PhD thesis, ETH Zürich, PhD Thesis No. 12743, 1998.

464. Rowe DM, editor. *Thermoelectrics Handbook - Macro to Nano.* CRC, Tayler&Francis, Boca Raton, 2006.

465. Joffe AF. The revival of thermoelectricity. *Scientific American*, 199:31–37, 1958.

466. Kishi M, Nemoto H, Yamamoto M, Sudou S, Mandai M, and Yamamoto S. Microthermoelectric modules and their application to wrist-watches as an energy source. In *IEEE Proc. ICT*, pages 301–307, 1999.

467. Böttner H. Thermoelectric Micro Devices: Current State, Recent Developments and Future Aspects for Technological Progress and Applications. In *IEEE Proc. ICT*, pages 511–518, 2002.

468. Böttner H, Nurnus J, Gavrikov A, Kühner G, Jägle M, Künzel C, Eberhard D, Plescher G, Schubert A, and Schlereth KH. New thermoelectric components using microsystem technologies. *J. MEMS*, 13:414–420, 2004.

469. Fleurial JP, Snyder GJ, Herman JA, Giauque PH, Philips WM, Ryan MA, Shakkot-tai P, Kolawa EA, and Nicolet MA. Thick-film thermoelectric microdevices. In *IEEE 18th ICT, 294-300*, 1999.

470. Stordeur M and Stark I. Low power thermoelectric generator: A self-sufficient energy source for sensor systems and microsystems. *MST news*, 21:31–32, 1997.

471. Shiozaki M, Sugiyama S, Watanabe N, Ueno H, and Itoigawa K. Flexible thin-film BiTe thermopile for room temperature power generation. In *MEMS Conference, 946-949*, 2006.

472. Böttner H, Chen G, and Venkatasubramanian R. Aspects of Thin-Film Superlattice Thermoelectric Materials, Devices, and Applications. *MRS bulletin*, 2006:211–217, 31.

473. Nolas GS, Poon J, and Kanatzidis M. Recent Developments in Bulk Thermoelectric Materials. *MRS bulletin*, 31:199–205, 2006.

474. Paul O and Ruther P. *CMOS - MEMS, Vol. 2 of Advanced Micro and Nanosystems*, chapter Material Characterization. Wiley-VCH, 2005.

475. Strasser M, Aigner R, Franosch M, and Wachutka G. Miniaturized thermoelectric generators based on poly-Si and poly-SiGe surface micro-machining. *Sensors and Actuators A*, 97-98:535–542, 2002.

476. Glosch H, Ashauer M, Pfeiffer U, and Lang W. A thermoelectric converter for energy supply. *Sensors and Actuators A*, 74:246–250, 1999.

477. Antes J, Schifferdecker D, Krause H, and Löbbecke S. Ein neues μ-Kalorimeter zur isothermen Bestimmung von thermodynamischen und kinetischen Kenngrößen stark exothermer Reaktionen. *Chemie Ingenieur Technik*, 76:1332–1333, 2004.

478. Schifferdecker D, Antes J, and Löbbecke S. A New Concept for the Measurement of Strong Exothermicities in Microreactors. In *IMRET8, 134b*, 2005.

479. Rencz M, Kollár, and V. Székely. Heat-flux sensor to support transient thermal characterisation of IC packages. *Sensors and Actuators A*, 116:284–292, 2004.

480. Shen CH and Gau C. Heat exchanger fabrication with arrays of sensors and heaters with its micro-scale impingement cooling process analysis and measurements. *Sensors and Actuators A*, 114:154–162, 2004.

481. Baier V, Födisch R, Ihring A, Kessler E, Lerchner J, Wolf G, Köhler JM, Nietzsch M, and Krügel M. Highly sensitive thermopile heat power sensor for micro-fluid calorimetry of biochemical processes. *Sensors and Actuators A*, 123-124:354–359, 2005.

482. Kasap S. *Thermoelectric effects in metals: Thermocouples*. An e-booklet, 1997.

483. Khandurina J and Guttman A. Microchip-based high-throughput screening analysis of combinatorial libraries. *Curr. Opin. Chem. Biol.*, 6:359–366, 2002.

484. Pfeiffer AJ, Mukherjee T, and Hauan S. Design and Optimization of Compact Microscale Electrophoretic Separation Systems. *Ind. Eng. Chem. Res.*, 43:3539–3553, 2004.

485. Kirby J and Hasselbrink EF. Zeta potential of microfluidic substrates: 1. Theory, experimental techniques, and effects on separations. *Electrophoresis*, 25:187–202, 2004.

486. Brunet E and Ajdari A. Generalized Onsager relations for electrokinetic effects in anisotropic and heterogeneous geometries. *Phys. Rev. E*, 69:016306, 2004.

487. Min JY, Hasselbrink EF, and Kim SJ. On the efficiency of electrokinetic pumping of liquids through nanoscale channels. *Sensors & Actuators B*, 98:368–377, 2004.

488. van der Heyden FHJ, Bonthuis DJ, Stein D, Meyer C, and Dekker C. Electrokinetic Energy Conversion Efficiency in Nanofluidic Channels. *Nano Letters*, 6:2232–2237, 2006.

489. Schmidt EF. *Unkonventionelle Energiewandler*. Elitera, Berlin, 1975.

490. *CFD-ACE+, CFDRC documentation, Chap 4 Chemistry module*. CFDRC, ESI Group, 2002.

491. Wacogne B, Sadani Z, and Gharbi T. Compensation structures for V-grooves connected to square apertures in KOH-etched (1 0 0) silicon: theory, simulation and experimentation. *Sensors and Actuators A*, 112:328–339, 2004.

492. McNamara S and Gianchandani Y. A Micromachined Knudsen Pump for On-Chip Vacuum. In *Transducers '03*, pages 1919–1922, 2003.

493. Alexeenko AA, Gimelshein SF, Muntz EP, and Ketsdever AD. Modeling of Thermal Transpiration Flows for Knudsen Compressor Optimization. In *AIAA, paper 963*, 2005.

494. Becker EW. Entwicklung des Trenndüsenverfahrens zur Uran-Anreicherung. *Chemie Ingenieur Technik*, 58:1–5, 1986.

495. Menz W and Mohr J. *Mikrosystemtechnik für Ingenieure*. VCH, Weinheim, 1997.

Index

Printing: Krips bv, Meppel
Binding: Stürtz, Würzburg